Fat Production and Consumption

Technologies and
Nutritional Implications

NATO ASI Series

Advanced Science Institutes Series

A series presenting the results of activities sponsored by the NATO Science Committee, which aims at the dissemination of advanced scientific and technological knowledge, with a view to strengthening links between scientific communities.

The series is published by an international board of publishers in conjunction with the NATO Scientific Affairs Division

A	**Life Sciences**	Plenum Publishing Corporation
B	**Physics**	New York and London
C	**Mathematical and Physical Sciences**	D. Reidel Publishing Company Dordrecht, Boston, and Lancaster
D	**Behavioral and Social Sciences**	Martinus Nijhoff Publishers
E	**Engineering and Materials Sciences**	The Hague, Boston, Dordrecht, and Lancaster
F	**Computer and Systems Sciences**	Springer-Verlag
G	**Ecological Sciences**	Berlin, Heidelberg, New York, London,
H	**Cell Biology**	Paris, and Tokyo

Recent Volumes in this Series

Series A: Life Sciences

Fat Production and Consumption

Technologies and Nutritional Implications

Edited by

C. Galli

Institute of Pharmacological Sciences
University of Milan
Milan, Italy

and

E. Fedeli

Experimental Station for the Industry of Oil and Fat
Milan, Italy

Plenum Press
New York and London
Published in cooperation with NATO Scientific Affairs Division

Proceedings of a NATO Advanced Research Workshop on
Advanced Technologies and Their Nutritional Implications in the
Production of Edible Fats,
held March 17–21, 1986,
in Selvino, Italy

Library of Congress Cataloging in Publication Data

NATO Advanced Research Workshop on Advanced Technologies and Their Nutritional Implications in the Production of Edible Fats (1986: Selvino, Italy)
Fat production and consumption.

(NATO ASI series. Series A, Life sciences; v. 131)
"Proceedings of a NATO Advanced Research Workshop on Advanced Tech-
nologies and Their Nutritional Implications in the Production of Edible Fats, held
March 17–21, 1986, in Selvino, Italy"—T.p. verso.
"Published in cooperation with NATO Scientific Affairs Division."
Includes bibliographies and index.
1. Cardiovascular system—Diseases—Nutritional aspects—Congresses. 2.
Lipids in human nutrition—Congresses. 3. Nutritionally induced diseases—Con-
gresses. 4. Oils and fats, Edible—Congresses. I. Galli, Claudio. II. Fedeli, E. III.
North Atlantic Treaty Organization. Scientific Affairs Division. IV. Title. V. Series.
[DNLM: 1. Dietary Fats—adverse effects—congresses. 2. Diseases—etiology—
congresses. 3. Fatty Acids—adverse effects—congresses. 4. Food Technology—
congresses. 5. Nutrition—congresses. QU 85 N277f 1986]
RC669.N275   1986   613.2'8   87-14124
ISBN 978-1-4615-9497-0   ISBN 978-1-4615-9495-6 (eBook)
DOI 10.1007/ 978-1-4615-9495-6

© 1987 Plenum Press, New York
Softcover reprint of the hardcover 1st edition 1987
A Division of Plenum Publishing Corporation
233 Spring Street, New York, N.Y. 10013

PREFACE

Among the major components of human diet, edible fats and oils are typically produced through various forms of technological manipulation of naturally available starting material. It is also generally recognized that dietary fat, namely the type and amount, play a significant role in modulating the health status of large population groups in economically advanced countries, and that, more specifically, the onset and progression of a number of diseases of wide incidence and large socio-economical relevance, such as hyperlipidaemias, diabetes, hypertension, are significantly affected by dietary fats.

Associations operating in public health preventive programs and clinical associations in affluent countries have recommended changes of dietary habits and especially of fat consumption in the whole community. Among the parameters to be modified of special relevance are the reduction of total fat, of saturated fatty acids and cholesterol intake and to increase the amount of unsaturated fatty acids. It is, however, becoming more and more evident that each member of the complex fatty acid moiety of our diet may play a different role from a nutritional point of view. This is particularly true for the highly unsaturated compounds belonging to the two metabolic series n-6 and n-3, contained in high concentrations in vegetable oils and in lipids from marine animals respectively, as well as for oleic acid, contained in high concentrations in olive oil.

Industrial interventions play a major role in devising strategies aimed to improve the nutritional properties of fat produced for human consumption, with the goals of optimizing its fatty acid composition, of enhancing its stability, and of controlling the processes leading to formation of undesired (isomers) or noxious (peroxides) products. Development, standardization and applications of technologies for large scale preparations of edible fat with controlled chemical characteristics and adequate nutritional properties are, thus, essential.

Also, since in large areas of the world, inadequate dietary caloric intake, especially in newborns and children, is still a major nutritional problem, another important goal is to provide appropriate fats and oils technologies.

For the above reasons, it appeared highly needed and timely to bring together experts from both the fields of nutrition and fat technology in order to develop an interdisciplinary approach towards the optimization of edible fat production for general and specialized nutritional use.

This volume represents the Proceedings of a meeting devoted to the analysis and discussion of major nutritional and technological aspects of edible fat production, held in Selvino, Italy, March 17-20, 1986, under the auspices of NATO as an Advanced Research Workshop, with the participation of 70 specialists in the biochemistry and the processing of fats and oils.

In addition to formal presentations there were two round table discussions centering on the biochemistry and two round tables centering on technology. The combined round tables at the end of the meeting were held with the aim to prepare general recommendations, included in this volume, concerning the technology and nutrition of edible fats and oils, to be addressed to Associations, Organizations and Institutions devoted to provide guide-lines for fat production and consumption in human diets.

The editors feel that the general aims of the Workshop were fulfilled and that the active involvement of the participants in the discussion of the various topics at the meeting is a good indication of the great interest for further research in these areas.

E. Fedeli

C. Galli

CONTENTS

NUTRITION – General Effects

Cardiovascular System

Thrombosis

Selected Fatty Acids

Minor Components

TECHNOLOGY

CONDITIONS INFLUENCING ESSENTIALITY OF POLYUNSATURATED FATTY ACIDS

Ralph T. Holman

Hormel Institute
University of Minnesota
Austin, MN 55912

STRUCTURES REQUIRED FOR ESSENTIALITY

The polyunsaturated fatty acids were shown to have special nutritional value in studies by Burr and Burr, who found that dietary fats contain an essential growth-promoting factor not due to fat-soluble vitamins, but which occurs in the fatty acid fraction. They demonstrated that linoleic acid, and perhaps linolenic acid, had this activity[1]. By the criteria of growth and dermatitis, linoleic acid was found to be more active than linolenic acid, and the essentiality of linolenic acid was not seriously considered. Only recently has the evidence come forth to indicate that linolenic acid is essential for other functions.

Linoleic acid was shown to be the precursor of a tetraenoic acid[2], the conversion of linoleic acid to arachidonic acid was shown[3], and the preferred pathway is via α-linolenic acid[4]. Linoleic acid is converted by 6-desaturation, elongation to C_{20}, 5-desaturation, elongation to C_{22}, and 4-desaturation to a series of longer chain, more highly unsaturated polyunsaturated fatty acids (PUFA), all having the same 6-carbon terminal structure (ω6). Linolenic acid is similarly converted through the same steps to a series of PUFA with a 3-carbon terminal structure, to constitute the ω3 family of PUFA. These two families of PUFA are known to have functions related to their occurrence in the essential structural lipids in biologically active membranes, and to their conversion by enzymatic oxidation reactions to highly biologically-active prostaglandins, prostacyclins, thromboxanes, leukotrienes and other active products of oxidation. These substances are all essential regulators of metabolism. The metabolism of the two families of essential PUFA is shown in Figure 1, with that of oleic acid.

Oleic acid (9-18:1 or 18:1ω9) can be desaturated and chain-elongated by the same series of reactions to form a series of ω9 PUFA[3]. This family of PUFA is normally present as minor components in animal tissue lipids, but when dietary supply of essential fatty acids (EFA) is low, these nonessential ω9 PUFA increase in proportion, and may become major components in overt EFA deficiency. The 20:3ω9 is thus a very useful indicator of the severity of EFA deficiency. Palmitoleic acid also undergoes the same series of reactions to produce a family of ω7 PUFA, which are of lower abundance. Because these metabolic sequences are performed by the same enzymatic systems, the several substrates are

1

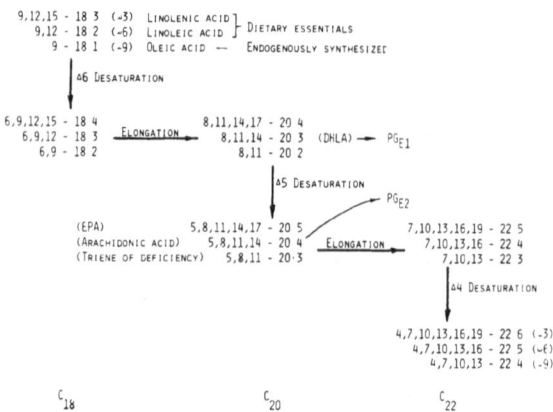

Fig. 1. Metabolism of Polyunsaturated Fatty acids in Liver.

competitive. The ability to compete is ω3 > ω6 > ω9.[5] Saturated fatty acids, naturally-occurring monoenoic acids and isomeric 18:1 acids likewise are competitive in these reactions, and the latter inhibit the metabolism of linoleic acid to arachidonic acid[6]. At low intake of essential PUFA, the nonessential PUFA are competitive with the diminished supplies of essential ω6 and ω3 PUFA and are incorporated into membrane phospholipids (PL) in substitution for the ω6 and ω3 PUFA normally found.

Membrane properties are modulated by the kind and proportions of fatty acids they contain. Isomeric 18:1 acids occurring in partially hydrogenated vegetable oils are likewise competitive with essential PUFA and are incorporated into membranes. They also may be converted to yet other PUFA of unnatural structure[7]. Odd chain[8] and unusual isomers and homologs of PUFA have been synthesized, tested for essentiality and found to have no activity, or less than that of the ω6 and ω3 acids[9]. Essentiality seems confined to even-chained homologs and those isomers in which the unsaturation is all-cis, methylene-interrupted and with either ω6 or ω3 terminal structure[9].

Unique PUFA products are formed from linolenic (ω3), linoleic (ω6) and oleic (ω9) acids, as is shown in Figure 1. There is no metabolic crossover among the three families of acids in animals. The biologically-active oxidative products formed from individual PUFA of different structures are likewise all unique, and their biological functions are also modified by differences in their structures. From only these three common precursors, 15 unique PUFA are formed (Figure 1). If each of these formed only one prostaglandin, a thromboxane, a prostacyclin and a leukotriene, 60 possible biologically-active regulators of metabolism could be formed, and the composite effects would bear a complex relationship to subtle modifications in the proportions of their precursor fatty acids.

NUTRITIONAL DEFICIENCIES OF PUFA

The prevalence of deficiencies of EFA is becoming increasingly evident. EFA deficiency is not a laboratory phenomenon only, but it occurs among humans more frequently, and from a greater variety of causes, than originally thought. The classic work of Hansen demonstrated the dietary need in infants for linoleic acid, and he found that with low intake, infants developed dermatitis which could be cured by linoleate or arachidonate[10]. The induction of dermatitis in humans given total parenteral nutrition without fat has been studied thoroughly and the changes in PUFA of tissue lipids found are parallel to those found in experimental animals[11]. In retrospect, we now know that linoleic acid deficiency develops more rapidly and more easily than linolenic acid deficiency, and most nutritional EFA deficiencies are largely linoleic acid deficiency. Chronic malnutrition, endemic in many parts of the world, is accompanied by the distorted PUFA profile of EFA deficiency[12], and anorexia nervosa also involves EFA deficiency. In nutritional EFA deficiency, PUFA profiles of tissue PL are similarly distorted from normal, and fortunately, the fatty acid pattern of serum PL indicates the PUFA status of the individual[11].

The essentiality of linolenic acid has been suspected since the discovery of EFA, but only recently has its possible function emerged. Many attempts have been made to induce linolenate deficiency and to precipitate abnormalities[13]. The abundance of ω3 PUFA in lipids of eye and nerve tissues has suggested that they must have a function in these tissues. Deficiency of ω3 acids induced nutritionally in monkeys is associated with loss of visual function[14]. In biochemical studies of the retina, the association of ω3 acids with function has been made[15]. One case of human linolenic acid deficiency has been described, in which neuropathy was associated with ω3 deficiency measured analytically[16]. When linolenate was provided, the neuropathy and the biochemical deficiency were eliminated.

QUANTITATIVE ASPECTS OF PUFA REQUIREMENT

The minimum nutrient requirement for linoleic acid[9] and linolenic acid[17] have been determined for rats in studies of the PUFA response in tissues to the oral dose of individual fatty acids. Maximum growth was attained by intake at about 1% of energy, but the maximum rate of growth was in the order arachidonate > linoleate > linolenate[18]. With respect to prevention of dermatitis, the efficacies were in the same order, but linolenate did not completely eliminate the dermatitis even at high intakes[18]. Weight gain and dermatitis are unspecific measures of EFA, and the most direct and specific measures of these two acids are the longer chain PUFA produced from them in normal metabolism. The best assay of linoleate intake is the measurement of the ω6 acids formed from it, and the best assay of linolenate intake is the ω3 acids formed from it.

Dose-response studies were made of linoleate, linolenate and arachidonate esters singly[18], and were extended to α-linolenic acid (18:3ω6)[19]. Pairs of pure dietary PUFA have also been studied with variable 18:3ω3 and constant 18:2ω6[20], variable 18:2ω6 and constant 18:3ω3[21] and variable 18:3ω3 and constant 20:4ω6[6] to assess interactions among PUFA. The effects of triglycerides of pure monoenoic acids[22] and of saturated fatty acids[23] upon the metabolism of linoleic acid have also been determined and found to be less than the effects of the ω3 and ω6 PUFA upon each other[24].

Interactions of fatty acids fed as mixtures of fats, and studied by analytical and statistical approaches[24], revealed high direct diet-tissue correlations among the ω6 acids and among the ω3 acids. The levels of saturated and monounsaturated fatty acids in the diet had measurable effects upon those correlations. Equations were derived which permit estimates of intake of 18:3ω3, 18:2ω6 and saturated acids from analysis of the tissue lipids. This study showed that, when mixed fats containing many fatty acids are ingested, the interactions between the individual fatty acids of the acyl pool mentioned above still operate and affect the

3

composition of tissue lipids in predictable fashion.

Criteria for PUFA adequacy are numerous. The assay of dermatitis is inadequate because it is affected by several other factors, including humidity. Growth is also affected by many factors other than PUFA. Only analytical measurements of tissue or blood lipids can give a direct index of PUFA status. In the first dose-response study, using alkaline isomerization for PUFA analysis, the triene/tetraene ratio was constructed to express the effects of dietary intake of linoleic acid in a single term[25]. When gas chromatography was introduced, many of the individual PUFA became measurable, and the 20:3ω9/20:4ω6 ratio was used to express more precisely the EFA status. For humans, the upper limit of normalcy for this new ratio needed revision downward to 0.2. The ratio probably now should be abandoned in favor of better measures of response, rather than tuning the ratio more finely. Many PUFA affect the biosynthesis of 20:3ω9 and 20:4ω6, and the ratio is now too simplistic for use as more than a rule of thumb. We prefer to consider the total chromatogram and examine the data for total ω6 acids, total metabolites of 18:2ω6, total ω3 acids, total metabolites of 18:3ω3, products of 6-desaturation, products of 5-desaturation, products of 4-desaturation, products of elongation to C_{20}, products of elongation to C_{22}, and product/precursor ratios for key metabolic reactions. The complex and voluminous data are now evaluated in preliminary fashion by computer. Consideration of the pattern of PUFA permits not only a measure of the dietary intake, but reveals metabolic abnormalities.

Quantitative nutrient requirements are derived from the dose-response curves, obtained by measured doses of pure fatty acid esters or by feeding mixtures of natural fats and oils. These curves are all exponential, approaching limiting values. Just as maximum weight gain (obesity) is a poor criterion of EFA requirement, so is maximum effect upon the composition of tissue lipids, for infinite dose would be required to induce maximum change. Minimum nutrient requirement (MNR) must be less than that which will induce the extreme change in the homeostasis existing among the PUFA. Any point on an exponential curve may be expressed in terms of the "half-change" value (I) for intake. We have chosen to express the MNR as that intake which will induce 70% of maximum change:

MNR = 1.7 I

The value of 70% of maximum change, which occurs at 1.7 I was chosen by analogy to the requirements for thiamin and vitamin C which were also set from measurable, exponential biochemical responses for which the factor was found to be near this value. The equation gives estimates of MNR very close to those deduced from the triene/tetraene ratio and from physical changes in EFA deficiency.

Determination of nutritive status is possible once the minimum nutrient requirement is set and the relationship between dose and response has been defined. The same relationships may be used to solve for dietary intake when tissue PUFA values are known. Appropriate equations for estimating linoleate intake for many species have been derived[26]. There is presently a need to re-investigate these relationships for humans using modern capillary gas chromatography which has improved accuracy and reproducibility.

Recently, we have begun a data base obtained by the more discriminating and precise capillary GLC analysis[27] for normal and diseased humans. Analyses are done upon four gross lipid classes of serum, phospholipids (PL), cholesteryl esters (CE), triglycerides (TG) and free fatty acids (FA). The best single analysis of EFA status is the PUFA pattern of serum PL, for this lipid class is richest in PUFA, it is the principal lipid component of membranes, and it responds to changes in dietary EFA most dramatically. The analyses of the other lipid classes provide additional and confirmatory information. The data are preferably expressed as % of total fatty acids of the lipid. The reason for this is that the properties of tissue membranes are influenced by the proportions of various types of fatty acids in their lipids. The proportions of PUFA within lipids therefore are more indicative of function of membranes than are the concentrations of PUFA or of the membranes in the total aqueous space.

OTHER NUTRITIONAL FACTORS AFFECTING PUFA STATUS

Caloric deficiency and limited intake of EFA have a bearing on the utilization of PUFA. Chronic malnutrition, involving protein and caloric deficiencies, has been found to involve a deficiency of EFA as well[12]. This is likewise true in anorexia nervosa, which is a form of starvation[28]. Long-term total parenteral nutrition without fat[11], in which the major energy source is glucose, induces rapid EFA deficiency through the inhibition by glucose of the release of fat and it's EFA from adipose tissue. Long-term intravenous nutrition without a source of EFA throws the patient into a serious blockage of his reserves of PUFA, and overt EFA deficiency is induced. This form of total parenteral nutrition has no place in modern clinical nutrition, except as a very short term expedient.

Previous history of EFA nutrition has a bearing upon the response of an individual to change in EFA intake. Although this is generally assumed from logic and experience of workers in the field, no systematic study has been made of the subject until recently. Two diets were prepared containing equal levels of dietary fat, the normal control containing linoleic acid, and the other not, to induce EFA deficiency. A large group of rats of equal age and size was divided into two groups which were fed the two diets. At intervals of one week, subgroups were separated from the main group and were fed the opposite diet. Thus, there were several groups which had been on linoleate-adequate diet for differing lengths of time before they were given an EFA deficient diet, and there were several which had been in deficiency for different periods of time before they were provided linoleate. The longer the period of adequate linoleate, the slower was the onset of deficiency. Conversely, recovery from deficiencies of different severities seemed to be equally rapid (Holman and Hill, unpublished). Perhaps it would be wise to fortify patients with EFA prior to surgery or other anticipated traumatic procedures.

Hormonal imbalances may disturb PUFA metabolism. Alloxan diabetes[29] and thiouracil-induced hypothyroidism[30] were observed to intensify the dermatitis of EFA deficiency. Diabetes induced by streptozotocin[31] was found to be associated with gross changes in PUFA pattern, notably a deficiency of arachidonic acid. Experimental diabetes and hypothyroidism are also known to be accompanied by hypercholesterolemia.

Diseases of genetic origin may involve abnormalities of PUFA pattern. Achrodermatitis enteropathica[32], Sjogren-Larsson Syndrome[33], multisystem neuronal degeneration[34], cystic fibrosis[35], a hepato-pancreato-renal syndrome (Lindahl et al., unpublished) and Prader-Willi syndrome of obesity[36] are all genetically induced diseases in which disturbed PUFA patterns occur. The defects in PUFA metabolism are probably secondary to defects in enzyme synthesis. Nevertheless, correction of the PUFA deficiency may improve metabolic capabilities of the subcellular organelles.

Impaired liver function occurs in three diseases which have been studied thus far. Cirrhosis was found to be associated with a highly significant arachidonic acid deficiency in serum PL[27]. Alcoholism without cirrhosis did not display abnormality of PUFA pattern, but alcoholism with cirrhosis was associated with more severe deficiency than was cirrhosis without alcoholism. A new syndrome (HPR Syndrome), involving abnormalities of liver, pancreas and kidney, has been found to be associated with serious inability to metabolize linoleic and linolenic acids to longer chain $\omega6$ and $\omega3$ acids (Lindahl et al., unpublished). The serum PL was found to have linoleic acid at three times normal level, but the $\omega6$ acids derived from it were hardly detectable, indicating serious defects in PUFA metabolism. In Reye's Syndrome, which involves malfunction of the liver, a deficiency of PUFA in serum PL and a ten-fold enhancement of PUFA in non-esterified FA have been observed, suggesting that viral infection may trigger uncontrolled lipolysis by phospholipase A_2[37].

Nutritional imbalances involving lipids are well known to affect the PUFA of tissues. Low intakes of linoleic and linolenic acids lead to gross tissue deficiencies of the $\omega6$ and $\omega3$ acids derived from them. Excesses of non-EFA such as saturated acids[23], monoenoic acids[22], and isomeric unsaturated acids[6] suppress the metabolism of PUFA.

In an effort to accelerate EFA deficiency, means were sought to increase lipid turnover. Experimental diabetes is accompanied by hypercholesterolemia and enhanced lipid transport, so alloxan-treated weanling rats were fed an EFA deficient diet. They were found to develop severe symptoms of EFA deficiency within a month rather than the three or more months required in normal rats fed the deficient diet[29]. It is now known that experimental diabetes induced by streptozotocin is accompanied by highly significant abnormalities in PUFA pattern, notably an arachidonic acid deficiency in the phospholipids[31].

Experimental hypothyroidism induced by thiouracil is also known to be accompanied by hypercholesterolemia. Weanling rats were fed 0.02% thiouracil and either 1% saturated fat or 1% linoleate, and the dermatitis of EFA deficiency developed severely within a month in the EFA deficient group, but only marginally in the linoleate group. The control group fed 0.5% linoleate and 0.5% saturated fat but no thiouracil did not develop significant dermatitis[30]. These results led to the suspicion that hypercholesterolemia may accelerate EFA deficiency, and so excess dietary cholesterol was tested with rats[38]. This treatment too accelerated EFA deficiency. The injection of aminonucleoside into weanling rats given an EFA deficient diet, which induced a hypercholesterolemia of 450 mg/dl, also induced dermatitis of EFA deficiency in two to three weeks[26,30]. Recently we observed that two hypolipidemic drugs induced abnormal PUFA patterns in liver PL of rats. We postulate that excessive intake of nonessential lipid, or the induced excessive transport of lipids requires mobilization of PUFA from tissues for synthesis of the PL necessary for their transport, leading to a relative deficiency of EFA. If the dietary fat does not provide sufficient EFA for its own transport, EFA must be mobilized from tissue reserves for synthesis of lipoproteins.

Vitamin E, a biological antioxidant, bears a relationship to the metabolism of the easily peroxidizable PUFA. The requirement for tocopherol increases with increases in the dietary PUFA. This complex phenomenon is the subject of a voluminous literature which has been reviewed[39].

Nutrients affecting growth exert a strong influence upon EFA utilization and PUFA pattern. We have observed in a group of 75 EFA-deficient, cholesterol-fed rats that those which grew most developed the most severe dermatitis. The dermatitis was scored, the rats were weighed, and the correlation between growth and dermatitis was r = 0.38 with p < 0.01[26]. That is, if animals fed equal EFA grow differently for some unknown reason, those that grow most are required to mobilize PUFA from tissues for the proliferation of new cells, therefore precipitating a more severe deficiency.

To induce different rates of growth, different levels of protein intake were provided to groups of EFA-deficient rats. The growth rate and dermal score increased with the protein intake, up to 30% protein. The content of 20:3ω9 in liver PL decreased, and the 20:4 6 increased with increasing protein level. When the PUFA profiles of a protein deficient group (10% protein) were compared with the optimum group (26% protein) as control, several significant differences were observed in both liver and heart PL. The 18:2ω6, 18:3ω6, 20:2ω6, 20:4ω6, 22:5ω6 and 22:6ω3 were significantly less, but the 22:4ω6, 20:5ω3 and 20:3ω9 were significantly higher in the protein deficient group than in the control group[40].

This phenomenon is probably a secondary response to the differences in growth induced by different protein levels. It is quite parallel with the results of two other recent studies with other nutrient deficiencies. The effect of zinc deficiency upon PUFA metabolism was studied with groups given low zinc diet fed ad libitum, normal zinc diet fed ad libitum and the normal zinc diet pair-fed to the low zinc group. Significant differences in profile were found between the low and normal zinc groups, but not between the low zinc group and the normal zinc group pair-fed to it. Thus, the difference in profile induced by zinc deficiency is really not traceable to lack of zinc, but to lack of growth[41]. A very similar phenomenon was found with biotin deficiency, which was associated with an altered PUFA pattern, but not if the control group was pair-fed to it, restricting the two groups to the same rate of growth[42]. Obviously, assessment of effects of other nutrients upon PUFA patterns should be done with pair-fed controls to keep compared groups at equal rates of growth.

Dietary carbohydrates also have an influence upon the fatty acid pattern of tissue lipids. Carbohydrate metabolism produces acetate which can be converted to long chain saturated acids, which may be desaturated at position 9 to form monoenoic acids. Thus, a high intake of carbohydrate tends to form relatively saturated and monounsaturated storage fat and tissue lipids. The extreme case of this is the high sucrose diet which has been traditionally used to develop EFA deficiency[26]. Starch as isolated from nature has sufficient adsorbed linoleic acid-containing lipids to prevent induction of overt symptoms of EFA deficiency when it is used in semi-synthetic diets. The nature of the sugar used in diets to induce EFA deficiency seems to have little effect[43], but a high starch diet induces an unusual amount of several odd-chain fatty acids. Dietary carbohydrate calories are the equivalent of saturated fat calories with respect to their influence upon PUFA status.

SIMILARITIES IN THE ABNORMAL PUFA PROFILES IN HUMAN DISEASES

Our program of study of PUFA profiles in several human diseases has revealed that many diseases do involve abnormal PUFA metabolism. Abnormalities may occur in the elongation and desaturation steps leading to long chain PUFA, or the selectivities in the transacylations of the many PUFA involved in their transport from the intestine to the destination in a tissue lipid may be skewed[44]. Each disease profile is unique, but

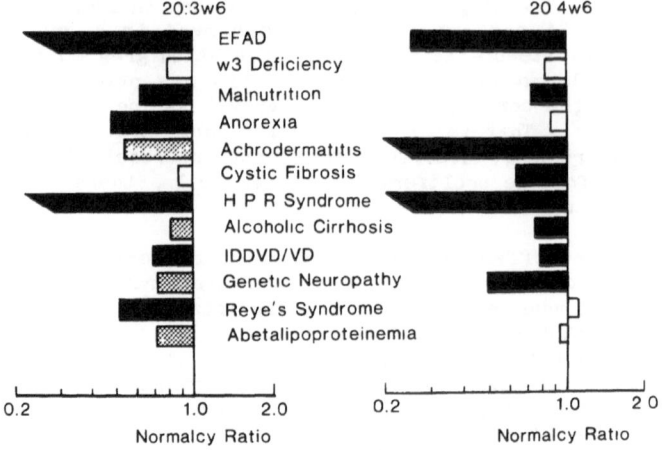

Fig. 2. Normalcy ratios for 20:3ω6 and 20:4ω6 in human serum PL in several diseases.

several have important features in common. The abnormal PUFA profiles have been published in detail in the original references, but some simi-larities are worth summarizing here. Almost all the diseases studied had some statistically significant abnormalities in individual PUFA when compared with normal controls. Of these, deficiencies of 20:3ω6, 20:4ω6, 20:5ω6 and 22:6ω3 occurred frequently and were common to several diseases. In Figure 2 the bars indicate the normalcy ratios (observed value/control value) for dihomogammalinolenic (20:3ω6) and arachidonic (20:4ω6) acids in serum PL in several physiological states, nutritional treatments and diseases. Arachidonic acid deficiency is the most common deficiency, and it is frequently accompanied by deficiency of 20:3ω6. These two prostaglandin precursors are often diminished in PL, and their ratios are often grossly altered, suggesting that prostaglandin patterns may also be skewed.

Two ω3 PUFA are also often abnormal in serum PL in disease states, as is shown in Figure 3. Chronic malnutrition is accompanied by statis-tically significant elevation of 20:5ω6 and decreased 22:6ω3. It appears that diseases often disturb metabolism of ω3 PUFA, and the attendant effects upon production of prostanoids and other autocoids from the ω3 acids may explain some of the features of the syndromes involved.

CONCLUSIONS

The importance of PUFA in nutrition and health is just now being realized. Much of the older literature may need revision, and phenomena studied with obsolete methods may need further study using current methods. This is especially true for the quantitive relationships deduced prior to the advent of modern analytical gas chromatography using capillary columns. A data base of fatty acid compositions of human serum lipids is needed for normal and diseased individuals against which patients can be compared for diagnosis. Similarly, a data base is needed from humans receiving known levels of linoleic and linolenic acids, from which quantitative relations can be derived, permitting assessment of EFA intake in patients and populations.

Caloric, protein, mineral and vitamin deficiencies are being found to significantly alter PUFA patterns in animal studies. From these, the importance of growth and stress are becoming recognized as strongly

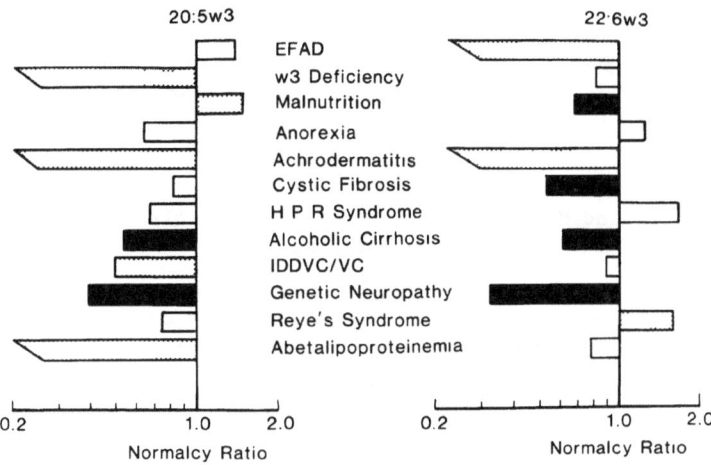

Fig. 3. Normalcy ratios for 20:5 3 and 22:6 3
in human serum PL in several diseases.

affecting PUFA metabolism. This area of nutritional interactions is one which will bear much study and will come to be a major thrust in understanding the metabolism and nutrition of PUFA. Applications to human nutritional and medical problems will certainly follow.

EFA deficiency has been unknowingly induced in countless humans by the life-saving use of total parenteral nutrition without EFA. Now that this has been amply demonstrated, the continued long-term use of preparations without fat emulsions cannot be justified. The composition of such emulsions and the dosage should be tuned more closely to provide for the full requirement of both families of essential PUFA. Future research may indicate the need for specialized emulsions to compensate for abnormal patterns of PUFA in patients under special stress.

The involvement of abnormalities in PUFA metabolism and pattern in tissue lipids in disease processes is slowly becoming apparent. The list of diseases which are accompanied by abnormal patterns of PUFA in tissue lipids is growing steadily. Consideration of nutritional means to restore these essential lipid constituents to normal will also be the business of the nutritional sciences in the near future.

The importance of the PUFA in nutrition is now firmly established, and the essentiality of both the ω6 and the ω3 PUFA is accepted. The provision of both groups of PUFA in adequate amounts should be an objective of the food scientists. Unfortunately, the PUFA present problems in storage as oils, for they are highly subject to autoxidation. The edible oils industry has solved this problem by lowering the proportions of PUFA in vegetable and fish oils through partial hydrogenation, thereby providing mankind with a relatively stable and abundant supply of dietary fats, cheaply meeting the needs of many populations for fat calories. This process has been of great benefit to man. Since this process was developed, we have become aware of the nutritional need for the PUFA which are removed by hydrogenation, and of the possibly adverse effects of the isomeric unsaturated acids and the saturated acids produced by hydrogenation. It would appear that the mandate for the next generation of food scientists would be to find better ways of preserving the dietary essential PUFA in vegetable and fish oils, making these essentials available in unaltered condition. Perhaps this problem is as much a matter of education as it is of technology, and for that reason this review is written.

ACKNOWLEDGEMENTS

Research reported from the author's laboratory was supported in part by NIH Grant 04514, by Program Project Grant HL 08214, by NIH Grant NS 14304, and by The Hormel Foundation.

REFERENCES

1. G. O. Burr and M. M. Burr, On the Nature and Role of the Fatty Acids Essential in Nutrition, J. Biol. Chem. 86:587-621 (1930).

2. I. Rieckehoff, R. T. Holman and G. O. Burr, Polyethenoid Fatty Acid Metabolism. Effect of Dietary Fat on Polyethenoid Fatty Acids of Rat Tissues, Arch. Biochem. 20:331-340 (1949).

3. J. F. Mead, The Metabolism of the Polyunsaturated Fatty Acids, Prog. Chem. Fats Other Lipids 9:159-192 (1971).

4. Y. L. Marcel, K. Christiansen and R. T. Holman, The Preferred Pathway from Linoleic Acid to Arachidonic Acid, Biochim. Biophys. Acta 164:25-34 (1968).

5. R. T. Holman, Nutritional and Metabolic Interrelationships Between Fatty Acids, Fed. Proc. 23:1062-1067 (1964).

6. E. G. Hill, S. B. Johnson, L. D. Lawson, M. M. Mahfouz and R. T. Holman, Perturbation of the Metabolism of Essential Fatty Acids by Dietary Partially Hydrogenated Vegetable Oil, Proc. Natl. Acad. Sci. USA 79:953-957 (1982).

7. R. T. Holman and M. M. Mahfouz, Cis and Trans Octadecenoic Acids as Precursors of Polyunsaturated Acids, Prog. Lipid Res. 20:151-156 (1981).

8. H. Schlenk, Odd Numbered Polyunsaturated Fatty Acids, Prog. Chem. Fats Other Lipids 9:587-606 (1971).

9. R. T. Holman, Biological Activities of and Requirements for Polyunsaturated Acids, Prog. Chem. Fats Other Lipids 9:611-682 (1971a).

10. A. E. Hansen, H. F. Wiese, A. N. Boelsche, M. E. Haggard, D. J. D. Adam and H. Davis, Role of Linoleic Acid in Infant Nutrition Clinical and Chemical Study of 428 Infants Fed on Milk Mixtures Varying in Kind and Amount of Fat, Pediatrics 31:171-192 (1963).

11. J. R. Paulsrud, L. Pensler, C. F. Whitten, S. Stewart and R. T. Holman, Essential Fatty Acid Deficiency in Infants Induced by Fat-Free Intravenous Feeding, Am. J. Clin. Nutr. 25:897-904 (1972).

12. R. T. Holman, S. B. Johnson, O. Mercuri, H. J. Itarte, M. A. Rodrigo and M. E. DeTomas, Essential Fatty Acid Deficiency in Malnourished Children, Am. J. Clin. Nutr. 34:1534-1539 (1981).

13. J. Tinoco, Dietary Requirements and Functions of α-Linolenic Acid in Animals, Prog. Lipid Res. 21:1-45 (1982).

14. M. Neuringer, W. E. Connor, D. S. Lin, L. Barstad and S. Luck, Biochemical and Functional Effects of Prenatal and Postnatal   3

Fatty Acid Deficiency on Retina and Brain in Rhesus Monkey, Proc. Natl. Acad. Sci. USA 83:(in press) (1986).

15. N. G. Bazan, T. S. Reddy, H. E. P. Bazan and D. L. Birkle, Metabolism of Arachidonic and Docosahexaenoic Acids in the Retina, Prog. Lipid Res. 25:(in press) (1986).

16. R. T. Holman, S. B. Johnson and T. F. Hatch, A Case of Human Linolenic Acid Deficiency Involving Neurological Abnormalities, Am. J. Clin. Nutr. 35:617-623 (1982).

17. C. Pudelkewics, J. Seufert and R. T. Holman, Requirement of the Female Rat for Linoleic and Linolenic Acids, J. Nutr. 94:138-146 (1968).

18. H. Mohrhauer and R. T. Holman, The Effect of Dose Level of Essential Fatty Acids upon Fatty Acid Composition of the Rat Liver, J. Lipid Res. 4:151-159 (1963).

19. P. T. Garcia and R. T. Holman, Competitive Inhibitions in the Metabolism of the Phospholipids, Triglycerides and Cholesteryl Esters of Rat Tissues, J. Am. Oil Chem. Soc. 42:1137-1141 (1965).

20. H. Mohrhauer and R. T. Holman, Effect of Linolenic Acid upon the Metabolism of Linoleic Acid, J. Nutr. 81:67-74 (1963a).

21. J. J. Rahm and R. T. Holman, Effect of Linoleic Acid upon the Metabolism of Linolenic Acid, J. Nutr. 84:15-19 (1964).

22. H. Mohrhauer, J. J. Rahm, J. Seufert and R. T. Holman, Metabolism of Linoleic Acid in Relation to Dietary Monoenoic Fatty Acids in the Rat, J. Nutr. 91:521-527 (1967).

23. H. Mohrhauer and R. T. Holman, Metabolism of Linoleic Acid in Relation to Dietary Saturated Fatty Acids in the Rat, J. Nutr. 91:528-534 (1967).

24. W. O. Caster, H. Mohrhauer and R. T. Holman, Effects of Twelve Common Fatty Acids in the Diet Upon the Composition of Liver Lipid in the Rat, J. Nutr. 89:217-225 (1966).

25. R. T. Holman, The Ratio of Trienoic:Tetraenoic Acids in Tissue Lipids as a Measure of Essential Fatty Acid Requirement, J. Nutr. 405-410 (1960).

26. R. T. Holman, Essential Fatty Acid Deficiency, Prog. Chem. Fats Other Lipids 9:279-348 (1971).

27. S. B. Johnson, E. Gordon, C. McClain, G. Low and R. T. Holman, Abnormal Polyunsaturated Fatty Acid Patterns of Serum Lipids in Alcoholism and Cirrhosis: Arachidonic Acid Deficiency in Cirrhosis, Proc. Natl. Acad. Sci. USA 82:1815-1818 (1985).

28. C. E. Adams, R. T. Holman, J. W. Erdman, R. A. Nelson, J. A. Jaskiewicz, S. B. Johnson and S. J. E. Grater, Plasma Fatty Acid Profile in Patients with Anorexia Nervosa, Am. J. Clin. Nutr. (in press) (1986).

29. J. J. Peifer and R. T. Holman, Essential Fatty Acids, Diabetes and Cholesterol, Arch. Biochem. 57:520 (1955).

30. R. T. Holman, The Lipids in Relation to Atherosclerosis, Am. J. Clin. Nutr. 8:95-103 (1960a).

31. R. T. Holman, S. B. Johnson, J. M. Gerrard, S. M. Mauer, S. Kupcho-Sandberg and D. M. Brown, Arachidonic Acid Deficiency in Streptozotocin Induced Diabetes, Proc. Natl. Acid. Sci. USA 80:2375-2379 (1983).

32. R. Cash and C. K. Berger, Acrodermatitis Enteropathica: Defective Metabolism of Unsaturated Fatty Acids, J. Pediatr. 74:717-729 (1969).

33. O. Hernell, G. Holmgren, S. F. Jagell, S. B. Johnson and R. T. Holman, Suspected Faulty Essential Fatty Acid Metabolism in Sjogren-Larsson Syndrome, Pediatr. Res. 16:45-49 (1982).

34. P. J. Dyck, J. K. Yao, D. E. Knickerbocker, R. T. Holman, M. R. Gomez, A. B. Hayles and E. H. Lambert, Multisystem Neuronal Degeneration Hepatosplenomegaly and Adrenocortical Deficiency Associated with Reduced Tissue Arachidonic Acid, Neurology 31:925-934 (1981).

35. J. D. Lloyd-Still, S. B. Johnson and R. T. Holman, Essential Fatty Acid Status in Cystic Fibrosis and the Effects of Safflower Oil Supplementation, Am. J. Clin. Nutr. 34:1-7 (1981).

36. R. A. Nelson, D. M. Huse, R. T. Holman, B. O. Kimbrough, H. W. Wahner, C. W. Callaway and A. B. Hayles, Nutrition, Metabolism, Body Composition and Response to the Ketogenic Diet in Prader-Willi Syndrome, in: "The Prader-Willi Syndrome," V. A. Holm and P. L. Pipes, eds., University Park Press, Baltimore (1981).

37. P. L. Ogburn, H. Sharp, J. D. Lloyd-Still, S. B. Johnson and R. T. Holman, Abnormal Polyunsaturated Fatty Acid Patterns of Serum Lipids in Reye's Syndrome, Proc. Natl. Acad. Sci. USA 79:908-911 (1982).

38. R. T. Holman and J. J. Peifer, Acceleration of Essential Fatty Acid Deficiency by Dietary Cholesterol, J. Nutr. 70:411-417 (1960).

39. L. A. Witting, The Interrelationship of Polyunsaturated Fatty Acids and Antioxidants In Vivo, Prog. Chem. Fats Other Lipids 9:517-554 (1960).

40. E. G. Hill and R. T. Holman, Effect of Dietary Protein Level Upon Essential Fatty Acid Deficiency, J. Nutr. 110:1057-1060 (1980).

41. T. R. Kramer, M. Briske-Anderson, S. B. Johnson and R. T. Holman, Influence of Reduced Food Intake on Polyunsaturated Fatty Acid Metabolism in Zinc-Deficient Rats, J. Nutr. 114:1224-1230 (1984).

42. T. R. Kramer, M. Briske-Anderson, S. B. Johnson and R. T. Holman, Effects of Biotin Deficiency on Polyunsaturated Fatty Acid Metabolism in Rats, J. Nutr. 114:2047-2052 (1984a).

43. J. J. Casal and R. T. Holman, The Effect of Kind of Dietary Carbohydrate Upon the Composition of Liver Fatty Acids of the Rat, J. Am. Chem. Soc. 42:1134-1137 (1965).

44. R. T. Holman, Control of Polyunsaturated Acids in Tissue Lipids, J. Am. Coll. Nutr. 5:236-265 (1986).

DIETARY FATS DURING EARLY DEVELOPMENT

M.A. Crwaford, W. Doyle and P.J. Drury

Nuffield Laboratory of Comparative Medicine
The Institute of Zoology, London, NW1 4RY
England, UK

INTERNATIONAL RECOMMENDATIONS ON DIETARY FATS

The paper will follow the approach used by the first part of the
Expert Committee published conjointly by the Food and Agricultural
Organisation and the World Health Organisation (FAO/WHO, 1978) in 1978
on the Role of Dietary Fats in Human Nutrition. The first section dealt
specifically with the role of dietary fats in early development. The
report made the important point that there was a need to reduce the intake
of saturated fats in countries at high risk to cardio-vascular disease,
obesity and maturity onset diabetes but was unique in that it also
referred to the need for additional fat in developing countries.

Since the 1978 FAO/WHO publications, there have been a number of more
recent reports including the DHSS COMA Report on the prevention of
cardiovascular disease (DHSS, 1984).

The National Advisory Council on Nutrition Education (NACNE, 1983)
and the Joint Advisory Committee on Nutrition Education (JACNE,1985) of
the U.K. were in line with previous recommendations (DHSS, 1974), the
Royal College of Physicians and British Cardiac Society (1976) and WHO,
(1982). There has also been increasing interest in the role of nutrition in
cancer (Doll et al., 1966) where dietary fats again appear relevant
(Carroll et al., 1981). The evidence led to a report by the National
Research Council of the USA, Committee on Diet, Nutrition and Cancer in
1982 (Committee on Diet, Nutrition and Cancer, 1982).

1.0 EARLY GENESIS OF HEART DISEASE IN CHILDHOOD

Two factors in relation to the metabolism of fats and fatty acids
seem to be directly relevant to heart disease.

1. The present diet generally used in Western communities is a high
   fat, high saturated fat diet. Consequently, the conversion
   processes for the essential fatty acids will be inhibited.

2. It is clear from our own studies and recent Australian work, that
   the hunting and gathering diet which pertained throughout human

13

evolution was relatively low in total fat and especially saturated fat but rich in essential fatty acids.

These two simple facts combined with the weight of experimental and contemporary epidemiological evidence identifying saturated fats with atherogenesis, thrombogenesis and the incidence of heart disease may well explain the contemporary high death rate from cardiovascular and related disease in Western communities. Understanding the involvement of the essential fatty acids as building materials for growing cell membranes offers an insight into the likelihood that the adult form of coronary heart disease begins in childhood.

2.0 THE CONVERSE NEED FOR MORE FATS & OILS IN DEVELOPING COUNTRIES

In developing countries the opposite problem exists. 70% of the malnutrition is linked to a deficit in dietary energy rather than protein (FAO, 1978). Some tens of millions of children suffer from malnutrition. It can be associated with severe infection and a retardation of growth and development if not death. In the worst regions as many as fifty percent of the children born may die from the combined effects of malnutrition and infection in early life but the effects of malnutrition are not just confined to loss of life. They may also remain in the shape of permanent disability and physiological or morphological changes, such as blindness or skeletal deformities. About sixty percent of the world's blindness is in India: the bulk of it is nutritional in origin and therefore preventable.

In infants and children perhaps the most important developmental aspect is that of the brain which is intimately related to intelligence and performance. Most of the work which has evaluated infant nutrition has been concerned with physical growth whereas what really matters in the human species is development of the brain and central nervous system. It is more difficult to assess nutritional effects on the brain as little attempt has yet been made to define and apply nutritional parameters for brain development. The approach in use is the assessment of body growth: we have no parallel measurement of brain development despite the fact that it is this aspect of biology rather than body growth which makes man different from other animals.

It is during early growth that the nutritional demands are most critical. About seventy percent of the brain cells divide during fetal growth and most of the remainder of brain development takes place in the first two postnatal years. Children who become blind from vitamin A deficiency cannot have their sight restored by feeding vitamin A at a later date. The effects of retardation of body growth and specifically brain growth, separate from body growth during fetal and early postnatal life, has been demonstrated experimentally in laboratory animals to lead to irreversible retardation of brain development. Children who have died from kwashiorkor have been found to have significantly smaller brain cell populations.

3.0 THE SAME PRIORITY OF THE MOTHER AND CHILD

The view point that we wish to express is that the priority for nutritional programmes should be "development". Secondly, that infant malnutrition should not be considered as an entity on its own or of the time it occurs. The origin of infant malnutrition probably lies in the ecological background of the mother.

The thesis to be presented is that the reproductive process builds in safeguards by requiring the transfer of energy from the food to the mother before pregnancy can occur. During and after pregnancy, the transfer of dietary energy from maternal stores to the fetus occurs via the placenta and to the infant via the milk. The successful transfer of energy to the mother, fetus and infant, guarantees the critical phases such as conception, the fetal growth thrust, birth, its perinatal period, and weaning. These guarantees are achieved by the transfer of energy stores as fat and so the focus of this paper will be on the role that lipids or fats play in early development. In particular, the evidence pointing to the likelihood that the principles of maternal nutrition which need to be considered in the outcome of pregnancy and brain development, stretch back before pregnancy at least to puberty. This argument has large scale implications for food and agricultural policy as well as for educational policy.

4.0   THE SPECIAL SIGNIFICANCE OF DIETARY ENERGY

The FAO/WHO report wrote that until relatively recently, "It was widely believed that infant malnutrition was due primarily to an insufficiency of protein in the diet. There is now adequate evidence to show that the most limiting factor is not protein, but energy. In a sizeable proportion of the children, however, both energy and protein are inadequate, because of the very small amounts of food consumed. But in these same children, the extent of the energy deficit is greater than that of protein. This finding has practical implications in the control and prevention of the problem".

5.0   THE CONTRIBUTION OF FAT AND OILS TO THE ENERGY DENSITY OF FOOD

Dietary energy is obtained from three major sources:

1.   Protein at 4 kcal (16.7 kJ)/g protein.

2.   Carbohydrate at 4 kcal (16.7kJ)/g carbohydrate.

3.   Fats or oils at 9 kcal (37.7 kJ)/g fat.

In most parts of the world 80 - 90% of the dietary energy comes from carbohydrates and fats. In Europe and North America about 40% of the dietary energy comes from dietary fat but in many parts of developing countries the fat intake may only account for 10% of the dietary energy.

The problem that was recognised is that in developing countries the energy density of the food is low because most of it comes from carbohydrate sources. Carbohydrate foods are bulky and may need to be boiled in water to make them edible. In the process, they swell as they absorb water further reducing the energy density of the food. It can be physically difficult for an infant or child to have enough food in his stomach to provide enough energy.

The converse occurs with fat which has the highest energy density of 9 kcal/g compared to 4 kcal for sugars and proteins. Fat is eaten as fat or oil and may be absorbed by food during cooking which increases the energy density of the food without increasing its volume. The energy value of 100 g boiled rice may be doubled by simply adding 10 g of fat or oil. Unfortunately, bottle feeding or early weaning of infants in developing countries has often been coupled with the dilution of the milk or food with water to make a thinner consistency. This is usually prompted by the high cost of the food but a more appropriate action would be to add fat or oil to the food which would increase its energy density.

In looking for a source of fats it is important to remember that animal production is expensive and excessive amounts of animal fat consumption has been associated with heart disease. This means that a solution to the problem of malnutrition could lie in the increased production and consumption of vegetable oils in developing countries. Indeed the FAO/WHO report makes the recommendation that "Edible oil production in developing countries should be increased". It also refers to the need to maintain the levels of essential fatty acids in the diet. This attitude brings us closer to considering fats in a qualitative sense in a manner analogous to the qualitative considerations of pregnancy.

6.0 FATS AND OILS - QUALITY

a) Structural and storage fats

Considerations of quality in fats and oils is needed because not only are there two different types of fats with different compositions and functions but, analogous with proteins we also have non-essential and essential fatty acids. Furthermore, in protein chemistry we have physiologically active derivatives of the essential amino-acids which act as hormones. For example, tryptophan is converted to 5-hydroxytryptamine and tyrosine to adrenalin. We have a precisely analogous situation on the essential fatty acids with dihomo-gamma-linolenic, arachidonic and eicosapentaenoic acids being converted to prostaglandins, lipoxins and leukotrienes.

Descriptions of the fatty acid content of natural products and foods have demonstrated that seed material is generally a rich source of linoleic acid whereas the lipid of green leafy vegetables contains alpha-linolenic acid. It used to be thought that tissue fats simply reflected diet. We now know that there are two types of fat in the body, i.e.

1. <u>Storage</u> - mostly adipose tissue rich in triglycerides, concentrated in non-essential fatty acids and used for energy.

2. <u>Structural</u> - phosphoglycerides and cholesterol; only two positions are available on the phosphoglyceride for fatty acids and as the 2-position incorporates polyunsaturated fatty acids, the structural lipids are rich in essential polyunsaturated fatty acids.

The storage fats largely reflect the diet whilst the structural fats are built to fairly tight fatty acid specifications depending on the specialized function of the cell membrane. The phosphoglycerides in the structural fats contain the parent essential fatty acids (EFA) found in plant life (linoleic, 18:2,n-6 and alpha-linolenic, 18:3,n-3 acids), and their derivatives (FAO/WHO 1978). The structural fats are in high demand for membrane rich systems, of which a good example is the brain where they account for 60% of the solid matter. The highly unsaturated fatty acids are particularly concentrated where there is a requirement for rapid

16

movement at a cellular level such as may be required in transport mechanisms in, for example, brain, its synaptic junctions and the retina, where only the long chain derivatives of the EFA are found and not the parent linoleic and alpha-linolenic acids (Sun & Sun, 1974; Crawford et al., 1976; Anderson & Benolken, R.M., 1977).

The linoleic series can produce a docosapentaenoic acid (C22:5,n-6) but the main component used in cell membranes is AA (C20:4,n-6). In the alpha-linolenic series the main metabolite is DXA (22:6,n-3). These long chain derivatives are the principal components of cell structural lipids and also include the direct precursors for PGs and LTs.

b) <u>Metabolic pathways of the essential fatty acids</u>

The derivation of PGs and LTs from EFAs is shown schematically in fig. 1.

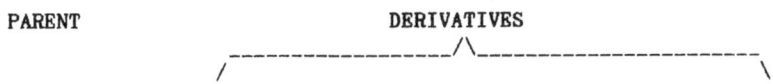

```
PARENT                          DERIVATIVES
                    -------------------/\----------------------
                   /                                           \

    6 desaturase    elongase      5 desaturase    elongase
18:2,n-6 --> 18:3,n-6 --> 20:3,n-6 --> 20:4,n-6 --> C22:4,n-6
Linoleic   -linolenic    dihomo- -   arachidonic   docosatetraenoic
                         linolenic

                    cyclo-oxygenase              lipoxygenase

        PGE1          PGE2 I2 TXA2   5-HPETE -> LTA4 B4 C4 D4

        LIPOXINS        PROSTAGLANDINS         LEUKOTRIENES
                           PGE3   I3


18:3,n-3 -> 18:4,n-3 -> 20:4,n-3 -> 20:5,n-3 -> 22:5,n-3 -> 22:6,n-3
  -Linolenic                     eicosapentaenoic      docosahexaenoic
```

Essential fatty acids are used for cell structures and the oxidative products of the 20 carbon chain length derivatives as hormone-like regulators of cell function.

c) <u>Requirements for Essential Fatty Acids</u>

The fatty acids share the same enzyme systems. This means that the different fatty acids will compete with each other. It is known, for example, that saturated fats suppress EFA activity (Holman, 1977). From studies on wild and modern domestic animals and foods we came to the conclusion that the food which was relevant to the period of human evolution, and man as a hunter-gatherer, would have been low in fat but the fat would have been rich in linoleic, alpha-linolenic acids and even in AA and DXA (Crawford, 1968, Crawford, 1974). The explanation for these observations is that the fat in the meat of wild animals or fish is predominantly structural and hence rich in phosphoglycerides with their component EFAs and their long chain derivatives. The lipid from plant foods is also, in general, rich in one or other of the parent EFAs. The fat in contemporary food, on the other hand, is mainly triglyceride with a high proportion of saturated or non-essential fatty acids. We have previously made the case that a major difference between man's food as a hunter and gatherer, with today's food, is that the balance of the fatty acids has been dramatically changed from a high proportion of EFA to one in which the non-essential saturated fatty acids dominate.

It is estimated that in populations not exposed to the contemporary Western, high fat diet, the requirement for EFA, is about 3% of the dietary energy. In populations, such as ours, where 40% of the dietary energy comes from fat, a high proportion of which is saturated or non-essential, about 4% of the dietary energy is derived from EFAs; it is likely that under these circumstances, the EFAs encounter significant competition in their metabolism. This would mean that the requirement for EFA is probably dependant on the amounts of other fatty acids present in the diet and in our case, is probably higher than 3% of the dietary energy.

A physiological guide can be obtained from consideration of human milk; where the essential fatty acid component occupies about 4-6% of the dietary energy which is about the same level as the protein, or rather its essential amino acid, component. It could be argued that because of the very high demand for cell membrane development, particularly in the growing brain, nervous and vascular systems, that the requirement for EFA is at its highest in infancy and childhood. In the adult the maintenance requirements might be less were it not for the evidence that saturated fats are both atherogenic and thrombogenic (Vergroesen, 1975). In Northern Europe and North America, there is a high incidence of both atherosclerosis and thrombosis. Some hold the view that this is because of an imbalance between the EFA needed for the integrity of the vascular endothelium and for prevention of platelet aggregation on the one hand and on the other, the high dietary levels of saturated fats which induce a relative EFA deficit, making endothelial cells more permeable and encouraging thrombus formation. In this case the requirement for EFA in the adult eating a Western high saturated fat diet could well be higher than expected if vascular health is taken as a criterion.

The above considerations now provides nutritional science with ample energy to justify a new attitude to fats i.e. the nutritional quality of fats and oils.

7.0 ENERGY REQUIREMENTS IN PREGNANCY

During early development, both the energy and qualitative considerations of fats and oils are especially important. The significance of dietary energy in the human species, is underlined in considering the development of the brain which reaches its peak in the human species. One fifth of the oxygen we breathe is used by the brain although the adult brain occupies only 2% of the body mass. By contrast, at birth, about 16% of the body mass is brain and the proportionate use of oxygen and energy by the neonate brain is very much higher being in the region of 60-70% of the total intake. Energetic considerations will be especially important to early brain growth in the fetus and infant.

In well-nourished mothers the normal gain in protein, including the fetus, expansion of maternal blood volume, uterus and mammary development is about 900g; the fetal gain accounting for 400g of the protein. The average gain in protein and fat over the whole pregnancy amounts to 20 Kcal (0.08 MJ) of protein and 120 Kcal (0.5 MJ) of lipid daily (Hytten & Leitch, 1971). However, the fat stores accumulated by the mother are built up early in pregnancy, ahead of the fetal growth spurt. These lipid stores provide for the energetic and essential fatty acid requirements when the fetal growth rate is maximal and the risk of inadequate maternal food intake could be high because of the volume considerations. After delivery, the residual maternal lipid deposits provide for about one third of the energy cost of milk production for the first 100 days of lactation. The well nourished mother will not only have stored fat to protect the growth of her fetus but also will have passed fat stores onto the fetus to provide the neonate with a reserve for the critical perinatal period. She will in addition retain sufficient fat to secure lactation success.

8.0 A STUDY ON MATERNAL NUTRITION IN PREGNANCY

Our own studies in the East End of London have illustrated that the problems of undernutrition are not confined to the developing countries. We studied the relative importance of protein, dietary fat and energy in relation to fetal development and low birth weight. Because of the higher incidence of low birth weight, perinatal mortality and handicap in the lower socio-economic groups, we also wished to examine the possibility that there is a greater nutritional problem amongst the lower income group.

Mrs. Doyle (Crawford et al., in press) studied the food intakes in groups of mothers from two contrasting socio-economic groups who were considered to be at high and low risk of producing a low birth weight baby. The dietary intakes of 100 mothers were assessed during one week in each trimester of their pregnancy; 76 were recruited from a relatively poor community who attended a maternity hospital in Hackney in the East End of London - the Salvation Army Mothers Hospital, Clapton and 24 were recruited from a relatively higher socio-economic community attending the Royal Free Hospital in Hampstead. Asian immigrants were excluded from this study on account of their smaller stature. We also excluded those who smoked more than 10 cigarettes a day because of the known deleterious effect of smoking on the outcome of pregnancy. In fact the majority did not smoke.

The mean birthweight of the low socio-economic group in Hackney infants was 3025 g while that of the Hampstead infants was 3313 g. Of the SAMH live births 11.8% were below 2500 g and 50% were at or below 3000 g. In contrast no RFH infants were below 2500 g and only 17% were below 3000 g.

A comparison of the nutrient intakes, within the Hackney group, of mothers of infants born below 2500 g and those above that figure, also showed a consistently lower intake of all nutrients in the low birth weight group. The Department of Health and Social Security in the U.K. recommends an energy intake during the first trimester as 2150 kcals and 2400 kcals in the second and third trimesters:

Table 1. MEAN CALORIE INTAKES, BIRTHWEIGHTS AND (S.E.) OF MOTHERS IF INFANTS

ABOVE AND BELOW 2500 g IN SAMH AND RFH

	1st trim.	2nd trim.	3rd trim.	Mean of 3 weeks	Mean BW g
SAMH <2500 g	1304 (193)	1504 (87)	1517 (180)	1445 (95)	1843 (174)
SAMH >2500 g	1662 (45)	1748 (48)	1790 (54)	1723 (40)	3201 (57)

RFH >2500 g 1961 (104) 2108 (86) 2065 (85) 2043 (78) 3313 (72)

Although there were differences in protein intakes between the high and low-income groups, both had a mean intake above the Recommended Daily Allowances (RDA) of 60 g during each trimester as was the mean intake for the LBW group over the three trimesters. These results emphasised that it was dietary energy that was limiting rather than protein; a breakdown of the energy deficit revealed that the major deficit was in the relatively low fat intake.

Whilst it is probable that the dietary recommendations may be set at too high a level, nonetheless the contrast in food intakes and birth weights between the two socio-economic groups in relation to birth weights is clear. It is also of interest that the estimation of protein requirements is probably the most precise of the assessments available to us and they were being met; the gap between the recommendation for dietary energy in the group with birth weights below 2500 g was greatest in the first trimester when maternal stores are being gathered and was too great for comfort.

These data emphasize the crucial role of dietary energy. If such severe deficits occur in the U.K. it is likely that the situation is no better in poor regions of the developing countries.

The largest component in the calorie deficit was in a deficit of dietary fats. As there was a difference in fats this would be predicted to be associated with lower intakes of EFA in the low birth weight group and this was found to be the case.

Table 2. FATTY ACID INTAKES (g/day) AND BIRTH WEIGHTS

	<2500 g n=9 Mean	S.E.	>2500 g n=62 Mean	S.E.
Birthweight g	1843	174	3201	57.5
Total fat g	62.1	5.0	72.6	1.75
18:2,n-6 g	5.45	0.62	8.19	0.42
20:4,n-6 g	0.088	0.02	0.103	0.005
18:3,n-3 g	0.870	0.145	1.152	0.057
Long chain n-3 g	0.183	0.046	0.242	0.025
Total EFA's g	6.59	0.8	9.69	0.5

The above differences were statistically significant and were obtained from the study of maternal food intakes in the East-End of London using weighed food intake data for seven days in each trimester giving a total of 21 days of weighed food intakes per mother throughout the pregnancy.

9.0 A CRITICAL ROLE FOR THE PLACENTA

Studies on maternal blood levels of EFA showed that there were low EFA levels in the low birth weight group. This factor might not simply be a result of low fat intakes but could also be exacerbated by the fact that a low calorie intake would impose a demand for the oxidation of fatty acids. Furthermore we also found the expected correlation with birth weight, placental weight and head circumference. We therefore examined the data on the basis of low placental weight, head circumference and birth weight and found that the blood EFA followed in each case. In addition, the lower levels of circulating EFAs reflected in the maternal blood, were also expressed in the fetal circulation obtained by sampling cord blood at the birth. In the fetal circulation, the differences were more pronounced with respect to arachidonic acid assumedly because the placenta concentrates arachidonic acid at the expense of linoleic acid for the fetus and returns linoleic acid to the maternal circulation.

Table 3.   MATERNAL PLASMA CPG WITH LOW AND HIGH PLACENTA WEIGHTS

FATTY ACID	MEAN	VARIANCE	S.E.
RANGE < 425 g	MEAN = 359.4	S.E. 19.2 n=14	
18:2w6	15.5	24	1.99
20:3w6	3.7	0.54	0.30
20:4w6	8.3	6.49	1.04
RANGE > 650 g	MEAN = 751.8	S.E. 15.8 n=10	
18:2w6	19.9	4.7	0.69
20:3w6	3.4	0.51	0.23
20:4w6	9.4	3.18	0.56

In the maternal data on plasma choline phosphoglycerides (CPG) shown above, the difference between low and high placental weights was found only in the linoleic acid content. However, for the same groups, the arachidonic acid was significantly lower in the cord plasma CPG which assumedly reflected placental function as is shown below.

Table 4   CORD PLASMA CPG WITH LOW AND HIGH RANGE PLACENTA WEIGHTS

FATTY ACID	MEAN	VARIANCE	S.E.
RANGE < 425 g	MEAN = 359.4	S.E. 19.2 n=14	
18:2w6	10.5	13.5	1.4
20:3w6	4.2	1.9	0.52
20:4w6	10.9	13.1	1.40
RANGE > 650 g	MEAN = 751.8	S.E. 15.8 n=10	
18:2w6	8.99	7.08	0.77
20:3w6	5.18	1.43	0.34
20:4w6	15.6	9.10	0.87

The data on maternal red cells in relation to birth weight were of special interest.   In blood, the plasma polar phosphoglycerides are present as the choline phosphoglyceride which has a relatively high linoleic and low arachidonic acid content. The red cell, by contrast, contains membrane systems which means that it contains the ethanolamine phosphoglycerides which are the converse of the choline fraction, being rich in the long chain derivatives and containing relatively low levels of linoleic acid. The red cells are also of interest because they tend to reflect a historical picture of EFA status whereas plasma is more influenced by daily events. There were significantly lower levels of arachidonic acid in the ethanolamine phosphoglycerides from maternal (and cord blood) red cells as shown below.

TABLE 5   MATERNAL RED CELL ETHANOLAMINE PHOSPHOGLYCERIDES AND BIRTH WEIGHT

FATTY ACID	MEAN	VARIANCE	S.E.
LOW BIRTH WEIGHT < 2500 g			
18:2w6	3.84	2.10	0.65
20:3w6	1.515	0.69	0.37
20:4w6	14.1	10.1	1.42
BIRTH WEIGHT > 2501 g			
18:2w6	4.64	1.28	0.24
20:3w6	1.34	0.11	0.07
20:4w6	19.0	3.19	0.39

TABLE 6   CORD RED CELL ETHANOLAMINE PHOSPHOGLYCERIDES AND BIRTH WEIGHT

FATTY ACID	MEAN	VARIANCE	S.E.
LOW BIRTH WEIGHT < 2500g			
18:2w6	4.65	16.1	1.79
20:3w6	2.11	0.12	0.16
20:4w6	15.1	14.9	1.00
BIRTH WEIGHT > 2501g			
18:2w6	2.13	0.71	0.23
20:3w6	2.00	0.42	0.18
20:4w6	20.7	37.1	1.69

Similar data was obtained for the n-3 essential fatty acids. In view of the fact that the placenta grows in advance of the fetal growth thrust in the last trimester, a relationship between placental weight and the outcome of pregnancy is to be expected. In view of the heavy investment in membranes by the growing placenta, which handles great lakes of blood to nourish the fetus, considerations of blood flow will also be essential ingredients of placental development.   Both the structural and flow requirements demand an input of essential fatty acids for membrane structures and the regulatory prostaglandins. Consequently, placental development is likely to be an important determinant of fetal growth and by the nature of its development, it is not difficult to understand how essential fatty acids will be required to play a major part in its own growth and function.

10.0   UNDERNUTRITION IN PREGNANCY AND LACTATION IN DEVELOPING COUNTRIES

In countries where undernutrition is common, maternal weight gain in pregnancy would be less and maternal fat deposition less.   There is no estimate of the amount of fat deposition which might represent a safeguard for optimum fetal development. However, in Maharrastra, India, there is evidence that mean birth weight at term is less than in the USA and U.K.. In India the incidence of perinatal mortality, failure to thrive and meet the expected growth velocities, despite being breast fed, is much greater. Data collected by the Shahs in Maharrastra (Shah & Shah, 1972, Shah & Shah, 1975) describes how maternal fat deposition during pregnancy appears to be about half that of U.K. ; women and birth weights are disproportionately low even taking into account the smaller heights and weights of the Indian mothers.

11.0 CONSIDERATIONS IN LACTATION

Comparative Observations

    Milk lipid accounts for 50-60% of the dietary energy for the newborn infant. In human milk the EFA content approximates the protein and occupies   about 6% of the dietary energy.   In our own studies on the fatty acid composition of milks from seven different countries, we have found that in Uganda, Tanzania and Sri Lanka, where the food is low in fat and high in carbohydrate, the shorter chain fatty acids of 10 - 14 carbon chain lengths were higher in proportion and the oleic acid lower, compared to the milk from mothers living on Western high fat diets.   These shorter chain fatty acids are synthesised from carbohydrates and proteins and act as efficient energy sources because their absorption is relatively rapid.

    Milk composition changes during the lactation with time.   In European mothers there is a tendency for the milk fat to increase during the first four months of lactation but in developing countries this increase is much less noticeable.   It is not unlikely that in parts of the world where malnutrition is endemic, the malnutrition is not just a function of infant malnutrition but is an ecological problem, intimately associated with maternal nutrition during pregnancy and lactation.   If the mother does not pass on the fat reserves to the fetus, and can do little better during lactation, then the infant will be at high risk when an attempt is made to wean it onto the low energy dense foods commonly used in areas where undernutrition is endemic.

    In Maharrastra, more than fifty percent of the babies did not grow as they should have done in the first six months despite being breast fed.

    Our evidence from East African and U.K. lactations, suggests that milk production is influenced by maternal nutrition. We found the mean total milk fat from mothers with low birth weight babies was less than those with normal weight babies and that the mothers with the low birth weight babies often failed to lactate.   In East Africa we found that the milk of four mothers whose babies were failing to thrive at the breast was as low as 1.5 - 2.0%. Although one cannot be certain, it is likely that the low energy density of the milk resulted from maternal undernutrition.

    To meet an infants energy requirement of about 600-800 cals/day, the mother would normally need to secrete about 800ml of milk; if the milk fat is less than 100ml, the baby would need to take about 2.5 litres which is unlikely to happen! To solve this problem, it will probably be necessary to adopt the policy to feed the mother to feed the child.

12.0 COMPARISON OF MILK FROM THAILAND AND HUNGARY

    In collaboration with the World Health Organisation we have studied the fat and fatty acid composition in mothers from Thailand and Hungary from milk samples taken at three weekly intervals over the first six months of lactation. There was no statistically significant differences in total milk fat (lipid) at the first three week collection and even at six weeks (second collection) any difference was marginal.   However, in all four collections after six weeks the Hungarian milks had a richer fat content. The lipid in the Hungarian mother's milks rose to over 7% and was significantly greater than the Thailand lipids from the third collection to the seventh. The calories delivered by milk lipid was similar in the first collection (435 cfd 405) but at later time points, the Hungarian milk lipid contributed more to the energy density of the milk; e.g. at 18 weeks the Hungarian milk lipid provided 633 calories per litre and the Thailand milk lipid 428 calories per litre. These results illustrate the inadequacy of data collected at a single time point

Expressed as a proportion of the energy the lipid was between 62.3-68.1% in the Hungarian milks and 58.0-59.7% in the Thailand milks (P<0.001). The milk protein, between 5.2 to 7.0%, was not significantly different between the two centres. The lactose was higher in the Thailand milks (34.1-35.7% compared to 26.7-30.1%).

13.0 COMPARISON OF FATTY ACIDS

It was of interest that in neither centre did the fatty acid profile of the milk lipid change significantly over the seven collections. There were striking differences in the milk fatty acids between the two centres. The proportions of lauric (C12:0) and myristic (C14:0) acids were much higher, whereas the oleic acid (C18:1) was much lower in the milks from the Thailand mothers. The mothers in Thailand would be expected to be eating much less fat than those in Hungary. It is likely that about 38-42% of the dietary energy might be coming from fat in Hungary but only perhaps 10-15% in Thailand. Fat constitutes a major energy component of human milk. It is probable that much of the fat in the milk of the Thailand mothers will need to be synthesised whereas in Hungary much of the milk fat will come directly from the diet. This argument is consistent with the much higher concentration of 10-14 carbon chain length fatty acids in the milks of the Thailand mothers as these fatty acids are derived mainly from biosynthesis in the mammary gland whereas the 16 and 18 carbon chain length fatty acids can be incorporated into milk lipid directly from the diet.

The lower levels of oleic acid (18:1,n-9) in the Thailand milks also indicates that a higher proportion of the lipid in the Hungarian milks is coming directly from the diet. In these circumstances the amounts of oleic acid presented in the Thailand milks do not match the amounts achieved in the milks of the Hungarian mothers nor that of the milks reported previously for well nourished mothers in the U.K., Denmark, or Finland. The conclusion to be drawn is that the rate of synthesis of oleic acid for milk is limited. This may mean that for optimum milk, a source of fat in the diet, rich in oleic acid could be important.

The fatty acid differences between the two centres are consistent with the interpretation that the physiology is attempting to counteract the lower energy availability as fat in the Thailand milks. The short chain fatty acids (lauric and myristic) are the most efficient energy substrates of the milk and high levels of these associated with a lower milk fat content could be a compensation mechanism. Indeed, the very much higher proportions of the short chain energy precursors in the Thailand milks compared to the Hungarians may mean that the efficiency of utilising the Thailand milks for energy may be substantially better than would be expected from direct analysis. These short chain fatty acids are the major fatty acid components of the so-called medium chain triglycerides that are used in clinical practice where there is an emergency demand for a readily absorbable high energy food.

14.0 MALNUTRITION AND LACTATION

The mothers in the Hungarian and Thailand groups were selected to have no health problems. The birth weights were all above 2600 g and there was no significant difference in the mean birth weights of the two groups. The growth rates of the infants were also similar. Although these parameters have only limited significance to human perfomance, there is no suggestion that there was any health difference between the two populations. However, there is a high incidence of infant malnutrition in Thailand which is relatively unknown in Hungary. We have found that mothers whose babies were failing to gain weight at the breast had a depressed level of milk fat (implying a deficit of calories) and of oleic acid. That is, the milk composition in maternal undernourishment was an exaggerated expression of the difference between the Hungarian and Thailand milks. The babies in the Thailand group grew well and one therefore concludes that the composition was satisfactory. However, it is possible that the relatively low fat and oleic acid content of the Thailand milks, although adequate, is near a border-line of safety. Whilst the short chain fatty acids could provide for immediate energy demands, the oleic acid would provide for storage fats (triglycerides of adipose fat) to be accumulated by the infant during lactation and to act as a buffer during the period of weaning onto foods of low energy density.

In cases which we have studied in which the infant was failing to thrive at the breast because of maternal undernourishment, the milk composition was radically altered. The milk fat was low, oleic acid content low and the levels of lauric and myristic elevated. However, protein and lactose contents of the milks were relatively little affected (Crawford et al., 1976).

The conclusion which flows from this discussion is that the role of dietary fats and oils in early development is of special relevance to the human species both quantitatively for energy and qualitatively for essential fatty acids. The most important aspects of human development, namely those of the brain and the vascular systems, are closely dependent on dietary lipids both for energy and specialist structural considerations. In both the Western world and the developing countries, the most serious health distortions are associated either with lipids in the qualitative sense (the West) or energetic sense (developing countries). Consequently the implications of this new knowledge to food and agricultural policies world wide are immense.

15.0 REFERENCES

Anderson, R.E., Benolken, R.M., Jackson, M.B. & Maude, M.B. (1977)
The relationship between membrane fatty acids and the development of the rat retina, in 'Functions and Bioshynthesis of Lipids'.
N.G. Bazan, R.R. Brenner, & N.M. Guisto, eds., Plenum Press, New York and London, pp. 547-559.

Carroll et al, (1981). Progress in Lipid Research. 20, pp. 685-690

Committee on Diet, Nutrition & Cancer. (1982). Assembly of Life Sciences, National Research Council. Washington, D.C.: National Academy Press.

Crawford, M.A. (1968) Fatty acid ratios in free-living and domestic animals, Lancet (i), 1329.

Crawford,M.A. (1975) The Re-evaluation of the nutrient role of animal products. Plenary lecture in Proc.Third World Conf.Animal Product (Melbourne 1973): 21-35, Reid,R.L. ed.,Sydney University Press

Crawford,M.A., Doyle,W., Craft,I.L. and Laurance,B.M. Comparison of food intake during pregnancy and birthweight in high and low socio-economic groups, Progress in Lipid Research,vol.25 (in press)

Crawford,M.A. and Rivers,J.P.W. (1976) The Protein Myth, in "Man/Food Equation", Steel,F. & Bourne,A. eds.,pp. 235-245,Academic Press London

Department of Health & Social Security (1974) Diet and Coronary Heart Disease: report of the Advisory Panel of the Committee on Medical Aspects of Food Policy on diet in relation to cardiovascular disease and cerebrovascular disease (Chairman: prof. Sir Frank Young). Report on Health & Social Subjects 7. HMSO London

Department of Health & Social Security (1984) Report on Health, Diet and Cardiovascular Disease, Committee on Medical Aspects of Food Policy, Social Subjects 28. HMSO London

Doll,R., Muir.C. and Waterhouse,J. (1966) Cancer Incidence in Five Countries, Springer-Verlag, New York

FAO and WHO Report of an Expert Consultation "Dietary Fats and Oils in Human Nutrition", FAO, Rome 1978

Holman,R.T. (1977) Function and Biosynthesis of Lipids, in "Essential Fatty Acids in Human Nutrition", Bazan,N.G., Brenner,R.R. and Guisto,N.M. eds.,pp.515-534, Plenum Press, New York-London

Hytten,P. and Leicht,I. (1971) The Physiology of Human Pregnancy, pp.69-80, Blackwell Scientific Publication, Oxford & London

Joint Nutrition Advisory Committee on Nutrition Education (1985) Eating for a Healthier Heart. The British Nutrition Foundation & Health Education Council, London

National Advisory Council on Nutrition Education (1983) Proposals for Nutritional Guidelines for Health Education in Britain. Health Education Council, London

Royal College of Physicians of London and British Cardiac Society (1976) Prevention of Coronary Heart Disease, J. Royal College of Physicians of London 10, 213

Shah,P.K. and Shah,P.M. (1972) Indian Pediatrics 9

Shah,P.K. and Shah,P.M. (1975) Indian Pediatrics 12, 64

Sun,G.Y. and Sun,A.Y. (1974) Synaptosomal plasma membrane: acyl group composition of phosphoglycerides and (Na and K)-ATPase activity during fatty acid deficiency, J.Neurochem. 22, 15

Vergroesen,A.J. (1975) The Role of Fats in Human Nutrition, Academic Press, London

WHO Expert Committee Prevention of Coronary Heart Disease (1982), Technical Report Series 678.

THE SIGNIFICANCE OF DIETARY FAT FOR METABOLIC DISEASES AND ATHEROSCLEROSIS IN PARTICULAR

Alessandro Menotti and Fulvia Seccareccia

Laboratory of Epidemiology and Biostatistics
Istituto Superiore di Sanità
Rome, Italy

During the last 40 years the research on the relationships between dietary fats and health, and more specifically fats and disease, have been mainly focused on the problem of atherosclerosis, cardiovascular diseases and coronary heart disease in particular (CHD).

Only more recently there has been an increasing interest toward the relationships of fats to cancer with still unsatisfactory results.

An easy and simplified way to look at the present knowledges is to consider the chain fat-serum lipids-atherosclerosis and heart disease.

From another point of view the following relationships are to be independently investigated i.e. (a) blood lipids to heart disease; (b) dietary fats to blood lipids and (c) dietary fats to heart disease.

Point (a), i.e. the relationship between blood lipids and heart disease, is that more deeply investigated and providing the largest scientific evidence. The positive relationships between the levels of serum cholesterol, and mainly LDL cholesterol, and the risk of future development of CHD have been shown in a large number of population samples all over the world, both between and within populations (1,2,3,4,5). Studies like the Seven Countries (3,4), the Ni-Hon-San (6) and the Framingham-Honolulu-Puerto Rico (7) have shown a direct relationship between the mean levels of total cholesterol of the studied population samples and the incidence and mortality from coronary heart disease in the subsequent years of observation (table 1). Conversely the same studies and many others (1,5) have shown, within single population samples, a similar relationship between individual levels of total serum cholesterol and of LDL cholesterol to the subsequent risk of CHD, and an inverse relationship of HDL-cholesterol to CHD (table 2).

This means that populations with higher mean levels of total and LDL cholesterol suffer a higher incidence and mortality from heart

Table 2. Standardized incidence ratio of first major
coronary event as a function of entry levels
of serum cholesterol. The pool 5 of the U.S.
Pooling Project. (ref. 5)

Serum cholesterol		Standardized
Quintile	mg/dl	incidence ratio
1	< 194	72
2	194–218	61
3	218–240	78
4	240–268	129
5	> 268	158

The mathematical background of this peculiar situation has been
developed only recently and shows the present limits in the conduction
of adeguate dietary surveys for this purpose. On the other hand the
relationship has been shown by intervention studies both in metabolic
wards and in free living individuals suggesting the possibility to
increase or decrease the blood lipid levels by modulating the amount and
types of fats in the diet. The classical equations by Keys (10) and by
Hegsted (11) give the mathematical explanations of such changes. Here
saturated fats have a hyper-cholesterolemic action, poly-fat a hypo-cholesterolemic action, mono-unsaturated fats seem neutral, whereas dietary
cholesterol has a hyper-cholesterolemic role (table 4).

Among the most recent contributions in this field there are two
specular studies conducted in Finland and in Italy. In particular a
mediterranean diet fed to Finnish families has been followed by a
substancial decrease of serum cholesterol, (12) whereas an Italian diet
enriched in butter, cream and cheese has induced, in Italian families,
an increase of cholesterol from 214 to 245 mg/dl (13).

Point (c) is essential for showing that dietary fats and heart
disease are correlated each others, through, at least partly, the blood
lipids and for exploring the existence of other fat induced mechanisms,
independent from the blood lipid changes and from their role in
initiating and favouring the progression of atherosclerosis.

When comparing different populations groups or samples it has been
shown that a direct correlation does exist between saturated fats
consumption and incidence-mortality from CHD (table 1).

The major studies in this field are again the Seven Countries
(3,4), the Ni-Hon-San (6) and, from the pathological point of view, the
Geographic Pathology of Atherosclerosis (14). Moreover similar correla-

Table 1. Some correlations between populations' dietary fat
intake, serum cholesterol and coronary heart disease
The Seven Countries Study (ref. 3,4)

	n of groups	correlation coefficient
saturated fats in the diet (% of calories) vs mean serum cholesterol (mg/dl)	14	0.89
median serum cholesterol (mg/dl) vs 10 year coronary mortality	16	0.80
saturated fats in the diet (% of calories) vs 10 year coronary mortality	16	0.84

disease. At the same time, within single populations samples, the individual risk of future coronary events is highly and directly correlated with casual levels of the same variables.

Much has been disputed about the shape of this relationship, some having identified a continous function (usually exponential), others having suggested a threshold beyond certain levels. The question seems still open.

Point (b) concerns the relationship of the amount of saturated fats to the levels of serum cholesterol and mainly of LDL cholesterol, the inverse relationship of poly-unsaturated fats to the same variables and the still uncertain and debated relationship of mono-unsaturated fats again to those variables. In this case the evidence is supported by the comparison of population groups while it is rather scanty when comparing single individuals (1,2).

Once more inter-population comparisons, like those available from the Seven Countries (3,4), the Ni-Hon-San Study (6) and a few others (1,2,7), support the idea that populations with a low intake of saturated fats have lower mean cholesterol levels than those characterized by high consumption of saturated fats (table 1). The evidence is definitely much more limited when comparing single individuals and this is likely due to the difficulties of classifying people in terms of dietary habits and fat consumption in particular, in connection with the fact that intra-individual variance is frequently greater than inter-individual variance (table 3). As a consequence only many repeated independent measurements (from 7 to 9 or more) can adequately classify or characterize an individual (8,9).

Table 3. Linear regression of serum cholesterol on dietary
variables in 1900 men.
The Western Electric Study (ref. 17)

Dietary variables	coefficient	p
saturated fats (% total calories)	1.104	0.014
poly-unsaturated fats (% total calories)	-2.096	0.064
cholesterol (mg/1000 calories)	0.034	0.044
constant	229.405	-

tions have been found combining the WHO mortality data and the FAO food
balance sheets of 20 countries (15).

In the Seven Countries (3,4), p.e., the correlations coefficient in
16 samples between saturated fats intake and CHD mortality was of
0.84.On the other hand, when comparing single individuals within
populations, it happens once again that intra-individual variance
overwhelms inter-individual variance obscuring in this way the relationship although a few accurate studies have documented the same kind of
facts observed between populations (16,17).

At least in the Western Electric Study (17) the relationships
between the individual fats consumption (and their different types) and
the risk of heart disease seem clear but they are not completely
mediated by the blood lipid levels and therefore other mechanisms,
likely bound to platelets aggregation, to the coagulation system and to
the myocardial metabolism might be involved (table 5). Similar results
have been obtained in the Seventh-Day Adventists Study where different
food items, instead of nutrients, have been considered for the analysis
(16).

The above picture is completed by some positive results of primary
prevention trials in population groups where, by dietary changes and/or
by the use of drugs, the blood lipid pattern has been favourably changed
and then followed, after some years, by a reduction of CHD incidence and
mortality, as compared to control groups.

In spite of some limitation and criticism the Helsinki Mental
Hospital trial (18), a similar study conducted in Minnesota (2), the
Veterans Administration Study (18) and the Oslo Study in Norway (19),
all support the hypothesis of a causal role of saturated fats in the
progression of atherosclerosis and in the occurence of organ complications. The amount of the reduction in CHD incidence and mortality has
usually been found proportional to the amount of reduction of mean serum

Table 4. The Keys et al. (ref. 10) and Hegsted et al. (ref. 11)
equations predicting the changes of serum cholesterol
as a function of fats' changes in the diet.
S= percentage of dietary calories from saturated fats;
P= percentage of dietary calories from poly-unsatura-
ted fats;
C= dietary cholesterol in mg/per day;
E= daily energy intake in kilocalories.

Keys	$1.26 \ (2S - P) + 1.5 \ \sqrt{1000 \ C/E}$
Hegsted	$2.16S - 1.65P + 0.677C$

cholesterol, as a conseguence of the dietary changes consisting in a
decrease of saturated fats and cholesterol and in a slight increase of
poly-fats.

Similar conclusions have been obtained by experimental work on
animals, including non-human primates, (21,22) and by the observation of
spontaneous changes of dietary habits in large population groups as
induced by war situations or by a new life style (1,2).

The changes in food consumption occurred in the United States along
the last 30 years, characterized by a reduction of animal fats, is
consistent with the observed reduction of mean serum cholesterol in the
general population and with the decline of CHD mortality.

In this framework the atherogenic role of saturated fats and mainly
of those with long carbon chains is largely documented. Beyond their
atherogenic action, through and increase of the endogenous synthesis of
cholesterol and of LDL, saturated fats worsen the catabolic rate of
cholesterol, seem to favour the development of hypertension, the
tendency of toward thrombosis and the cellular growth which characteri-
zes the arterial atherosclerotic lesions (1,2). However a number of
problems are still open and need to be mentioned

The role of mono-unsaturated fats on blood lipids seems neutral
along with most of the experimental studies (for isocaloric substitution
with starch) but perhaps they are a little hypercholesterolemic in very
high amounts. Their role, anyhow, is again under scrutiny due to the
possible protective activity of oleic acid (and/or of other components
of olive oil). In fact populations using olive oil as the basic fat seem
to be protected from CHD (3,4) and perhaps from cancer. In any case
olive oil, beyond its content in oleic acid has a marked choleretic and
cholagoghe function.

It has been repeatedly shown that poly-fat and linoleic acid in
particular have a hypo-cholesterolemic action and do favour a slow-down
of atherosclerosis progression and a protective action on the heart
muscle against ischaemia. However in order to contrast the hyperchole-

Table 5. 19 year risk of fatal CHD in 1900 men as a function of base-line dietary variables adjusted, in a logistic regression, for seven other possible confounding variables. The Western Electric Study. (ref. 17)

Dietary variables	coefficient	p
Hegsted diet score	0.029	0.004
Keys diet score	0.027	0.010
saturated fats (% of total calories)	0.031	0.144
poly-unsaturated fats (% of total calories)	-0.258	0.010
cholesterol (mg/1000 calories)	0.003	0.008

sterolemic action of saturated fats a double amount of poly-fats is needed (10).

Anxiety about the possible hazards of the intake of large amounts of poly-fats, mainly if cooked for long time and at high temperatures, has been risen several times. The production of peroxides, the breaking of carbon chains and the production of free radicals might in fact favour carcinogenesis (1,2). Moreover an excess of dietary poly-fats, independently from cooking at high temperatures, might favour the production of gallstones. On the other hand the possible favourable role of some long chain poly-fats, like icosapentanoic acid, in limiting platelets aggregation and coagulation factors activity is a recent accomplishment which will be discussed deeper by others.

Finally the role of dietary cholesterol has to be briefly considered. Opinions and data are contrasting but, apart from different enphasis given to the experimental or observational data, it is common agreement that its role on serum cholesterol is secondary to that of saturated fats except in the case of extremely high intake (1,2).

In considering the whole set of evidence on dietary fats as a possible cause of heart disease it should be recalled that their action is conditioned by individual genetical reactivity and by a number of other contingent factors which act, at least, on their intestinal absorbtion. It is known, p.e., that fat absorbtion is reduced with ageing, when the meal is accompained by alcoholic beverages and when the diet is rich in dietary fibres; moreover some vegetable proteins seems to contribute to a reduction of serum cholesterol through mechanisms involving the intestinal absorbtion of fats and/or the liver synthesis of cholesterol (1,2).

A peculiar problem concerns the effect of alcohol in increasing the levels of serum HDL-cholesterol, i.e. the protective fraction of cholesterol. A substancial increase of alcohol intake has however never been suggested for this purpose.

Recent reports from several Committees (1,2,23,24) suggest that total fats should not exceed 30 per cent of total caloric intake (and even less in sedentary adults); saturated fats should represent no more than 7-8 per cent of total calories; poly-fats 4-5 per cent but anyhow no more than 10 per cent, whereas mono-unsaturated fats should cover the difference; dietary cholesterol should not exceed 300 mg per day. Here it should be recalled that small amounts of saturated fats, like 7-8 per cent of total calories, do correspond to the intake of some free living non-malnourished populations which are characterized by a freedom from the epidemic of early CHD and by long expetancy of life (3,4,6,7). In general a diet with such a low content of saturated fats followed for long time is usually accompained by optimal mean levels of serum cholesterol in adult populations, i.e. around 170-180 mg/dl.

Since saturated fats seem to represent also an epidemic factor for breast cancer, perhaps ovarian cancer and with some uncertainties colon cancer, such suggested reccomandations could be accepted also for those purposes (1,2).

In most industrializes countries, including Italy, the amount of total fats, saturated fats and cholesterol in the diet, definitely exceeds those levels and therefore a sharp change in the eating habits is needed in order to reach the above mentioned goals.

In Italy, for istance, estimates from food balance sheets give a 32 per cent of total fat, a 10-12 per cent of saturated fats whereas a clear separation of mono-unsaturated fats from poly-unsaturated fats seems difficult. In this country the major sources of saturated fats are the dairy products, mainly cheese, meat and then vegetables oils.

The implementation of such reccomandations, as expressed in nutrients, into the choice of food is complex and must keep into account also the use of other nutrients and food items.

When reccomending this or that type of food and mainly of shortening, it should not be forgotten that the mix of different fatty acids in the several food items depends on the agricoltural production, on industrial processing and storing and on ways of cooking. Moreover part of dietary fats is invisible and this usually represents a large proportion which may exceed 50 per cent. Finally part of shortenings is waisted and therefore the role of fats in relation to the health effects is likely underestimated.

The most common advice in order to reach the above mentioned dietary goals calls for de-enphasizing the use of dairy products like whole milk, cream, butter and cheese, of meat and mainly fat meat and sausages, of eggs and of deep fried food. In terms of shortenings preference should be given to vegetable oils and to soft margarines, with the reccomandation, anyhow, to avoid high temperatures and long cooking. Caution should be put in the choice of industrial precooked food which can include large amounts of fats and in the use of "junk" food in general.

Large amount of starchy food items, of legumes, vegetables and fruits plus preference of fish and poultry over meat are other complementary choices.

Such guide lines are usually reccomended to all people aged 2 and above also considering that the nutritional adheguacy and safety of such approach are largely documented.

The help of the medical profession, of the health services, of agricolture and food industry in favouring such changes will be essential in the attempt to reach those goals.

REFERENCES

1. Report of Inter-Society Commission for Heart Disease Resources, Optimal Resources for Primary Prevention of Atherosclerotic Diseases, Circulation 70:155A (1984).
2. Report of Nutrition Committee, Rationale of the Diet-Heart Statement of the American Heart Association, Circulation 65:839A (1982).
3. A. Keys, C. Aravanis, H. Blackburn, R. Buzina, B.S. Djordjevic, A.S. Dontas, F. Fidanza, M.J. Karvonen, N. Kimura, A. Menotti, I. Mohacek, S. Nedeljkovic, V. Puddu, S. Punsar, M.L. Taylor and F.S. Van Buchen, "Seven Countries. A multivariate analysis of death and coronary heart disease", Harvard Univ. Press, Cambridge Mass and London, England (1980).
4. A. Keys, C. Aravanis, F.S.P. Van Buchen, H. Blackburn, R. Buzina, B.S. Djordjevic, F. Fidanza, M.J. Karvonen, N. Kimura, A. Menotti, J. Nedeljkovic, V. Puddu and H.L. Taylor, The diet and all causes death rate in the Seven Countries Study, Lancet 2:58 (1981).
5. Pooling Project Research Group, Relationship of blood pressure, serum cholesterol, smoking habits, relative weight and ECG abnormalities to incidence of major coronary events: final report of the Pooling Project, J. Chron. Dis. 31:201 (1978).
6. T.L. Robertson, H. Kato, G.G. Rhoads, A. Kagan, M. Marmot, S.L. Syme, T. Gordon, R.M. Worth, J.L. Belsky, D.S. Dock, M. Miyanishi and S. Kawamoto, Epidemiological studies of coronary heart disease and stroke in Japanese men living in Japan, Hawaii and California. Incidence of myocardial infarction and death from coronary heart disease, Am. J. Cardiol. 39:239 (1977).
7. T. Gordon, M.R. Garcia-Palmieri, A. Kagan, W.B. Kannel and J. Schiffman, Differences in coronary heart disease in Framingham, Honolulu and Puerto Rico, J. Chron. Dis. 27:329 (1974).
8. K. Liu, J. Stamler, J. Mc Keever and P. Mc Keever, Statistical methods to assess and minimize the role of intra-individual

variability in obscuring the relationship between dietary lipids and serum cholesterol, J. Chron. Dis. 31:399 (1978).

9. D.R. Jacobs, J.T. Anderson and H. Blackburn, Diet and serum cholesterol. Do zero correlation negate the relationship ? Am. J. Epidem. 110:77 (1979).

10. A. Keys, J.T. Anderson, and F. Grande, Serum cholesterol response to changes in the diet 1-5, Metabolism 14:747 (1965).

11. D.M. Hegsted, R.B. Mc Gandy, M.L. Luyers and F.J. Stare, Quantitative effects of dietary fats on serum cholesterol in man. Am. J. Clin. Nutr. 17:281 (1965).

12. C. Enholm, J.K. Huttunen, P. Pietinen, U. Leino, M. Mutanen, E. Kostiainen, J. Pikkarainen, R. Dougherty, J. Iacono and P. Puska, Effect of diet on serum lipoproteins in a population with a high risk of coronary heart disease, N. Engl. J. Med. 307:850 (1982).

13. A. Ferro-Luzzi, P. Strazzullo, C. Scaccini, A. Siani, S. Sette, M.A. Mariani, P. Mastranzo, R.M. Dougherty, J.M. Iacono, M. Mancini, Changing the Mediterranean diet: effects on blood lipids, Am. J. Clin. Nutr. 40:1027 (1984).

14. H.C. Mc Gill (ed.), "The geographical pathology of atherosclerosis", Williams and Wilkins Co., Baltimore (1968).

15. J. Stamler, Population studies, in: "Nutrition Lipids and Coronary Heart Disease. A global view", R. Levy, B. Rifkind, B. Dennis, N. Ernst, ed., Raven Press, New York (1979).

16. E.L. Wynder, F.R. Lemon, and U. Brass, Cancer and coronary artery disease among Seventh-Day Adventist, Cancer 20:1016 (1959).

17. R.B. Shekelle, A. Mac Millan Shryoch, O. Paul, M. Lepper, J. Stamler, S. Liu, and W. J. Raynor, Diet, serum cholesterol and death from coronary heart disease. The Western Electric Study, New Engl. J. Med. 304:65 (1981).

18. M. Miettinen, O. Turpeinen, M.J. Karvonen, R. Elosuo, and E. Paavilainen, Effect of cholesterol-lowering diet on mortality from coronary heart disease and other causes. A twelve-year clinical trial in men and women, Lancet 2:835 (1972).

19. S. Dayton, M.L. Pearce, S. Hashimoto, W.J. Dixon, U. Tomiyasu, A controlled clinical trial of a diet high in unsaturated fat in preventing complications of atherosclerosis, Circulation 40:suppl.II (1969).

20. I. Hjerman, K. Velve Byre, and P. Leren, Effect of diet and smoking on the incidence of coronary heart disease. Report from the Oslo study group of a randomized trial in healthy men, Lancet 2:1303 (1981).

21. R.W. Wissler, and D. Vesselinovitch, Studies of regression of advanced atherosclerosis in experimental animals and man, Ann. N.Y. Acad. Sci. 275:363 (1976).

22. R.W. Wissler, and D. Vesselinovitch, Atherosclerosis in non-human primates, Adv. Vet. Sc. Comp. Med. 21:351 (1977).

23. Consensus Conference, Lowering blood cholesterol to prevent heart disease, JAMA 253:2080 (1985).

24. Gruppo Italiano per l'Epidemiologia e la Prevenzione delle malattie Cardiovascolari e Degenerative. Prevenzione Primaria della Cardiopatia Coronarica, Rapporto Rimini 84, Giorn. Arterioscl. 9:249 (1984).

25. A. Mariani-Costantini, Evoluzione del comportamento alimentare in

Italia, Aggiornamento del medico 3:69 (1983).

26. E. Cialfa, I consumi alimentari, Agricoltura 30:15 (1982).

DIETARY AND PHARMACOLOGICAL CONTROL OF HUMAN AND EXPERIMENTAL

HYPERCHOLESTEROLEMIA

Andrea Poli, Elena Tremoli, Claudio Galli
and Rodolfo Paoletti

Institute of Pharmacological Sciences
Via Balzaretti, 9, Milan, Italy

Elevated levels of plasma cholesterol are known to increase the incidence of clinical manifestations of atherosclerosis in man. A favourable effect of plasma cholesterol reduction, in terms of a diminution of the clinical complications of atherosclerosis has, on the other hand, only recently been shown. The Lipid Research Clinics Coronary Primary Prevention Trial (LRC-CPPT), in fact, which will be examined with more details later, has demonstrated that a reduction (even as little as 8%) of the plasma levels of total cholesterol results in the decrease of the risk of fatal or non fatal myocardial infarctions by 19%, and that the total death rate (death for all causes) is also positively affected, even if less strikingly, by this pharmacological treatment.

The LRC-CPPT study (1,2) was conducted on a population of 3806 patients, selected among 480000, meeting the following criteria: plasma total cholesterol >265 mg/dl; LDL cholesterol >190 mg/dl; age between 35 and 59 years; plasma triglycerides <300 mg/dl and absence of clinically overt coronary heart disease (CHD) (patients with only positive exercise test were admitted). Patients with hypertension or conditions associated with secondary hyperlipoproteinemia were also excluded. The recruited patients were randomly assigned, according to a double blind design, to an active treatment with cholestyramine (24 g/day) or to placebo; both groups were placed on a hypolipidic diet. After a follow-up averaging 7.4 years, patients on cholestyramine showed a significant reduction of total and LDL cholesterol levels, which was greater (8.5 and 12.6% for total and LDL cholesterol respectively) than that obtained in the placebo group. Over the same time period, the actively treated group experimented a 19% reduction of major CHD events, and a similar reduction of the incidence of new cases of angina, positive exercise tests or by-pass surgery procedures. Even more interestingly, the subpopulation whose response in terms of total cholesterol reduction was greater than 25%, and greater than 35% for what concerns LDL cholesterol reduction (most likely the subpopulation more adherent to the prescribed dosage of 24 g/day of cholestyramine), experienced a 50% reduction of the incidence of the major CHD endpoints when compared to patients remaining at cholesterol pretreatment levels.

A critical point is now to establish if these data may be extended to any other intervention able to decrease with similar potency plasma cholesterol levels. The experimental evidences are in this respect rather inconclusive.

A reduction of the incidence of cardiovascular events has been previously demonstrated after hypolipidemic treatments performed with different drugs (usually clofibrate) (3,4,5). The total mortality, however, was in some of these trials unaffected (or even increased), due to an unexpected and, as yet, unexplained increase in non-atherosclerotic deaths, the largest part of them being due to cancer.

The results of the LRC-CPPT study have been of borderline statistical significance for what concerns non vascular mortality, because of an increase in violent deaths (suicide and accidents), while the data on coronary events have been definitively positive.

It is thus possible to hypothesize that the reduction of coronary events is related to the decrease in cholesterol levels, while the increase in non cardiovascular deaths, not shared by the intervention trial conducted with cholestyramine, represents a side effect of the drug treatment adopted in the previous trials.

This indicates that reduction in plasma cholesterol levels is "per se" a positive modification. Dietary manipulations, in this perspective, would be of extreme importance, since they can easily be suggested to the general population, through educational campaigns. The impact of a small decrease in the cholesterol levels, but extended to a large part of the population, could be in fact even more important than that of a steeper decrease of this parameter, but restricted to a limited subpopulation (a reduction of 1% of the plasma cholesterol levels, according to the epidemiological evidence, is likely to reduce the incidence of CHD by about 2%).

Another interesting point, is that increased levels of plasma cholesterol induce an increase in the response of platelets to aggregating stimuli (6,7,8). This is of particular interest since platelet hyperaggregability might be related to the genesis of thrombus, which is often the direct cause of the occlusion of a diseased but patent arterial vessel: increased levels of plasma cholesterol, thus, can not only promote the formation of atheromas, but also facilitate the transformation by which a previously asymptomatic plaque causes a clinical event. This would, on the other hand, mean that from a reduction of plasma cholesterol levels one could expect a double order of advantages: short term advantages, due to the reduction of platelet hyperaggregability, to be obtained as soon as the platelet-wall lipids have equilibrated with the new (lower) cholesterol levels, and long term advantages, due to the reduction of ateromas size and degeneration.

Data on the interrelationship between hypercholesterolemia and platelet hyperaggregability have been obtained with different experimental approaches.

Studies on human platelets obtained from hypercholesterolemic subjects (type IIA according to Fredrickson's classification of hyperlipoproteinemias) have shown that the whole mechanism of aggregation control is in these patients shifted toward hyperreactivity. Thromboxane B2 production is increased in platelets from these patients after addition of labelled arachidonic acid. A linear correlation (r=0.97) may be evidenziated between production of thromboxane B2 and levels of total cholesterol in plasma (7).

More detailed studies proved that platelets from type II hyperlipidemic patients produce significantly more malonyldialdehyde than normocholesterolemic subjects (60% of the patients exceeding the 95th percentile of a control population) when stimuled with collagen or thrombin. The production of TxB2 is also increased after collagen stimulation, but no correlation could be identified between plasma cholesterol and TxB2 production. Plasma triglycerides apparently play no relevant role in this sense, since no significant difference between platelets from type IIA and IIB patients could be identified (9).

On the other hand, the response of platelets from hypercholesterolemic subjects to the antiaggregatory effect of PGI2 is significantly decreased as compared to platelets from control patients. Higher concentrations of PGI2 are in fact required in vitro by platelets from patients with type IIA hyperlipoproteinemia to inhibit platelet aggregation induced by ADP, collagen or epinephrine. The activity of PGI2 stimulated adenylate cyclase, on the other hand, measured in washed membranes of platelets from these patients, is not significantly different from that measured in platelets of controls (10).

Studies on the interrelationship between arachidonic acid metabolites and hyperlipidemia have been conducted also in experimental animals, yielding results consistent with those obtained in vitro with human platelets. Experiments on animals, in particular, have allowed to study the biological activity of the endothelial cell layer of arteries, focusing on the production of active metabolites by these cells.

Platelets from rabbits fed diets enriched in cholesterol show hyperaggregation in response to different stimuli. TxB2 is produced in excess from the same platelets (8,11), while the arterial wall of rabbits with advanced atherosclerotic lesions shows a decreased production of prostacyclin. Aortas from hypercholesterolemic rabbits with less advanced atherosclerotic lesions, on the other hand, show increased production of PgI2 as compared to normolipidemic animals (12): this is likely to indicate an attempt of the arterial wall to compensate the hyperreactivity of the platelets of these animals. It is interesting to underline that the platelet hyperaggregabiliy which can be observed in hypercholesterolemic animals can be reversed by different approaches. Metformin, a biguanide compound already known to reduce atheroma formation in hypercholesterolemic rabbits without affecting their plasma cholesterol levels (13), has been shown to decrease platelet sensitiity to aggregating agents, also in the absence of any significant modification of the lipidic parameter of the animal studied (14). It is interesting to note that Metformin has no direct effects on the platelet cyclooxygenase pathway.

Data supporting the interrelationships between cholesterol and platelet aggregation have been obtained also from an interesting trial conducted in two populations with greatly different prevalence of hyperlipoproteinemia, and thus of incidence of myocardial infarction, and with strongly different dietary habits (15).

Middle aged couples free from any clinical evidence of disease were selected in North Karelia (Finland) and in Cilento (south Italy) in the age range between 35 and 49 years. After a two week baseline period, the italian couples were switched for 6 weeks to a diet with a high content in animal fats, and a low P/S ratio, returning in the following 6 weeks to their standard diet rich in vegetable oils and with a high P/S ratio. Finnish couples followed a similar but inverse experimental protocol: two weeks of baseline at their usual high fat diet (low P/S ratio), 6 weeks of intervention on a low fat, high P/S ratio, return to standard finnish diet.

Lipids and lipoprotein plasma levels were modified in the two populations by the dietary intervention as expected. Total cholesterol decreased by 24% in men and 21% in women during the high P/S diet in Finland; levels returned to pretreatment levels after reassumption of the high fat diet. HDL cholesterol also decreased significantly (17%). Triglycerides were unaffected. In Italy opposite modifications of cholesterol levels were described (total cholesterol rose by 19% in men and 16% in women). Triglycerides were only marginally modified also in Italy. These changes in the lipid and lipoprotein pattern were associated to interesting modifications of platelet behaviour. Levels of TxB2 in collagen stimulated platelet rich plasma, which were at the baseline determination significantly higher in the North Karelia men, as compared to the subjects living in South Italy, decreased during the low-fat intervention diet. In men living in South Italy, a similar decrease could be observed after the switch back from the intervention period (high fat diet) to their standard diet (low fat, high P/S). No significant modification of platelet thromboxane formation, in the experimental system tested, could be shown in women, both in Finland and in Italy.

Data obtained with hypolipidemic agents in patients are equivocal. While some authors have described a reduction of platelet aggregability in response to hypocholesterolemic theraphy with clofibrate (16), other groups have obtained, with the same drug, inconsistent results (17). Tiadenol, a hypolipidemic agent with a mechanism of action completely different from that of the fibrates family (that is affecting synthesis of lipoproteins instead of their metabolism) has also been described to reduce platelet aggregability in type IIA patients (18), but also these observations have been challenged (17). Short term modifications of plasma cholesterol and LDL levels, however, may not be reflected by concomitant changes in platelet lipid composition. This latter effect could require longer periods of pharmacological control of hyperlipidemia.

References

1. Lipid Research Clinics Program: The LRC-CPPT results. I. Reduction in the incidence of Coronary Heart Disease, JAMA 251: 351 (1984).
2. Lipid Research Clinics Program: The LPC-CPPT results. II. The relationship of reduction in incidence of coronary heart disease to cholesterol lowering, JAMA 251: 365 (1984).
3. Committee of the principal investigators, WHO Clofibrate Trial: A cooperative trial in the primary prevention of ischemic heart disease using clofibrate, Br Heart J 40: 1069 (1978).
4. Coronary Drug Project Research Group: Clofibrate and Niacine in coronary heart disease, JAMA 231: 360 (1975).
5. Davis CE, Havlik RJ: Clinical trials of Lipid lowering drugs and coronary artery disease prevention, in : "Hyperlipidemia - Diagnosis and Therapy", Rifkind B, Levy R, eds, Grune and Stratton inc, New York, pp 79-92 (1977).
6. Carvalho A, Colman RW, Lees RS: Platelet function in hyperlipidemia, N Engl J Med 290: 434 (1974).
7. Tremoli E, Folco GC, Agradi E, Galli C: Platelet thromboxanes and serum cholesterol, Lancet i: 107 (1979).
8. Tremoli E, Sirtori CR, Maderna P and Paoletti R: Enhanced platelet sensitivity and arachidonic acid metabolism associated with animal and human hyperlipidemias, in :"New Trends in Nutrition, Lipid Research and Cardiovascular disease" Alan R Riss, New York, pp 151-163 (1981).
9. Tremoli E, Maderna P, Colli S, Morazzoni G, Sirtori M and Sirtori CR: Increased platelet sensitivity and thromboxane B2 formation in type II hyperlipoproteinemic patients, Eur J Clin Invest 14: 329 (1984).
10. Colli S, Lombroso M, Maderna P, Tremoli E and Nicosia S: Effects of PGI2 on platelet aggregation and adenylate cyclase activity in human type IIA hypercholesterolemia, Biochem Pharm 32: 1989 (1983).
11. Zmuda A, Dembinska-Kiec A, Chytkowski A, Griglewski RJ: Experimental atherosclerosis in rabbits. Platelet aggregation, thromboxane A2 generation and antiaggregatory potency of Prostacyclin, Prostaglandins 14: 1035 (1977).
12. Tremoli E, Socini A, Petroni A and Galli C: Increased platelet aggregability is associated with increased prostacyclin production by vessel wall in hypercholesterolemic rabbits, Prostaglandins 24: 397 (1982).
13. Sirtori CR, Catapano A, Ghiselli GC, Innocenti AL and Rodriguez J: Metformin - an antiatherosclerotic agent modifying very low density

lipoproteins in rabbits, <u>Atherosclerosis</u> 23: 73 (1976).

14. Tremoli E, Ghiselli GC, Maderna P, Colli S and Sirtori CR: Metformin reduces platelet hypersensitivity in hypercholesterolemic rabbits, <u>Atherosclerosis</u> 41: 53 (1982).

15. Tremoli E, Petroni A, Socini A, Maderna P, Colli S, Paoletti R, Galli C, Ferro-Luzzi A, Strazzullo P, Mancini M, Puska P, Iacono J and Dougherty R: Dietary interventions in North Karelia, Finland and South Italy. Modification of Thromboxane B2 formation in platelets of male subjects only, <u>Atherosclerosis</u> 59: 101 (1986).

16. De Carvalho ACA, Colman RW and Lees RS: Clofibrate reversal of platelet hypersensitivity in hyperbetalipoproteinemia, <u>Circulation</u> 50: 590 (1974).

17. Sirtori M, Montanari G, Gianfranceschi G, Malacrida MG, Battistin P, Morazzoni G, Tremoli E, Colli S, Maderna P and Sirtori CR: Clofibrate and Tiadenol treatment in hyperlipoproteinemias, <u>Atherosclerosis</u> 49: 149 (1983).

18. Vergani C and D'Angelo A: Effet de Tiadenol sur les lipides seriques et sur le function plaquettaire, <u>La Vie Med</u> 2: 1 (1980).

DIETARY FATTY ACIDS AND VASCULAR EICOSANOIDS

C. Galli, C. Mosconi, L. Medini and E. Tremoli

Institute of Pharmacological Sciences
University of Milan
Via Andrea del Sarto,21
20129 Milan, Italy

INTRODUCTION

The precursor-product relationships between 20 carbon polyunsaturated fatty acids (PUFA) and eicosanoids (E) suggest that modifications of the dietary intake of the 18 carbon PUFA - the short chain polyunsaturated fatty acids (SCP) 18:2 w6 linoleic (LA) and 18:3 w3 alpha linolenic - which are converted in biological systems to the E - precursor(s) fatty acids through desaturation and elongation reactions, or the administration of preformed E precursors may affect E production in the body. This is of special interest in the vascular compartment, where E exert diverse biological activities and may modulate the progression of pathological states. Experimental studies have indeed demonstrated that changes of dietary fatty acids affect E formation in several biological systems and especially in the vascular compartment (1,2). Most of the information obtained, however, concerns changes of the stimulated production of E by specific cell types (e.g. platelets, leukocytes, etc.), whereas, on the other side, evaluation of the urinary excretion of E metabolites does not provide information on changes in selected biological compartments. In addition, variations of E production may not closely relate' to functional modifications in a given cell type.

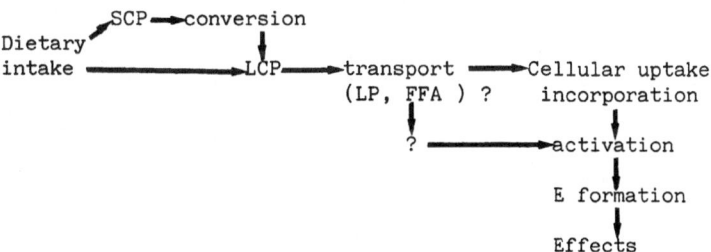

Figure 1. Relationships between dietary fatty acids and eicosanoid production

Legend:SCP,short chain polyunsaturated fatty acids, LCP,long chain polyunsaturated fatty acids, LP,lipoprotein ,FFA,free fatty acids.

Although correlations between E - precursor level and E formation after stimulation of specific cells have been observed, the overall effects of dietary fatty acids on a) the fatty acid composition of cellular phospholipids and b) E formation are difficult to predict. This is dependent upon the complexity of the sequence of steps interposed between the intake of dietary SCP, the accumulation of E precursor fatty acids and the formation of E, as shown in the scheme depicted in Fig.1.

The following aspects appear of relevance in the whole sequence between the intake of PUFA with the diet and E production in tissues:

a) sources of the E - precursor fatty acids for incorporation in tissue lipids (endogenous conversion of the SCP to the LCP in the liver vs other tissues, intake of preformed LCP), b) the mechanisms for transport of the LCP from the major site(s) of endogenous formation (e.g.the liver) to peripheral cells, and c) those dictating the incorporation of LCP in selected lipid classes in different cell types.

SPECIFIC ASPECTS OF LONG CHAIN POLYUNSATURATED FATTY ACID UTILIZATION

a) Sources of E - precursor fatty acids for incorporation in cellular lipids

The individual phospholipid classes present in cellular and subcellular membranes of different cell types contain quite different levels of LCP. High concentrations of LCP, and especially of the E - precursor arachidonic acid (AA), are found in membrane lipids even in cell types such as platelets which have a relatively short biological life and are not very active in converting SCP to LCP. It would appear that, in analogy with the metabolic processes regulating cholesterol synthesis in the body and utilization at various cellular sites, also LCP are mainly formed, and possibly stored, in the liver and subsequently transported to peripheral tissues. Animal studies have provided some evidence for this : e.g in EFA deficiency induced in the rat already at birth - when lipid and LCP deposition in brain are normally very active - and subsequently maintained throughout lactation and weaning, the concentration of AA progressively falls in liver during the first few weeks of extrauterine life, being replaced by 20:3 w9, the biochemical marker of EFA deficiency. At the same time AA continues to accumulate in brain. When levels of AA in liver reach a value of about ten percent of controls, remaining subsequently at a very low plateau, the accumulation of AA in brain ends, being replaced by that of 20:3 w9 (3). This suggests that the liver acts as a store of AA and that in EFA deficiency this fatty acid is released from the liver to be incorporated in tissues with high EFA requirement, such as the developing brain. More recently, significant depletion of AA in liver together with relative sparing of kidney and heart from depletion of LCP, and transport of labelled AA from liver to the above organs, during EFA deficiency in rats, have also been described (4). Limited information is however available on the mechanisms for LCP transport in the plasma compartment from the liver to peripheral tissues. In addition, the relative contribution of the local conversion of SCP to LCP vs that taking place in the liver, to the accumulation of LCP in membranes of different cell types, have not been adequately explored. Finally, some foods, such as to some extent human milk and lean meat for the w6 acids and more significantly fish products for the w3 acids, represent appreciable sources of preformed LCP. These may be more effectively utilized for incorporation in tissue structural lipids and for E synthesis than their SCP precursors. Information on the actual concentrations of AA and of other LCP in different foods are limited.

rabbit, as shown in table 3, in spite of an increase of this fatty acid in liver. This suggests that a) conversion of LA to AA does not occur to an appreciable extent in platelets, and that b) the fatty acid composition of platelet phospholipids reflects that of plasma lipids, where elevation of the LA/AA ratio occurs after dietary LA supplementation (see table 2).

Table 2. Levels (weight percentage) of LA and AA in plasma CE of rats and humans after increase of dietary LA levels.

Fatty Acid	Rat (10)		Human (11)	
	Basal Diet	+ 3 en% LA	Basal Diet	+ 5 en % LA
LA	30.1+1.9	18.3+0.5	47.0+1.0	56.3+1.3
AA	25.8+4.5	41.1+4.2	8.7+0.6	9.6+0.9
AA/LA	0.87	2.28	0.18	0.17

Values are the average + SEM of at least 10 determinations. Number in brackets indicate the reference paper.

Table 3. Levels of LA and AA in platelet phosphatidyl choline of rabbits and men after low and high dietary LA.

Fatty Acid	Rabbit (12)		Man (11)	
	Low LA	High LA	Low LA	High LA
LA	26.6+0.8 a	34.8+1.4 a	7.5+0.3 b	11.4+0.4 b
AA	6.9+0.5 c	5.3+0.3 c	17.0+0.4	15.7+0.9
AA/LA	0.23	0.15	2.27	1.38

Values are the average + SEM of at least 10 determinations. Number in brackets indicate the reference paper. Values with the same letter are significantly different from each other. In rabbits, the semisynthetic diets contained respectively 2.2 en% and 11.0 en % LA and in men, the two diets contained 3.2 and 8.7 en % LA.

The above data and relative considerations indicate that the pattern of utilization of dietary LA for conversion to LCP and especially to AA is rather complex, and that considerable species and tissue differences are present, which should be kept in mind when dietary studies in different conditions are compared. Generally speaking, a substantial increase of LA levels in the diet, as recommended by various commissions on dietary goals, does not result in elevation of AA in circulating cells such as platelets, suggesting that formation of LCP from SCP and their incorporation in cells is under tight control.

A number of studies, some of which will be presented and discussed by other authors of this volume, have shown that the administration of oils rich in 20:5 w3 (eicosapentaenoic acid or EPA) results in incorporation of this fatty acid in plasma and tissue lipids, thus providing a basis for the consequent modifications in the pattern of E formation in various types of cells and tissues. This fatty acid is present in significant concentrations in fish foods, which represent, thus, a source of preformed LCP of the w3 series not very common in western human diet. Incorporation of EPA in various cellular phospholipid classes, after dietary intake, presumably follows distribution patterns different from those of w3 LCP endogenously formed from the SCP precursor, and also from that of AA. It has in fact been reported that administration of EPA with the diet does not result in accumulation of this fatty acid in cellular phosphatidyl inositol (PI) (15), which, instead, incorporates selectively AA. This behaviour may explain some of peculiarities in the effects of EPA administration on eicosanoid production.

In addition to the accumulation in tissue phospholipids, dietary EPA affects the pattern of cellular LCP of the w6 series, interfering with the conversion of LA to AA. This effect was observed even with a relatively low supplementation of EPA (9), which, in the presence of an adequate supply of w6 fatty acids, did not result in the accumulation of w3 LCP, as shown in table 4.

Table 4. Percentage levels of LA , AA and EPA in rat platelet phospholipids after dietary supplementation of corn oil (CO) and fish
oil (FO), alone or in combination (7)

	5 en % CO	10 en % CO	5 en % FO	5 en % FO + 5 en % CO
LA	2.5	3.8	1.9	4.5
AA	10.9	10.2	6.7	7.0
EPA	n.d.	n.d.	2.4	n.d.
AA/LA	4.4	2.7	3.5	1.6

Values have been obtained by analyzing pools of platelet samples. CO (56 percent LA) and FO (8 percent EPA) were fed for a period of 8 weeks.

It appears that when EPA was fed together with LA, to rats, it did not accumulate in platelets, but it resulted in accumulation of LA and in reduction of AA, indicating that LA administered through the CO supplement was not effectively converted to AA. This effect, by itself, will affect eicosanoid production, in spite of lack of EPA accumulation.

b) Transport of LCP in the plasma compartment and c) incorporation of LCP in selected phospholipid classes at the cellular level

As already mentioned, these are points of great relevance in order to understand the regulation of the LCP balance in cells, also in relation to E synthesis. The relative roles of various lipoproteins or of the free fatty acid fraction in plasma, in the transport of LCP have not been clarified. Also, the mechanisms responsible for the selected uptake of LCP at the cellular level are just beginning to be explored in detail. Thus, specific binding proteins for AA have been described in rat heart (5) and a selective arachidonoyl CoA synthetase has been described in lymphocytes (6).

Considerable species differences appear to exist in LCP metabolism, possibly as a consequence of differences in one or more of the above points (a-c), and this is reflected in appreciable interspecies variations in the levels of LCP in the plasma compartment, as indicated in table 1. Levels of LA are similar in human and rabbit plasma phospholipids, but those of AA are quite lower in rabbits. Levels of AA, on the other side, are similar in rats and men, in spite of quite higher levels of the AA precursor LA in men. The product / precursor ratio AA/LA is highest in rats and very low in rabbits, suggesting that the overall conversion of LA to AA is more efficient in rats than in the other species, and that LCP metabolism in rabbits is not very active.

Table 1. Levels (weight percentages) of linoleic (LA) and arachidonic (AA) acids in plasma phospholipids of different species.

Fatty acid	Rat (7)	Rabbit (8)	Man (9)
LA	11.8+3.4	25.8+5.3	24.7+3.7
AA	8.9+2.8	2.2+0.2	12.5+2.3
AA/LA	0.75	0.08	0.51

Values are the average + SD of at least 10 determinations. Number in brackets indicate the reference paper. Rats and rabbits were fed semi-synthetic diets containing 10 % by weight corn oil, and human subjects consumed a diet containing 4.5 en% LA.

c) Dietary w6 and w3 PUFA and LCP in tissues

Dietary linoleic acid (LA)

The very efficient utilization of dietary LA for conversion to AA in the rat compared to man is shown also by data obtained after increasing dietary levels of this SCP. When LA levels in the diets are raised of about 3 to 5 en % in rats or humans, levels of AA are almost doubled in rat plasma cholesterol esters (CE), whereas they are not appreciably modified in human CE (table 2). This difference, however, may represent a combined effect of differences in PUFA metabolism and LCAT activity, responsible of the incorporation of AA in plasma CE, between the two species.

When fatty acids of platelet phospholipids are analyzed after administration of diets rich in LA, in animal species such as the rabbit and man, it appears that AA levels are not increased, especially in the

Dietary eicosapentaenoic acid (EPA)

Changes of specific polyunsaturated fatty acids in the diet result in modified E production in different cells and tissues.

Linoleic acid (LA)

It has been shown by a number of studies that intake of increasing amounts of linoleate with the diet affects E formation, e.g. in whole blood and platelets, although the effects depend upon the level of dietary supplementation. In fact, either stimulation, or no effect or depression have been described (2). Again, species and also sex differences may be important, in this respect. Limited modifications of urinary prostaglandin excretion after changes of dietary linoleic acid intake have been e.g. described in one human study (14).

In our own experience (12), the administration of high levels (more than 10 en %) of LA to rabbits for periods of 4 to 12 weeks resulted in accumulation of this fatty acid, replacing to some extent AA, in platelet phospholipids. The lower aggregability and reduced thromboxane production of these platelets was thus interpreted as a consequence of the reduced AA content, and also, possibly, of the inhibition of AA cyclooxygenase by the excess LA.

Reduction of thromboxane production in stimulated platelet, associated with decrease of plasma total and especially LDL cholesterol levels, was observed also in a study carried out on a set of healthy human couples in North Karelia, after a six week dietary intervention based on an increase of dietary LA from 3.7 to 8.2 en % and a rise of the P/S ratio from 0.15 to 1.3 (11). A small but significant fall of the AA /LA ratio in platelet phospholipids was also observed at the end of the dietary intervention, but it is difficult to attribute the reduction of platelet thromboxane production to this change in platelet fatty acids. The inhibiting effect of dietary LA on platelet thromboxane may be more easily observed in animal species such as rabbits and humans which do not appear to convert very efficiently this SCP to the E precursor AA, than e.g. in the rat. In fact, in this animal species supplementation of an additional 3 en % LA in respect of the standard dietary content for a period of 8 weeks resulted in elevation of platelet AA and in enhanced platelet aggregability (12).

The effects of dietary LA on prostacyclin production are more difficult to evaluate, since formation of this type of E cannot be adequately measured in "in vitro" preparations and data on the effects on the relese of urinary metabolites are not available. In our studies we have observed reduced release of the stable metabolite 6 keto PGF1alfa from perfused aortas obtained from LA supplemented animals (8). Studies carried out in cultured endothelial cells have shown an inhibiting effect of LA, upon incubation, on prostacyclin release (15).

Eicosapentaenoic acid (EPA)

Following the pioneer work of Dyerberg et al. (16), administration of 20:5 w3 (EPA) containing oils or of preparations enriched with EPA has been shown in several animal species and especially in humans, not only to affect platelet and leukocyte function, but also to result in formation of E derived from this fatty acid by the above circulating cells as well as by cells in the vessel walls (17,18). This occurs in association with reduced synthesis of E of the 2 series, due to both reduction of 20:4 w6 (AA) in tissues and inhibition of the enzymatic oxygenation of AA by the accumulated EPA.

These effects will be discussed in detail by various authors in this volume.

Table 5. TxB2 production by stimulated platelets from rats fed w6 and w3
        PUFA alone or in combination.

	5 en % CO	10 en % CO	5 en % FO	5 en % FO + 5 en % CO
TxB2 a)	420	195	180	145

a) Pg/ul PRP at 5 min after 10 I.U./ml thrombin. See table 4 for details
on feeding conditions.

The accumulation of EPA in cellular lipids, and the consequent effect
on E production is also affected by the balance between this fatty acid
and PUFA of the w6 series in the diet, at least in the rat (7), as already discussed (see table 4). Thus, the combined administration of w6
and w3 LCP resulted in a significant fall of the AA/LA ratio in platelet
phospholipids, in spite of lack of incorporation of EPA in platelets,
under these dietary conditions. Platelet thromboxane production,
however, was still significantly reduced by the combination of w6 and w3
fatty acids in the diet, as shown in table 5. Administration of fish
food rich in EPA has also been shown to reduce plasma cholesterol and
especially triglyceride levels in human normal (19) and hyperlipaemic
(20) subjects. The combined effects of fish foods containing EPA or of
preparations enriched with this fatty acid on platelet function and
blood lipids suggest that a substantial increase of fish consumption
should be encouraged. The use of preparations rich in LCP of the w3
series (EPA and docosahexaenoic acid, DHA) may also be very beneficial
in conditions of abnormal platelet reactivity, associated with hyper-
lipaemia. From a general point of view, however, the prolonged consumption of very high amounts of fatty acids which are highly unsaturated
and thus susceptible to autoxidation and peroxidation processes may not
be completely exempt from undesirable effects. Very little is known
about the physiological roles of the highly unsaturated members of the
w3 fatty acid series, but it seems reasonable to postulate that in
animal species living in conditions of low temperature, low oxygen
tension, low light exposure and high external pressure, such as marine
animals, they may represent a form of biochemical adaptation to specialized living conditions.

CONCLUSIONS

Correlations between dietary fatty acids, E precursor levels in
cellular lipids and E formation are complex to evaluate. Generally
speaking correlations have been observed between E precursor(s) levels
in different types of cell and E formation after stimulation, but the
relationships between dietary PUFA and tissue fatty acids are not always
predictable. The effects of dietary manipulations on E formation "in
vivo" are being explored, but conceptual limitations in the interpretation of the data are evident. It is also conceivable that
modulation of the release of E in the vascular compartment by dietary
fatty acids may be mediated by modifications of circulating factors
(free fatty acids, lipoproteins, lipid peroxidation products), in
addition to induced changes of precursor levels in vessel walls.
More generally speaking it is also conceivable that, to some extent,
the basal production of E in different tissues may be based on the
utilization of a small pool of free precursor(s), undergoing rapid

turnover and rapidly affected by dietary manipulations. This point requires, however, detailed investigations.

Dietary lipids may affect E formation in tissues also by ways which do not appear to be directly mediated by changes of E precursor levels. Thus, conditions favouring the accumulation of lipid peroxidation products, such as unbalances between PUFA and antioxidants in the diet or the administration of fats subjected to thermal deterioration, result in increased thromboxane and reduced prostacyclin formation (21).

In conclusion, dietary fatty acids (especially dietary PUFA which may affect cellular levels of E precursor fatty acids) and conditions affecting PUFA stability in the diet and/or in the body, appear to influence E formation in the vascular compartment.

The following general considerations deserve attention and further investigation : a) informations obtained in studies of the arachidonic acid cascade in experimental and human pathologies and of their pharmacological control suggest that manipulation of most E, with the possible exception of prostacyclin, should be aimed mainly to reduction of their formation or to reduction of their biological effects (formation of products with modified activity). b) In mammals,incorporation of E precursor PUFA in tissues is controlled by complex mechanisms and conditions generally limiting and protecting their accumulation appear to prevail (e.g. very low levels of E precursor fatty acids in diets, rate limiting steps in the conversion of 18 carbon EFA to long chain PUFA, selected incorporation in specific pools, presence of natural antioxidants, etc.).

As a consequence it appears that manipulation of dietary lipids should aim to : a) reduce the intake of compounds which have been shown to favour the endogenous synthesis of E such as thromboxane and leukotrienes, b) favour the accumulation in tissues of PUFA which are precursors of less active E (e.g. EPA) mainly through the processes operating in biological systems, i.e. administration of the short chain w3 precursor, followed by desaturation and elongation or administration of w3 fatty acids as components of tissue structural lipids, associated with their supply of natural antioxidants, in the form of seafood, c) preserve the levels of PUFA and of natural antioxidants in dietary lipids. The above aspects demand further investigation on the processes controlling the formation, transport and selective incorporation of PUFA in human tissues, on one side, and on the other, technological interventions to cope with the preservation of PUFA and antioxidants in food.

REFERENCES

1. S.H.jr Goodnight, W.S. Harris, W.E. Connor and D.R.Illingworth, Polyunsaturated fatty acids , hyperlipidemia and thrombosis, Arteriosclerosis 2:87 (1982).
2. M.M.Mathias and J.Dupont, Quantitative relationships between dietary linoleate and prostaglandin (eicosanoid) biosynthesis, Lipids 20 :791 (1985).
3. C.Galli,E.Agradi and R.Paoletti, Accumulation of trienoic fatty acids in rat brain after depletion of liver n-6 polyunsaturated fatty acids, J.Neurochem. 24:1187 (1975).
4. J. B. Lefkowith, V. Flippo, H. Sprecher, and P.Needleman, Paradoxical conservation of cardiac and renal arachidonate content in essential fatty acid deficiency, J.Biol.Chem. 260 (29):5736 (1985).
5. P. Needleman, A. Wyche, H.Sprecher, W.J.Elliott and A.Evers, A unique cardiac cytosolic acyltransferase with preferential selectivity for fatty acids that form cycloxygenase / lipoxygenase metabolites and reverse essential fatty acid deficiency, Biochim.Biophys.Acta 836:267 (1985).
6. A.S. Taylor, H. Sprecher and J.H. Russell, Characterization of an arachidonic acid-selective acyl-CoA synthetase from murine lympho-

cytes, Biochim.Biophys Acta, 833:229 (1985).

7. A.Socini, C.Galli, C.Colombo, and E.Tremoli, Fish oil administration as a supplement to a corn oil containing diet affects arterial prostacyclin production more than platelet thromboxane formation in the rat, Prostaglandins 25: 693 (1983).

8. I. Masi, E. Giani, C. Galli, E.Tremoli and C.Sirtori, Diets rich in saturated, monounsaturated and polyunsaturated fatty acids differently affect plasma lipids, platelet and arterial wall eicosanoids in rabbits, Ann.Nutr.Metab.,30:66 (1986).

9. R.M. Dougherty, C. Galli, A. Ferro-Luzzi and J.M. Iacono, The lipid and phospholipid fatty acid composition of plasma, red blood cells and platelets and how they are affected by dietary lipids:a study of normal subjects from Italy, Finland and the U.S.A., Am.J.Clin. Nutr. in press.

10. E. Giani, I.Masi, C.Colombo and C.Galli, Sex differences in platelet thromboxane and arterial prostacyclin production in control and n-6 fatty acid supplemented rats, Prostaglandins, 28:573 (1984).

11. E. Tremoli, A. Petroni, A. Socini, P. Maderna, S. Colli, R.Paoletti, C. Galli, A. Ferro-Luzzi, P. Strazzullo, M. Mancini, P. Puska, J. Iacono, and R. Dougherty, Dietary interventions in North Karelia, Finland and South Italy. Modification of thromboxane B2 formation in platelets of male subjects only, Atherosclerosis 59:101 (1986).

12. C. Galli, E.Agradi, A.Petroni and E Tremoli, Differential effects of dietary fatty acids on the accumulation of arachidonic acid and its metabolic conversion through the cyclooxygenase and lipoxygenase in platelets and vascular tissue, Lipids 16:165 (1981).

13. B.J. Weaver and B. Holub, The relative incorporation of arachidonic acid and eicosapentaenoic acid into human platelet phospholipids, Lipids 20:773 (1985).

14. A. Ferretti, J. Judd, M.W. Marshall, V.P. Flanagan , J.M. Ramon and E.J. Matusik, Moderate changes in linoleate intake do not influence the systemic production of E prostaglandins, Lipids 20:268 (1985).

15. A.A. Spector, T.L. Kaduce, J.C. Hoak and G.L. Fry, Utilization of arachidonic and linoleic acids by cultured endothelial cells. J.Clin.Invest. 68:1003 (1981).

16. J. Dyerberg, and H.O.Bang, Haemostatic function and platelet polyunsaturated fatty acids in Eskimos, Lancet 2:43 (1979).

17. S. Fischer, and P.C. Weber, Thromboxane A3 (TxA3) is formed in human platelets after dietary eicosapentaenoic acid (C20:5 n-3),Biochem. Biophys.Res.Comm. 116:1091 (1983).

18. S. Fischer and P.C. Weber, Prostaglandin I3 is formed in vivo in man after dietary eicosapentaenoic acid, Nature(Lond.) 307:165 (1984).

19. W.S.Harris, W.E. Connor and M.P. McMurry, The comparative reductions of the plasma lipids and lipoproteins by dietary polyunsaturated fats : salmon oil versus vegetable oils. Metabolism 32:179 (1983).

22. B.E.Phillipson,D.W.Rothrock,W.E.Connor,W.S.Harris and D.R.Illingworth Reduction of plasma lipids,lipoproteins and apoproteins by dietary fish oils in patients with hyperglyceridemia. N.Engl.J.Med. 312: 1210 (1985).

21. E. Giani, I. Masi and C. Galli, Heated fat, vitamin E and vascular eicosanoids , Lipids 20:439 (1985).

EFFECTS OF SATURATED AND POLYUNSATURATED FATTY ACIDS ON

PLASMA LIPIDS, PLATELETS AND THE VASCULAR SYSTEM

Scott H. Goodnight, Jr., William E. Connor
and D. Roger Illingworth

Departments of Medicine and Clinical Pathology
The Oregon Health Sciences University
Portland, Oregon

INTRODUCTION

The hypothesis that alterations in dietary fat intake may influence
the development of atherosclerosis and thrombogenesis has been intensely
studied for several decades (1). For example, in the early 1950's, it
was shown that the feeding of a diet high in saturated fat to humans
would raise plasma cholesterol, whereas the substitution of
polyunsaturated fatty acids into the diet would reduce the plasma
cholesterol (2,3). Since atheromatous lipid deposits may occur in the
arterial vasculature when plasma levels of cholesterol are high, a
reduction in circulating cholesterol and low density lipoproteins may
result in a reduced incidence of occlusive vascular disease in man (4).
Diets rich in saturated fat may also predispose to vascular thrombosis
as evidenced by a series of studies in animals and man which have been
published in the last several decades (1). In one such study, the
intravenous infusion of saturated fatty acids into animals led to
thrombosis of the great vessels, whereas long-chain polyunsaturated
fatty acids did not (5). Further, Renaud showed that animals fed butter
or specific saturated fatty acids were more prone to venous thrombi
after the infusion of endotoxin (6). Another animal model was developed
by Hornstra, et al., who measured the number of hours required for the
thrombotic occlusion of a polyethylene cannula inserted into the aorta
of rats. Once again, dietary saturated fatty acids were more
thrombogenic and produced a marked shortening of the occlusion time
compared to the feeding of polyunsaturated fatty acids (7), Experiments
have also been carried out in man whereby circulating total or saturated
free fatty acids were increased by periods of fasting or occurred in
patients with acute myocardial infarction. The rise in free fatty acid
led to increased platelet reactivity (8,9). The increased platelet
reactivity in these conditions may be due to shortening of prostacyclin
survival secondary to displacement of PGI_2 from fatty acid binding sites
on albumin (10).

Since the evidence seems strong that polyunsaturated fatty acids
may have a beneficial effect in athersclerosis, more recent studies have
been directed at the effects of dietary polyunsaturated fatty acids on
lipids and lipoprotein metabolism as well as their influence on platelet
reactivity in vascular thrombosis. This research area has taken on a
aura of excitement following the observations of Dyerberg and Bang

concerning the effects of fish oil in humans (11). It is now quite clear that polyunsaturated fats may vary considerably in their effects on humans depending on whether they belong to the ω6 (e.g., linoleic acid) or the to ω3 (e.g., eicosapentaenoic) family of fatty acids. The following sections of this paper will focus on the effects of ω6 and ω3 fatty acids on lipids and lipoproteins as well as on platelet and vascular function in its relationship to thromboembolism.

EFFECT OF POLYUNSATURATED FATTY ACIDS ON PLASMA LIPIDS AND LIPOPROTEINS

The initial studies on the effects of polyunsaturated fatty acids on plasma cholesterol did not draw a major distinction between those fatty acids derived from vegetable oils (predominantly linoleic) and those from fish oils (eicosapentaenoic acid and docosahexaenoic acid). However, Dyerberg's studies on the Greenland Eskimos and the discovery that certain polyunsaturated fatty acids may alter cellular prostaglandin production spurred further study of the effects of the ω3 fatty acids from fish oil on plasma lipids and lipoproteins.

In one such study performed in our laboratory, we compared the hypolipidemic effects of a diet containing large amounts of salmon oil to two control diets, one containing a quantity of saturated fatty acids normally consumed by a typical American population, and another diet rich in the ω6 fatty acid, linoleic acid (12). In this study, 12 normal healthy adults were prospectively studied as outpatients in a Clinical Research Center where they obtained all of their meals. Three distinct diets, each containing 40% fat and 500 mg of cholesterol were given in a random order for four weeks. Each dietary regimen was followed by three weeks of a normal diet to allow for a "washout" period. One diet contained a high proportion of ω3 fatty acids derived from salmon and salmon oil. The second diet contained vegetable oil as a source of fat (a mixture of safflower and corn oil) and the third diet contained the saturated fatty acids. A fasting blood sample was obtained twice weekly and the plasma analyzed for total cholesterol, triglycerides, fatty acids and lipoprotein classes (HDL, LDL and VLDL).

The fatty acid composition in the plasma changed greatly during the salmon oil dietary period. Total ω3 fatty acids rose markedly in the cholesterol ester, triglyceride and phospholipid fractions and reached 25 to 30% of the total fatty acids. The increase was detectable after only one day of feeding and was maximal by day 10. Other fatty acid changes included a 50% reduction in linoleic acid in most lipid fractions and a 23% drop in arachidonic acid in the phospholipid fraction.

Total plasma cholesterol fell from 188 to 162 mg/dL (p < 0.001) during the salmon phase of the diet compared to the control (saturated) phase. LDL cholesterol decreased from 128 to 108 mg/dL (p < 0.005), but HDL cholesterol was unchanged. Of note, plasma triglycerides dropped from 74 to 48 mg/dL (p < 0.005). The vegetable oil diet also decreased cholesterol values in these volunteers to a similar extent (i.e., about 11%), but did not alter triglyceride levels.

Although previous studies had shown that fish oil could produce similar reductions in plasma cholesterol as could vegetable oils, it has not been appreciated that fish oil had such a clear cut effect of lowering plasma triglycerides. The mechanism of the hypotriglyceridemic effect of the fish oil was not clear from this study although it suggested that either the rate of synthesis of VLDL was reduced or the fractional rate of catabolism was increased.

The effect of the fish oil feeding on triglyceride levels was later extended to patients with hereditary causes of hyperlipidemia (13.) In this study, 20 patients with long-standing moderate to severe hypertriglyceridemia were given three sequential diets in the Clinical Research Center. The first was a low fat diet, the second a fish oil diet, and the third was a diet containing polyunsaturated vegetable oils. Ten of the patients studied had elevated levels of both LDL and VLDL (i.e., type IIb phenotype) with mean plasma cholesterol values of 337 mg/dL and mean triglyceride levels of 355 mg/dL. The other 10 patients had type V hyperlipidemia with chylomicrons present when fasting and exceedingly high chylomicron and VLDL levels (e.g., a mean of 2874 mg/dL). These patients were studied in a steady state and were not ingesting lipid-lowering drugs.

The type V patients were given a control diet that was exceedingly low in fat (5%) and cholesterol (< 100 mg/dL). During the fish oil and vegetable phases, fat content was increased to 20-30% and contained about 20-30 g of ω3 fatty acids or 47 g of ω6 fatty acids respectively. The type IIb patients had similar diets although the control period contained 20-30% total fat.

Consumption of the fish oil diet was associated with a marked hypolipidemic effect in all of the patients with type IIb or type V hyperlipidemia. In the type IIb group, plasma cholesterol fell by 27% and the triglyceride levels fell even further due to a decrease in VLDL levels (216 to 55 mg/dL). The vegetable oil diet did not decrease the VLDL triglycerides nearly so much (304 to 177 mg/dL).

The type V patients had an even more striking decrease in their plasma lipid levels. While ingesting the fish oil diet, chylomicrons cleared from the plasma while fasting and the chylomicron triglyceride content fell from 443 to 22 mg/dL. Further, the fish diet led to a fall of total plasma triglycerides from 1353 to 281 mg/dL, which is a drop of 79%. A similar fall was seen in VLDL triglycerides from 1087 to 167 mg/dL. Total cholesterol fell from 373 to 207 mg/dL which was accounted for mainly by a fall in VLDL cholesterol. LDL cholesterol rose from 84 to 125 mg/dL. These findings were in stark contrast to the vegetable oil diet which caused a striking increase in plasma triglyceride within 3 to 4 days. In fact, because of the risk that hypertriglyceridemia could lead to acute abdominal pain and pancreatitis, this dietary phase was prematurely discontinued in 10-14 days.

The dramatic reductions in plasma triglyceride levels in these patients with type V hyperlipidemia would suggest that a dietary change to include fish and fish oil would be beneficial in treating these patients and could reduce the complications of acute pancreatitis and acute abdominal pain often found in these subjects. Since most patients find that it is very difficult to comply with a very low fat diet on a long term basis, these and other patients with type IIb or type IV hyperlipidemia may find a diet containing fish and/or fish oil much more palatable.

The mechanisms responsible for the fall in LDL and VLDL induced by fish oil feeding has been studied (14-16). The most likely mechanism for reduced plasma VLDL is due to decreased synthesis of the lipoprotein by the liver. The hypertriglyceridemia induced by carbohydrate loading is blocked by fish oil feeding (14) and in turnover studies, VLDL triglyceride and apolipoprotein B synthesis was inhibited (15,16). Lastly, fish oil has been shown to lower LDL synthesis in normal individuals (17), and since in this population LDL apolipoprotein B is derived almost exclusively from VLDL apolipoprotein B, this finding is also consistent with a reduction in VLDL synthesis (18).

EFFECT OF POLYUNSATURATED FATTY ACIDS ON PLATELETS, VESSEL WALLS AND THROMBOSIS

The first studies examinining the effect of polyunsaturated fatty acids on platelet and vascular function were performed using vegetable oils containing linoleic acid. Galli, et al., observed that rabbits fed corn oil had a reproducible increase in linoleic acid in platelet membranes, whereas platelet arachidonic acid concentration was decreased (19,20). Similar findings were observed by McGregor and associates in experiments performed in rats (21,22). Corn oil has also been administered to human subjects with an increase in platelet linoleate but no change in the platelet content of arachidonic acid (23). *In vitro* data using human endothelial cells as well as platelets has shown that enrichment of the incubation medium with linoleic acid leads to an increase in the cellular content of linoleate by 5-25%, which is associated with a fall in arachidonic levels by 25-40% (24). In sum, in many experimental situations, feeding a diet enriched in linoleic acid leads to a decrease in cellular arachidonic acid, whereas diets that contain saturated fatty acids may have the opposite effect (1).

These changes in platelet composition may also be associated with a decrease in platelet function. For example, sensitivity to arachidonic acid and collagen-induced platelet aggregation was reduced in rabbits fed corn oil (25). In humans, Renaud, et al., reported that platelet aggretation in response to thrombin and ADP was decreased in French farmers eating a low rather than a high saturated fat diet (26). Similarly, decreases in platelet aggregation to collagen were found in Norwegian subjects given a supplement of corn oil (23). Studies examining prostaglandin production by platelets in animals or humans given linoleic acid as vegetable oil have generally shown reduced production of thromboxane B_2 (23,27,28). In spite of these findings however, it is generally agreed that the cutaneous bleeding time is not increased in subjects given a diet containing large amounts of vegetable oil.

Overall, these studies indicate that consumption of diets rich in ω6 polyunsaturated fatty acids by animals or man leads to a decreased arachidonic acid content of platelets, a mild decrease in platelet aggregation and reduced thromboxane A_2 production but that bleeding times are not significantly affected.

In contrast, the effects of dietary ω3 fatty acids in the form of fish oil have shown reproducible effects on bleeding time and platelet function in man. Dyerberg first demonstrated that platelet aggregation was decreased and bleeding times prolonged in Eskimos who were ingesting a high ω3 fatty acid containing diet (29). Several studies have now been reported showing that similar findings in Western man given fish oil supplements or by substituting fish or fish oil for other dietary fats (e.g., 1, 30,31).

In studies from our laboratory, we fed large amounts of salmon and salmon oil for four weeks to 11 normal, unmedicated volunteers who also ingested a control (typical American) diet prior to or after the fish oil diet (30). A three-week washout period was included between each dietary phase. The diets were carefully constructed to be identical in terms of calories, cholesterol, fat, carbohydrate and protein content, with the only difference being the source of fatty acids. Platelet phospholipid fatty acids showed an increase of the ω3 fatty acid eicosapentaenoic acid from 0.1 to 6.0% of the total fatty acids, whereas linoleic and arachidonic acids fell during the fish oil dietary period. Platelet function tests showed that the bleeding time was prolonged in 10 of 11 subjects, platelet retention on glass beads was reduced (89% to 78%: p < 0.001), and platelet aggregation to low doses of ADP was inhibited. Additional studies using another form of fish oil (Max Epa) showed a reduction of platelet thromboxane B_2 production by 60% as measured by radioimmunoassay (45). The reasons for the reduced production of thromboxane A_2 may include the decreased content of arachidonic acid in the platelet membrane phospholipids, competition by eicosapentaenoic acid with arachidonic acid for platelet cyclooxygenase, reduced biologic activity of thromboxane A_3 or, inhibition of platelet cyclooxygenase activity by docosahexaenoic acid (DHA) (32-35).

In contrast to platelets, the effects of fish oil on vascular endothelial cell production of PGI_2 has been challenging to study because endothelial cells are difficult to sample and culture and because much of the previous work using animal vessels may not be relevant to humans (36). However, several studies have shown decreases in arachidonic acid content and increases in eicosapentaenoic and docosahexaenoic acids in vessels of animals fed fish oil (37-39). Rapp, Connor, et al., have shown that $\omega 3$ fatty acid incorporation into atherosclerotic plaques in vessel walls occurs in patients who eventually had surgical resection of these vessels (unpublished observations).

The effect of fish oil diets on the vascular production of PGI_2 has been variable, in part due to the experimental model that was utilized. We and others (38, 40) have shown decreased production of 6 keto $F_{1\alpha}$ from vessel walls of animals, whereas von Schacky, et al., has elegantly shown that urinary metabolites of PGI_2 (PGI_2-M) are not decreased in humans ingesting $\omega 3$ fatty acid supplements (41). Additional studies are needed to determine the effect of fish oil feeding on the prostaglandins produced from specific vessels such as the coronary or cerebral arteries in man.

In sum, fish oil feeding to humans appears to influence platelet and endothelial cell function in a beneficial (i.e., antithrombotic) manner with increases in bleeding time and reductions in platelet reactivity and thromboxane synthesis. Vascular endothelial cell production of prostacyclin probably persists during fish oil feeding. Additional studies will be needed to show that these changes can reduce the incidence of arterial or venous thrombosis in man.

Studies examining the specific antithrombotic properties of fish oil in the diet are sparse and are limited to a small number of studies of experimental coronary or cerebral artery infarction in dogs and cats. In one such study, the volume of cerebral infarction was measured in a series of cats who were given a diet containing fish oil and who underwent surgical ligation of the middle cerebral artery. In the four surviving cats receiving menhaden oil, the volume of infarction was 7% of the brain compared to 19% in 5 control animals (42).

Another study by the same investigators utilized mongrel dogs who were given an electrically-induced coronary artery occlusion over a 24 hour time period. The fish oil treated animals may have had a smaller volume of myocardial infarction as compared to the control animals, although the incidence of early sudden death secondary to ventricular fibrilation was the same in both groups (43).

Hay, et al., have reported platelet survival, beta thromboglobulin and platelet factor 4 measurements in a group of human subjects with known athersclerotic vascular disease who were given dietary supplements of fish oil. Treatment with 3.5 g eicosapentaenoic acid per day led to a prolongation of platelet survival and a decrease in the plasma levels of the platelet specific proteins (44). Whether these changes will ultimately translate into a reduced propensity for vascular thromboembolism is not yet known.

One potentially deleterious effect of fish oil on human platelets which deserves mention is thrombocytopenia. Dyerberg, in his early studies of the Eskimos in Greenland, noted that his subjects' platelet counts were substantially lower than a European control population ($171,000$ vs $232,000/mm^3$) and two of his subjects had platelet counts of $<$ than $100,000$ mm^3 (29). When we performed our studies using salmon and salmon oil, we also noted a fall in the platelet count (i.e., $318,000$ to $209,000/mm^3$) in subjects on the fish oil diet (30). The platelet count of one of our subjects fell to $90,000/mm^3$ over a three week period of time. When the salmon oil feeding was stopped, the platelet count rapidly rose to normal levels. A number of other studies since then using varying kinds of fish oil and in different doses have also noted mild to moderate degrees of thrombocytopenia in their subjects.

However, in general, a positive correlation exists between the total dose of fish oil and the magnitude of thrombocytopenia. Lower doses of oil do not result in a decreased platelet count (45).

One of our subjects was studied in detail with bone marrow examination and [51]Chromium platelet survival platelet studies during fish oil feeding (45). Of interest, bone marrow megakaryocyte morphology and platelet survival remained normal even though the platelet count fell to < 100,000/mm^3. The size of the platelets as measured by the mean platelet volume increased dramatically. These findings would suggest that the fish oil or some component of it is altering the demarcation of platelets from the megakaryocyte cytoplasm resulting in fewer numbers of large platelets. Therefore, the circulating platelet mass remained in the normal range.

Although these findings are of theoretic interest, low doses of fish oil given for prolonged periods of time fail to induce clinically significant thrombocytopenia, so there is probably little reason for concern. However, there still may be the unusual subject who will develop an exaggerated fall in platelet count following fish oil feeding.

In conclusion, the study of the effects of polyunsaturated fatty acids from vegetable oils and more recently fish oils has reached an exciting stage. ω3 fatty acids derived from fish oil have been shown to reduce plasma triglycerides and may reduce plasma cholesterol as well. Further, these fatty acids inhibit platelet function while likely preserving the vascular endothelial production of prostacyclin. However, long term clinical studies in man using fish oil supplements will be required before definitive statements can be made about the ability of dietary fish oils to prevent or inhibit the generation of atherosclerosis or to prevent the formation of venous or arterial thromboembolism.

Acknowledgements

These studies were supported in part by research grants HL25687 and HL07295 from the National Heart, Lung and Blood Institute and a Clinical Research Center grant RR334 from the Division of Research Resources of the National Institutes of Health. Additional grant support was obtained from the Oregon Heart Association and the Medical Research Foundation of Oregon.

References

1. Goodnight SH Jr, Harris WS, Conner WE, et al. Polyunsaturated fatty acids, hyperlipidemia, and thrombosis. *Arteriosclerosis* 1982; 2:87-113.

2. Hegsted DM, McGrandy RB, Myers ML, et al. Quantitative effects of dietary fat on serum cholesterol in man. *Am J Clin Nutr* 1965; 17:281-295.

3. Keys A, Anderson JT, Grande F. Prediction of serum cholesterol responses of man to change in fats in the diet. *Lancet* 1957; 2:959-966.

4. Connor WE. The relationship of hyperlipoproteinemia to atherosclerosis: the decisive role of dietary cholesterol and fat, In: Scanu AM, ed. *The Biochemistry of Atherosclerosis*. New York: Marcel Dekker, Inc., 1979:371-418.

5. Hoak JC, Connor WE, Eckstein JW, et al. Fatty acid-induced thrombosis and death: mechanisms and prevention. *J Lab Clin Med* 1964; 3:791-800.

6. Renaud S. Endotoxin-induced hepatic vein thrombosis in
    hyperlipemic rats. Sequence of events. *Lab Invest* 1965; 14:424-
    430.

7. Hornstra G, Lussenburg RN. Relationship between the type of dietary
    fatty acid and arterial thrombosis tendency in rats.
    Atherosclerosis 1975; 22:499-516.

8. Gjesdal K, Nordøy A, Wang H, et al. Effects of fasting on plasma
    and platelet-free fatty acids and platelet function in healthy
    males. *Thromb Haemost* 1976; 36:325-333.

9. Gjesdal K. Platelet function and plasma free fatty acids during
    acute myocardial infarction and severe angina pectoris. Scan J
    Haematol 1976; 17:205-212.

10. Goodnight SH Jr, Inkeles SB, Kovach NL, et al. Reduced
    prostacyclin survival after fasting induced elevation of plasma free
    fatty acids. *Thromb Haemost* 1985; 54:418-421.

11. Dyerberg J, Bang HO, Stofferson E, et al. Eicosapentaenoic acid
    and prevention of thrombosis and atherosclerosis. *Lancet* 1978;
    2:117-119.

12. Harris WS, Connor WE, McMurry MP. The comparative reductions of
    the plasma lipids and lipoproteins by dietary polyunsaturated fats:
    salmon oil versus vegetable oils. *Metabolism* 1983; 32:179-184.

13. Phillipson BE, Rothrock DW, Connor WE, et al. Reduction of plasma
    lipids, lipoproteins, and apoproteins by dietary fish oils in
    patients with hypertriglyceridemia. *N Engl J Med* 1985; 312:1210-
    1216.

14. Harris WS, Connor WE, Inkeles SB, et al. Dietary omega-3 fatty
    acids prevent carbohydrate-induced hypertriglyceridemia.
    Metabolism 1984; 33:1016-1019.

15. Nestel PJ, Connor WE, Reardon MR, et al. Suppression by diets rich
    in fish oil of very low density lipoportein production in man. *J
    Clin Invest* 1984; 74:82-89.

16. Harris WS, Conner WE, Illingworth DR, et al. The mechanism of the
    hypotriglyceridemic effect of dietary omega-3 fatty acids in man.
    Clin Res 1984; 32:560a. abstract.

17. Illingworth DR, Harris WS, Connor WE. Inhibition of low density
    lipoprotein synthesis by dietary omega-3 fatty acids in humans.
    Arteriosclerosis 1984; 4:270-275.

18. Sigurdsson G, Nicoll A, Lewis B. Conversion of very low density
    lipoprotein to low density lipoprotein: a metabolic study of
    apolipoprotein B kinetics in human subjects. *J Clin Invest* 1975;
    56:1481-1490.

19. Galli C, Agradi E, Socini A, et al. Modulation of prostaglandin
    production in tissues by dietary essential fatty acids. *Acta Med
    Scan* 1980; 642 (Suppl):171-179.

20. Galli C, Agradi E, Petroni A, et al. Differential effects of
    dietary fatty acids on the accumulation of arachidonic acid and its

metabolic conversions through the cyclooxygenase and lipoxygenase in platelets and vascular tissue. *Lipids* 1981; 16:165-172.

21. McGregor L, Morazain R, Renaud S. A comparison of the effects of dietary short and long chain saturated fatty acids on platelet functions, platelet phospholipids, and blood coagulation in rats. *Lab Invest* 1980; 43:438-442.

22. McGregor L, Renaud S. Effect of dietary linoleic acid deficiency on platelet aggregation and phospholipid fatty acid of rats. *Thromb Res* 1978; 12:921-927.

23. Brox JH, Kille JE, Gunnes S, et al. The effect of cod liver oil and corn oil on platelets and vessel wall in man. *Thromb Haemost* 1981; 46:604-611.

24. Spector AA, Kaduce TL, Hoak JC, et al. Utilization of arachidonic acid and linoleic acids by cultured human endothelial cells. *J Clin Invest* 1981; 68:1003-1011.

25. Agradi E, Tremoll E, Colombo C, et al. Influence of short term dietary supplementation of different lipids on aggregation and arachidonic acid metabolism in rabbit platelets. *Prostaglandins* 1978; 16:973-984.

26. Renaud S, Dumont E, Godsey F, et al. Platelet functions in relation to dietary fats in farmers from two regions of France. *Thromb Haemost* 1978; 40:518-531.

27. Galli C, Agradi E, Petroni A, et al. Differential effects of dietary fatty acids on the accumulation of arachidonic acid and its metabolic conversions through the cyclooxygenase and lipoxygenase in platelets and vascular tissue. *Lipids* 1981; 16:165-172.

28. Hornstra G, Chalt A, Karvonen MJ, et al. Influence of dietary fat on platelet function in men. *Lancet* 1973; 1:1155-1157.

29. Dyerberg J, Bang HO. Hemostatic function and platelet polyunsaturated fatty acids in Eskimos. *Lancet* 1979; 2:433-435.

30. Goodnight SH Jr, Harris WS, Connor WE. The effects of dietary $\omega 3$ fatty acids on platelet composition and function in man: a prospective, controlled study. *Blood* 1981; 58:880-885.

31. Siess W, Scherer B, Bohlig B, et al. Platelet-membrane fatty acids, platelet aggregation, and thromboxane formation during a mackeral diet. *Lancet* 1980; 1:441-444.

32. Culp BR, Titus BG, Lands WEM. Inhibition of prostaglandin biosynthesis by eicosapentaenoic acid. *Prostaglandins Med* 1979; 3:269-278.

33. Needleman P, Raz A, Minkes MS, et al. Triene prostaglandins: prostacyclin and thromboxane biosynthesis and unique biological properties. *Proc Natl Acad Sci* 1979; 76:944-948.

34. Fischer S, Schacky CV, Siess W, et al. Uptake, release and metabolism of docosahexaenoic acid (DHA, $C22:\omega 3$) in human platelets and neutrophils. *Biochem Biophys Res Commun* 1984; 120:907-918.

35. Corey EJ, Shih C, Cashman JR. Docosahexaenoic acid is a strong inhibitor of prostaglandin but not leukotriene biosynthesis. *Proc Natl Acad Sci* 1983; 80:3581-3584.

36. Morita I, Takahashi R, Saito Y, et al. Effects of eicosapentaenoic acid on arachidonic acid metabolism in cultured vascular cells and platelets: species difference. *Thromb Res* 1983; 31:211-217.

37. Bruckner GG, Lokesh B, German B, et al. Biosynthesis of prostanoids, tissue fatty acid composition and thrombotic parameters in rats fed diets enriched with docosahexaenoic (22:6n3) or eicosapentaenoic (20:5n3) acids. *Thromb Res* 1984; 34:479-497.

38. Hornstra G, Christ-Hazelhof E, Haddeman E, et al. Fish oil feeding lowers thromboxane and prostacyclin production by rat platelets and aorta and does not result in the formation of prostaglandin I_3. *Prostaglandins* 1981; 21:727-738.

39. Vas Dias FW, Gibney MJ, Taylor TG. The effect of polyunsaturated fatty acids of the n-3 and n-6 series on platelet aggregation and platelet and aortic fatty acid composition in rabbits. *Atherosclerosis* 1982; 43:245-257.

40. Scherhag R, Kramer HJ, Dusing R. Dietary administration of eicosapentaenoic and linoleic acid increases arterial blood pressure and suppresses vascular prostacyclin synthesis in the rat. *Prostaglandins* 1982; 23:369-382.

41. von Schacky C, Fischer S, Weber PC: Long-term effects of dietary marine ω3 fatty acids upon plasma and cellular lipids, platelet function and eicosanoid formation in humans. *J Clin Invest* 1985; 76:1626-1631.

42. Black KL, Culp B, Madison D, et al. The protective effects of dietary fish oil on focal cerebral infarction. *Prostaglandins Med* 1979; 3:257-268.

43. Culp BR, Lands WEM, Lucchesi BR, et al. The effect of dietary supplementation of fish oil on experimental myocardial infarction. *Prostaglandins* 1980; 20:1021-1031.

44. Hay CRM, Durber AP, Raynor R. Effect of fish oil on platelet kinetics in patients with ischaemic heart disease. *Lancet* 1982; 1:1269-1272.

45. Goodnight SH Jr. The antithrombitic effects of fish oil. In: *Health Effects of Polyunsaturated Fatty Acids in Seafoods*. Academic Press (in press).

DIETARY FATTY ACIDS AND THROMBOSIS

Arne Nordøy

Dept. of Medicine, University Hospital
University of Tromsø
9012 Tromsø, Norway

The process of thrombosis is a multifactorial event involving the vessel wall, the cellular elements of the flowing blood and the plasma coagulation proteins. The process is like the tide; building up and breaking down, continuously, repeating itself. The forces contributing to the final occlusive event include both those involved in thrombogenesis and thrombolysis. In man, venous and arterial thrombosis are both, particular in the Western World, main causes of morbidity and mortality (1). In more than 90% of cases with acute myocardial infarction, arterial thrombosis has been localized in the coronary arteries during the early hours after debut of symptoms (2). The processes of atherosclerosis and thrombosis which act together are closely connected, both resulting in narrowing and occlusion of the artery. Even if the connections between these two processes are not fully elucidated, there is good evidence that an arterial thrombus may transfer into an atherosclerotic lesion and furthermore, an atherosclerotic lesion may be the base for a secondary thrombus which then represent the final occlusive event.

The dietary fat consumption show great geographical, cultural, economical and traditional variations throughout the world. This variation in dietary fat intake has been associated with the great geographical variations in the incidence of coronary heart disease. The main families of fatty acids are given in fig. 1.

MAJOR FAMILIES OF FATTY ACIDS (FA)

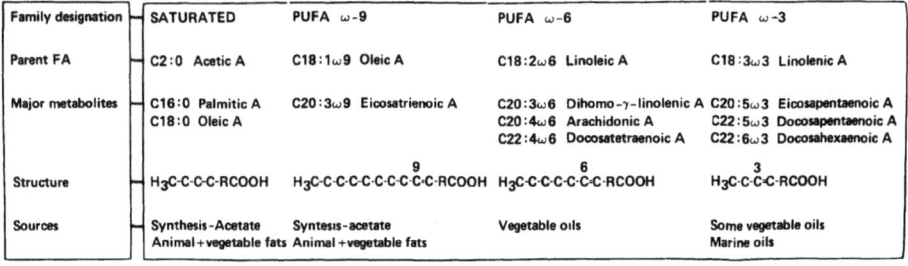

Family designation	SATURATED	PUFA ω-9	PUFA ω-6	PUFA ω-3
Parent FA	C2:0 Acetic A	C18:1ω9 Oleic A	C18:2ω6 Linoleic A	C18:3ω3 Linolenic A
Major metabolites	C16:0 Palmitic A C18:0 Oleic A	C20:3ω9 Eicosatrienoic A	C20:3ω6 Dihomo-γ-linolenic A C20:4ω6 Arachidonic A C22:4ω6 Docosatetraenoic A	C20:5ω3 Eicosapentaenoic A C22:5ω3 Docosapentaenoic A C22:6ω3 Docosahexaenoic A
Structure	$H_3C\text{-}C\text{-}C\text{-}C\text{-}RCOOH$	$H_3C\text{-}C\text{-}C\text{-}C\text{-}C\text{-}C\text{-}C\text{-}C\text{-}C\text{-}C\text{-}C\text{-}RCOOH$ (9)	$H_3C\text{-}C\text{-}C\text{-}C\text{-}C\text{-}C\text{-}C\text{-}RCOOH$ (6)	$H_3C\text{-}C\text{-}C\text{-}RCOOH$ (3)
Sources	Synthesis–Acetate Animal+vegetable fats	Syntesis–acetate Animal+vegetable fats	Vegetable oils	Some vegetable oils Marine oils

Fig. 1. The major families of fatty acids.

Of these, a high intake of saturated fatty acids has been associated with a high total serum cholesterol and a high incidence of CHD, whereas a high intake of polyunsaturated fatty acids (PUFA) from the n-6 and probably even more from the n-3 families has been observed in populations with low serum cholesterol values and a low incidence of coronary heart disease (1,3).

In the following I will try to review the relation between dietary fatty acids and thrombosis as observed in the experimental animal and in man. Possible associations are outlined in fig. 2.

EXPERIMENTAL THROMBOSIS

Since the late 1950's a series of different models of experimental thrombosis have basically confirmed that a diet rich in saturated fats increases the tendency to thrombosis (4). This increased thrombotic tendency seems to be present in all types of animals used and includes both venous and arterial thrombosis. Furthermore, both thrombosis occuring spontaneously and thrombosis occuring when a weak thrombogenic stimulus are added to dietary manipulations show the same relation to saturated fats. When the saturated fat intake increased from 10 to 40% of the total calories an increased incidence of thrombosis has been reported (5). Furthermore, when (n-6) PUFA, particularly linoleic acid (18:2 n-6) replaced a certain amount of the saturated fat, there was a tendency to reversal of the high trombotic tendency (5). However, the effects of n-6 fatty acids in this regard show conflicting results. Early studies (6) indicated that addition of linoleate to diets rich in saturated fats did not inhibit the occurrence of spontaneous thrombosis. However, most reports have reported a beneficial effect on thrombosis tendency when (n-6) PUFA have replaced saturated dietary fats (4).

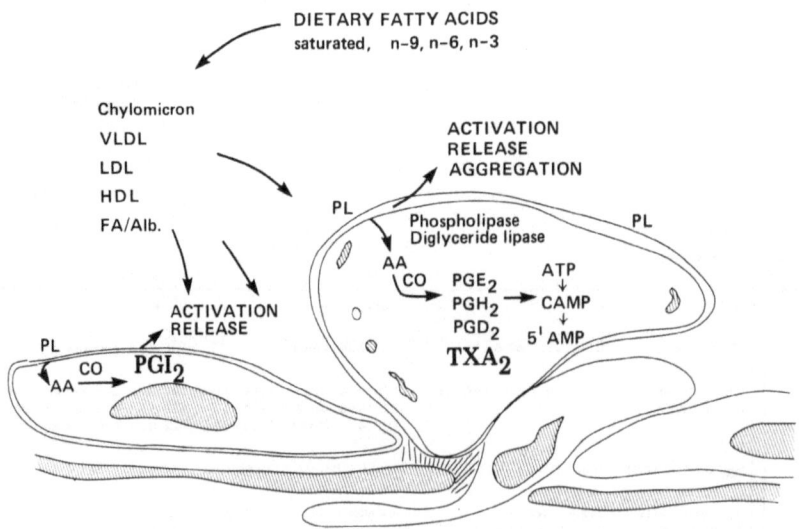

Fig. 2. Possible associations between dietary lipids, plasma lipids, platelets and the vessel wall which may be of significance for the process of thrombosis.

In a series of studies in rats, platelet thrombosis was induced in the pulmonary circulation by intravenous injection of ADP (5). The extent of thrombosis was correlated to the concentration of ADP. When a small dose of ADP (1mg/100 g.bwt.) was injected into rats fed a diet containing 40% fat the incidence of thrombosis was markedly increased compared to animals on a low fat diet (5).

Supplementing the diet with oils (8% of total lipids), rich in linoleic or linolenic acids reduced this increased tendency. When higher concentrations of ADP was used (5 mg/100 g.bwt.) only oils rich in linolenic acid (18:3 n-3) was able to counteract the effect of the saturated fat diet. By a further increase of the thrombotic stimulus (ADP-17 mg/100 g.bwt.) neither n-6 nor n-3 fatty acid supplement were able to counteract the combined thrombotic effects of ADP and saturated fats. We did not exclude, however, that a further increase in the n-6 or n-3 PUFA content alone or combined, of the diet could have accomplished such an effect. A similar tendency with regard to antithrombogenic effect of 18:3 (n-3) was observed in another experimental model in rats where the jugular veins were chemically damaged and the incidence of thrombosis was examined 24 hours later (7). In retrospective, it is of considerable interest to recognize that in those animals given supplementary 18:3 n-3 a significant increase of the very long chain PUFA of the n-3 family were found in platelet phospholipids with a concomitant reduction of the n-6/n-3 ratio (8). Recent dietary studies in rats have confirmed that increasing dietary linolenate reduces the synthesis of dienoic prostaglandins (9).

In a series of studies by Hornstra (4) and Renaud et al (10) using other experimental models similar results have been reported. The studies by Hornstra using rats and the filter loop technique showed a nearly linear correlation between the dietary linoleic acid content and the occlusion time, that is the time it takes for the loop to be occluded by a thrombus (4). Diets rich in cod liver oil with a high content of n-3 fatty acids was even more efficient than oils rich in linolenic acid as an antithrombotic substance (11). Recent studies in cats and dogs have shown that dietary supplement of fish oils in both these species reduced the infarction areas after occlusion of cerebral and coronary arteries, respectively (12,13). In another experimental model cod liver oil could prevent intimal hyperplasia in autogenous vein grafts used for arterial bypass, also in dogs (14). These studies indicate that supplement of dietary n-3 fatty acids by their effect on platelet and endothelial cell function is able to inhibit thrombosis and reduce the consequences of platelet activation on the vessel wall.

HUMAN THROMBOSIS

Comparative studies investigating the mortality pattern in different countries have repeatedly confirmed that a high dietary intake of saturated fats is positively correlated to the incidence of CHD (15). During the Second World War a dramatic reduction both in myocardial infarction and in postsurgical venous thrombosis in Norway was associated with a reduced intake of animal fats (16,17). During these five years fish and fish products substituted for meat and meat products. The 7-country study confirmed a close association between intake of saturated fat and incidence of CHD. This pattern is still existent in 1980, even if we have registered a reduced incidence of CHD in many of the western countries (18)(Fig. 2). On the basic of these epidemiological studies it should be repeated and concluded that a high intake of dietary saturated fatty acids represent a risk factor for development of arterial thrombosis.

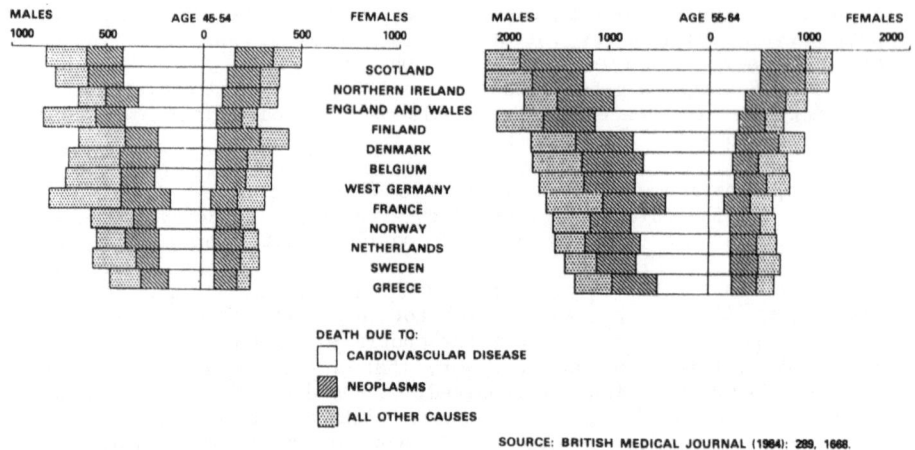

SOURCE: BRITISH MEDICAL JOURNAL (1984): 289, 1668.

Fig. 3. European mortality rates per 100000 for 1980.

Investigating the impact of dietary polyunsaturated fatty acids, it has been established that when n-6 PUFA replace saturated fatty acids and the p/s ratio in the diet is increased, the incidence of myocardial infarction is reduced (19,20). From the fascinating studies of Dyerberg et al (21) it is obvious that among Greenland Eskimoes there is an extremely low incidence of CHD (3) associated with a high intake of fat particularly rich in n-3 fatty acids coming from marine products. A similar tendency has been reported from comparative studies in Japan (22,23) where the incidence of coronary heart disease among fishermen with a daily intake of fish of about 250 g per day is extremely low compared with farmers having much lower intake of fish products, about 90 g per day (23). These studies have associated the effect of fish intake on coronary heart disease by changes in platelet thromboxane synthesis with reduced platelet aggregation and eventually prolonged primary bleeding time. However, it has been difficult to establish good correlations between the intake of eicosapentaenoic acid and the effects on platelet function. Furthermore, the recent report (24) from the Zutphen study, an ongoing longitudinal investigation of the relations of diet and other risk factors to chronic diseases, seems to establish a significant lower mortality from coronary heart disease among men who consume at least 30 g of fish per day than in men who did not eat fish. It should be stressed that men with a "high" (= 45 per day) fish consumption also had a significantly higher intake of monounsaturated and polyunsaturated fatty acids than the group not eating fish. Furthermore, the fish consumption of the men in the Dutch study consisted of 2/3 lean fish and only 1/3 fatty fish. The lean fish intake was also inversely related to mortality from CHD. The inverse relation between lean fish and coronary heart disease can not be explained by the intake of eicosapentaenoic acid. A review on the present studies relating fish intake to coronary heart disease is given in Table 1.

At present we do not know the mechanisms involved in the relation between fish intake and prevention of CHD. The epidemiological studies so far indicate that the effects of fish at least, only partly, may be explained by the intake of eicosapentaenoic fatty acids (20:5 n-3). More likely, the balance between saturated, monounsaturated and PUFA of the n-6 and n-3 families may be significant. Furthermore, is has not been excluded that other components (proteins) may be of significance.

In a recent not finished study comparing dietary habits and platelet function in two population groups living along the coastline of Norway and living up in the mountain areas, we have indications that even if the intake of fish is higher along the coastline, the platelet parameters probably reflecting an antithrombotic activity is superior among people living in the mountains. During the coming year we hopefully may explain these differences.

Table 1. The relation between fish intake and coronary heart disease.

POPULATION	FISH INTAKE (g/day)	C H D	REFERENCE
ESKIMOS	400*	Low	(21,25)
JAPANESE FISHERMEN	250	Low	(22)
JAPANESE FARMERS	100	Moderate	(22)
JAPANESE (HAWAI)	170	Unchanged	(26)
DUTCH (Zutphen)	>30	Low	(24)
AMERICANS (CHICAGO)	>35	Low	(27)
NORWEGIANS	Ca 100	Unchanged	(28)

*Including seal & whale

CONCLUSION

Avoiding parameters in the blood that may or may not be associated with an increased thrombotic tendency, at present, experimental and clinical studies exploring the direct effects of dietary lipid composition on the occurence of venous and arterial thrombosis, all indicate a positive correlation between a high saturated fat intake and thrombosis. Furthermore, when polyunsaturated fatty acids of the n-6 and n-3 families replaces the saturated fatty acids in the diet, the thrombotic tendency is reduced. There is indications that fish and fish products, rich in very long chain n-3 fatty acids is of particular significance as antithrombotic agents, however, further studies are needed to explore if other constituents of fish are significant and how the balance between the various dietary fatty acids should be in an ideal diet.

REFERENCES

1. Working group on arteriosclerosis of the National Heart, Lung and Blood Institute. Decline in coronary heart disease mortality, 1968-1978. Vol. 2.Bethesda Md. NIH. 157 (1981) (DHHs publ no) (NIH) 2035 (1982).
2. M.A. DeWood, J. Spores, R, Notske, L.T. Mouser, R, Buroughes, M.S. Golden, and H.T. Lang, Prevalence of total coronary occlusion during the early hours of transmural myocardial infarction. N. Engl.J.Med. 303:897 (1980).
3. N. Kromann, and A. Green, Epidemiological studies in the Upernavik District, Greenland. Acta Med. Scand. 208:401 (1980)
4. G. Hornstra, Dietary fats prostanoids and arterial thrombosis. Develop. Immunol. Hematol. J. 4:1 (1982).
5. A. Nordøy, Lipids and thrombogenesis. Annal. Clin. Res. 13:40 (1981)
6. A. N. Howard, and G.A. Gresham, The dietary induction of thrombosis and atherosclerosis. J. Atheroscl. Res. 4:40 (1964).

7. A. Nordøy, The influence of saturated fats, cholesterol, corn oil and linseed oil on experimental venous thrombosis in rats. Thrombos. Diathes. Haemorrh. 13:244 (1965).

8. A. Nordøy, J.T. Hamlin, A.B. Chandler, adn H. Newland, The influence on dietary fats on plasma and platelet lipids and ADP-induced platelet thrombosis in the rat. Scand. J. Haematol. 5:458 (1968).

9. D.A. Hwang, and A.E. Carrol, Decreased formation of prostaglandins derived from arachidonic acid by dietary linolenate in rats. Am. J. Nutr. 33:590 (1980).

10. S. Renaud, and J. Gautheron, Dietary fats and experimental (cardiac and venous) thrombosis. Haemostasis. 2:53 (1973/74).

11. G. Hornstra, E. Hadderman, and F. tenHoor, Fish oils, prostaglandins and arterial thrombosis. Lancet ii. 1080 (1979).

12. B.R. Culp, W.E.M. Lands, B.R. Lucahesi, B. Pitt, and J. Ronson, The effect of dietary supplementation of fish oil on experimental myocardial infarction. Prostaglandins 20:1021 (1980).

13. K.L. Black, B.R. Culp, D. Madison, O.S. Randall, and W.E.M. Lands, The protective effect of dietary fish oil on focal cerebral infarction. Prostaglandins Med. 5: (1979).

14. R.W. Landymore, C.E. Kunley, J.H. Cooper, M. MacAulay, B. Sheridan, and C. Cameron, Cod liver oil in the prevention of intimal hyperplasia in autogenous vein grafts used for arterial bypass. J. Thorac. Cardiovasc. Surg. 89:351 (1985).

15. A. Keys (ed), Coronary heart disease in seven countries. Circulation 41:1 (1970).

16. R.A. Jensen, Postoperative thrombosis - emboli: The frequency in the period 1940 to 1948. Acta Med. Scand. 103:263 (1952).

17. A. Strøm, and E.A. Jensen, Mortality from circulatory diseases in Norway 1940-1950. Lancet i:126 (1951).

18. J.C., Catford, and S. Ford, On the state of the public ill health: Premature mortality. Brit. Med. J. 289:1668 (1984).

19. P. Leren, The Oslo diet heart study. Circulation 42:935 (1970).

20. I. Hjerman, K.V. Byre, I. Holme and P. Leren, Effect of diet and smoking intervention on the incidence of coronary heart disease. Lancet ii:1303 (1981).

21. J. Dyerberg, H.O. Bang, E. Stoffersen, S. Moncada, and J.R. Vane, Eicosapentaenoic acid and prevention of thrombosis and atherosclerosis. Lancet ii:117 (1978).

22. Y. Kagawa, M. Nishizawa, M. Suzuki, T. Miyatake, T. Hamato, K. Goto, E. Motonaga, H. Izumikawa, H. Hirata and A. Ebihara, Eicosapolyenoic acid of serum lipids of Japanese islanders with low incidence of cardio-vascular diseases. J. Nutr. Sci, Vitaminol. 28:441 (1982).

23. G. Tamura, Hiraia, T. Terano, H. Saitoh, H. Tahara, A. Kumagai and S. Yoshida, Platelet function, vascular reactivity and blood viscosity in the Japanese. 2. Inter. Congr. Ess. Fatty Acids, Prostagl. Leukotrienes (Abstr.) (London) (1985).

24. D. Kromhout, E.B. Bosschicter, and C. de L. Coulader, The inverse relation between fish consumption and 20-year mortality from coronary heart disease. N. Engl. J. Med. 312:1205 (1985).

25. H.O. Bang, J. Dyerberg, and H.M. Sinclair, The composition of the Eskimo food in north western Greenland. Amer. J. Clin, Nutr. 33:2657 (1980).

26. J.D. Curb, and D.M. Reed, Fish consumption and mortality from coronary heart disease, N. Engl. J. Med. 313:821 (1985).

27. R.B. Shekelle, L.W. Misell, O. Paul, A.M. Shryock, and J. Stamler, Fish comsumption and mortality from coronary heart disease, N. Engl. J. Med. 313:820 (1985).

28. S.E. Vollset, J. Heuch, and E. Bjelke, Fish consumption and mortality from coronary heart dieseuse, N. Engl. J. Med. 313: 820 (1985).

UNEXPECTED EFFECTS OF DIETARY PALM OIL ON ARTERIAL THROMBOSIS (RAT)

AND ATHEROSCLEROSIS (RABBIT)   COMPARISON WITH OTHER VEGTABLE OILS

AND FISH OIL

G. Hornstra[1], A.A.H.M. Hennissen[2], D.T.S. Tan[3]
and R. Kalafusz[1]

Departments of Biochemistry (1) and Human Biology (2)
Limburg University, Maastricht, The Netherlands
(3) Palm Oil Research Institute, Kuala Lumpur, Malaysia

ABSTRACT

Dietary long chain saturated fatty acids promote arterial thrombosis
in rats. Palm oil (PO), however, containing 50% saturated fatty acids
(mainly palmitic acid), does not follow this general rule. As compared
with a control diet containing 5 energy % (en%) sunflowerseed oil
(SO), a diet containing 50 en% physically refined palm oil (PO-R)
inhibits the thrombotic obstruction of a loop -shaped cannula, inserted
into the lower aorta of male rats. Alkali-refined PO appeared somewhat
more active than PO-R. Rabbits , fed a 32 en% PO-containing diet
for more than 18 months, did not develope more atherosclerosis than
animals fed diets containing various unsaturated oils (fish oil, linseed
oil, olive oil, and sunflowerseed oil). These effects of PO are associated
with alterations in the formation of pro- and antithrombotic prostanoids
and in some functions of blood platelets. PO-feeding hardly affected
blood lipid levels in rabbits.

INTRODUCTION

For a long time already dietary lipids are known to affect experimental atherosclerosis in various animal models (1-6). Moreover,
epidemiological studies clearly indicate that in man the type of dietary
fat has a distinct influence on coronary artery disease, which is
one of the clinical manifestations of atherosclerosis (7-11).

The available data indicate that, in general, long chain saturated
fatty acids promote atherogenesis, whereas polyunsaturated fatty acids
present in vegetable- and fish oils, inhibit the formation of the
atherosclerotic lesion. These dietary fat effects have been thought
to be associated with their influence on cholesterol metabolism. However,
it is known now that also arterial thrombosis is implicated in formation
and complications of atherosclerosis. Consequently, dietary lipids

may influence atherosclerosis via the thrombotic pathway as well. We, therefore, are investigating the effect of type and amount of dietary fats on arterial thrombosis tendency, experimental atherosclerosis and some related parameters, e.g. platelet function, prostanoid metabolism and blood lipids.

EFFECT OF DIETARY LIPIDS ON ARTERIAL THROMBOSIS IN RATS

Experimental Model

Studies were carried out using a newly developed method for the measurements of the arterial thrombosis tendency (12). The method is based on the insertion of a loop-shaped polyethylene cannula into the abdominal aorta of male rats. At the cannula tips, vascular damage and bloodflow disturbances cause the formation of a platelet-rich, fibrin-poor, mural thrombus which, after about 4 days, obstructs the lumen of the vessel completely. Since the loop is made of translucent material and partly protrudes from the body, the moment of obstruction can be easily determined as it is accompanied by a change in colour of the blood from light red (loop open) to dark blue (loop obstructed).

The time elapsing between insertion of the loop and its complete obstruction is called the obstruction time (OT) and is longer, the lower the arterial thrombosis tendency.

Compounds influencing platelet aggregation and blood coagulation, affect the arterial thrombosis tendency as measured with this technique in a predictable manner. Moreover, known risk factors for atherosclerosis and thrombosis were shown to be associated with shortened obstruction times. Therefore, this so-called 'aorta-loop-technique' provides a useful tool for measuring arterial thrombosis tendency in rats (13).

Relation between Dietary Fat Type and Arterial Thrombosis Tendency

To investigate the dietary fat effect on arterial thrombosis tendency, five week old male Wistar rats were fed adequate diets, the carbohydrate moiety of which was partly replaced by various amounts of different oils and fats. Aorta loops were inserted after feeding for 8-10 weeks. The loops were checked for obturation twice daily. OT's were calculated in hours and because of skew distribution, transformed to their logarithmic values.

These studies, which have been described in detail elsewhere (14), demonstrated that :
- oils rich in polyunsaturated fatty acids of the (n-6) or (n-3) families and poor in saturated fatty acids, have an antithrombotic effect
- oleic acid is neutral as to arterial thrombus formation. It is antithrombotic only if it replaces prothrombotic dietary lipids
- long chain ($\geqslant$ 14 carbon atoms) saturated fatty acids promote arterial thrombus formation.

This latter conclusion is based on experiments with 12 different oils and fats. Using a computerized multiple regression analysis it appeared that OT is best described by the sum of half the dietary amount of myristic acid (14:0) plus all longer chain saturated fatty acids present in the diets. When this parameter is plotted against log OT (Fig.1) a perfect fit is obtained, except for medium chain triglycerides (MCT) and palm oil (PO).

It should be noted that almost all the fatty acids of MCT (mainly 8:0 and 10:0) are metabolized quite differently from the fatty acids with longer carbon chains. Absorption and metabolism of 8:0 and 10:0 very much resemble those of carbohydrates, so that MCT feeding can be considered a low fat feeding which is known to induce a short OT. This may explain the anomalous finding with MCT. However, we could not find any explanation for the exceptional behaviour of palm oil; which we confirmed in a further experiment (not shown).

These experiments were carried out almost 15 years ago, but in two recent studies we were able to confirm the antithrombotic effect of palm oil.

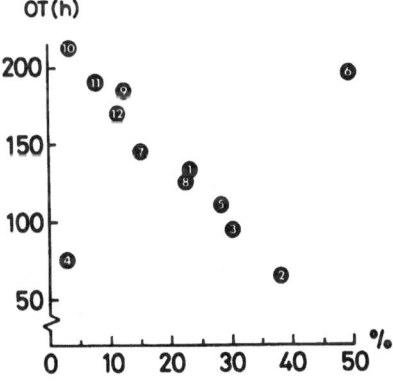

Fig. 1 Relationship between saturated fatty acid content and thrombo-genicity (OT) of dietary fats in rats. For full details: see ref. 14. All diets contained 50 en% fat.

1. Coconut oil
2. Triglyceride mixture
3. Hydrogenated coconut oil
4. Medium chain triglycerides
5. Whale oil
6. Palm oil
7. Olive oil
8. Hydrogenated soyabean oil
9. Linseed oil
10. Rapeseed oil, old
11. Rapeseed oil, new
12. Sunflowerseed oil.

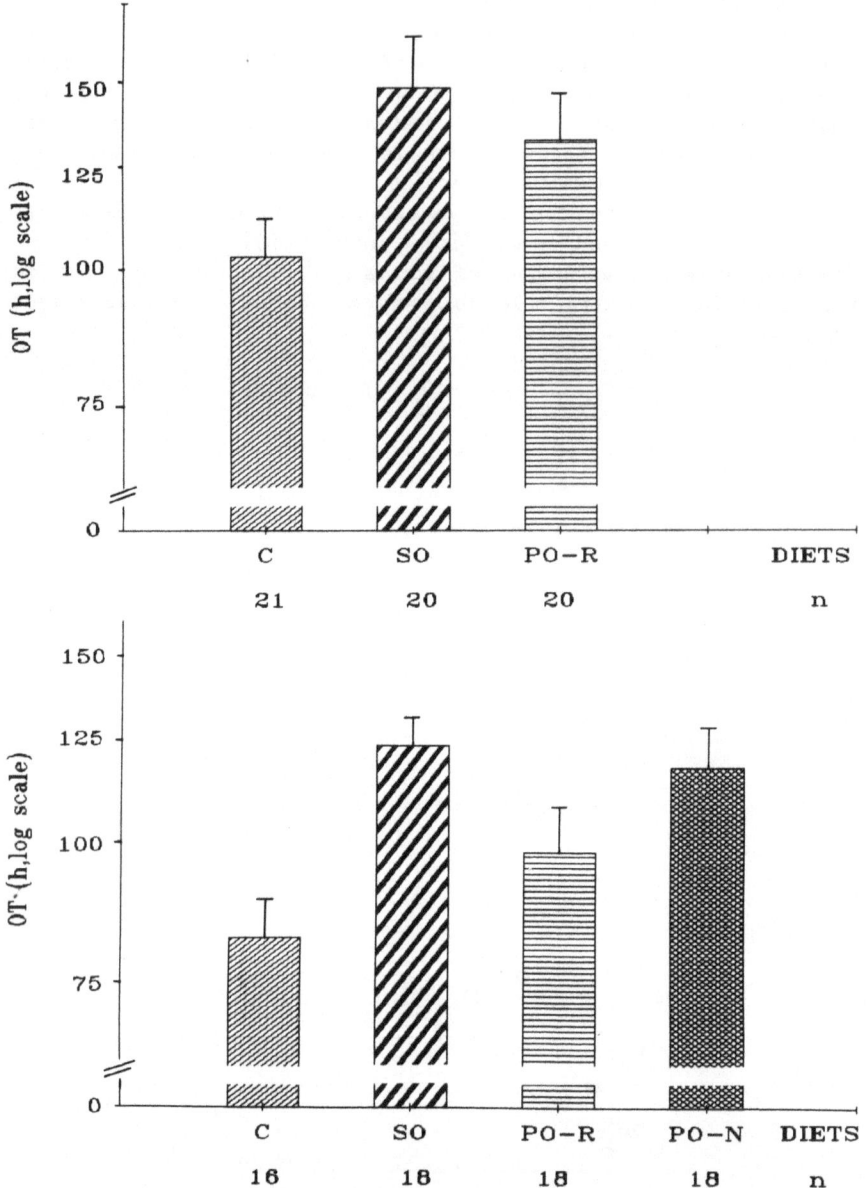

Fig.2. Effect of dietary palm oil on arterial thrombosis tendency in rats
(Obstruction time. h, log scale. mean + s.e.m.)
C   : negative control group (5 en% sunflowerseed oil)
SO   : positive control group (50 en% sunflowerseed oil)
PO-R   : 50 en% physically-refined palm oil
PO-N   : 50 en% alkali-refined palm oil
a. (upper panel) : First experiment
b. (lower panel) : Second experiment .

Recent studies on the antithrombotic effect of dietary palm oil

In the first of these studies, rats were fed a diet containing 50% of its digestible energy (en%) as physically refined palm oil (PO-R).

Two other groups of rats were given diets containing either 5 (negative control) or 50 (positive control en% sunflowerseed oil. After 8-10 weeks of feeding, arterial thrombosis tendency was measured using the aorta-loop-technique. As shown in Fig. 2a, the average obstruction time in the PO-R group is longer than in the negative control group (C). This difference is statistically significant. The OT in the PO-R group does not differ significantly from that in the positive control group (SO), which is also significantly longer than in the negative control group. This demonstrates that palm oil, notwithstanding its high content of saturated fatty acids, has an antithrombotic effect which is comparable to that of the highly unsaturated sunflowerseed oil.

In a second study, this antithrombotic effect of physically refined palm oil was compared to that of alkali-refined palm oil (PO-N). Again, palm-oil fed animals had longer obstruction times than the negative control group, which confirms the antithrombotic effect of palm oil. Detailed statistical analysis demonstrated that PO-N was significantly more antithrombotic than PO-R (Fig. 2b).

MECHANISM OF THE ANTI-THROMBOTIC EFFECT OF PALM OIL

The formation of an arterial thrombus is the result of at least two processes: the activation and aggregation of blood platelets and the process of blood clotting. Both processes are initiated by damaged vascular tissue and each of the two processes is promoted by the other. Blood coagulation, as measured by prothrombin- and activated partial thromboplastin times, is hardly affected by the dietary fat type (15). Therefore, our recent studies were concentrated on two important thrombotic functions of blood platelets: platelet aggregation and the platelet release reaction.

Platelet aggregation

When blood platelets become activated, for instance as a result of contact with a damaged blood vessel, they start to adhere to one another. This process is called platelet aggregation and is instrumental in the formation of an arterial thrombus (16). In our studies the aggregation of blood platelets was measured in blood taken from the animals while inserting the aorta loop for arterial thrombosis measurements. The blood samples were activated with collagen, thereby mimicking a damaged blood vessel, and platelet aggregation as a function of time was determined by following the changes in the electrical impedance of the blood sample (17), using a Chronolog whole blood lumi-aggregometer, model 500.

In both studies platelet aggregation was increased in the positive control group (SO) which, in fact, is in contrast to the diminished thrombosis tendency observed for this group. This anomaly has been observed before (18) using a different in vitro technique for measuring platelet aggregation, and is still unexplained.

Palm oil feeding does not give rise to an enhanced platelet aggregation reaction; in the second study it even tends to reduce it (Fig.3a). In both studies the aggregation response of platelets from palm-oil fed animals was significantly lower than that of platelets from the positive control group.

Platelet release reaction

Upon activation and following aggregation, blood platelets start to release the contents of their storage granules. Some of the released substances cause other blood platelets to aggregate; therefore the platelet release reaction strongly contributes to the formation of an arterial thrombus (16).

Since the platelet release reaction is accompanied by the liberation of adenosine triphosphate (ATP) from platelet granules, the ATP 'production' of activated platelets is a reliable marker of the platelet release reaction.

We determined the ATP release of platelets during the measurement of the collagen-induced aggregation in the second study, using the Chronolog whole blood lumi-aggregometer (17). As illustrated in fig.3b, platelets of palm oil fed animals produce less ATP upon activation with collagen than those obtained from animal of both control groups. The difference with the SO-group is statistically significant. This strongly indicates that platelets from palm oil fed animals are less thrombotic than those from the negative and positive control groups.

EFFECT OF PALM OIL ON EXPERIMENTAL ATHEROSCLEROSIS IN RABBITS

Since platelet aggregation and arterial thrombosis are implicated in atherogenesis, it is of extreme importance to investigate whether the antithrombotic effect of palm oil is also associated with an effect on atherosclerosis.

To find that out, we fed rabbits with cholesterol-free diets, containing 32 en% of palm oil, sunflowerseed oil, olive oil, linseed oil and fish oil. All diets contained, moreover, 8 en% sunflowerseed oil to prevent possible essential fatty acid deficiency. After 18 months of feeding, aortas were removed from the animals and after staining the atherosclerotic plaques with a lipophylic dye, the surface area of the thoracic aorta covered by the plaques was measured morphometrically.

This study, carried out in cooperation with Unilever Research, Vlaardingen,The Netherlands, demonstrated that palm oil, together with sunflowerseed oil, caused the lowest degree of atherosclerosis: aortas of palm-oil fed animals not only contained fewer lesions than those of the other groups (Fig. 4a) but the relative plaque surface was also smaller (Fig. 4b). It should be added that, due to the use of soyabean protein, the average degree of atherosclerosis was rather low in all groups.Moreover, the individual variability was very wide, as a result of which the differences did not reach statistical significance.However, it is clear that, although PO contains a large amount of saturated fatty acids, this is not associated with the expected (19) promoting effect on experimental atherosclerosis.

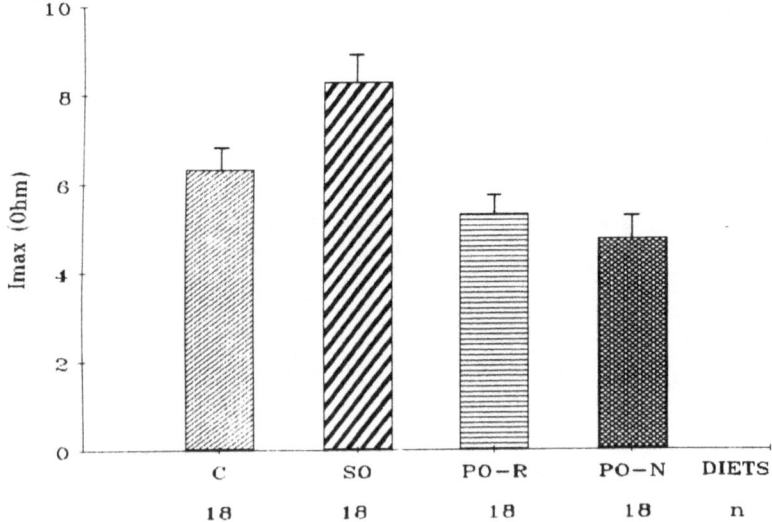

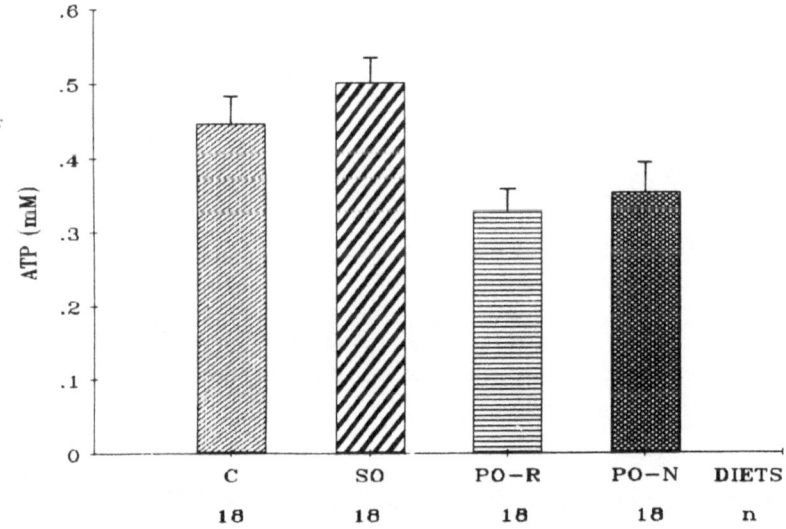

Fig.3. Effect of dietary palm oil on platelet aggregation and ATP-release
upon activation of whole blood with collagen.
a. (upper panel) platelet aggregation, Imax, mean + s.e.m.
b. (lower panel) ATP release, mM, mean + s.e.m.
For explanation of group code : see Fig.2 .

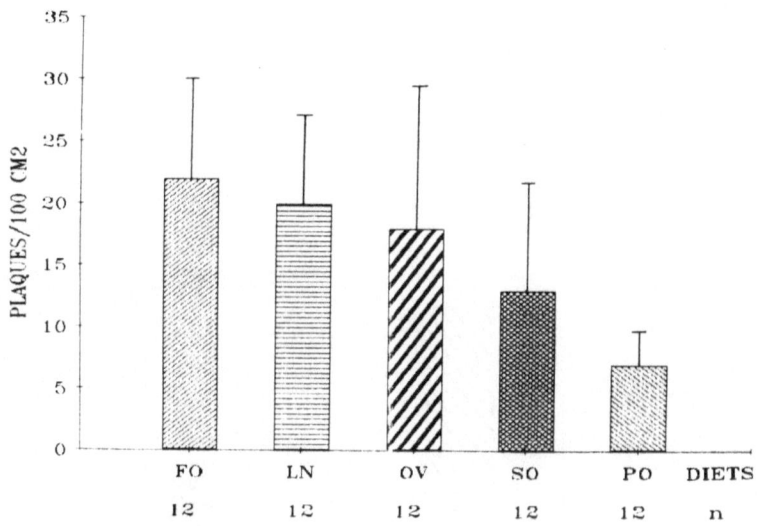

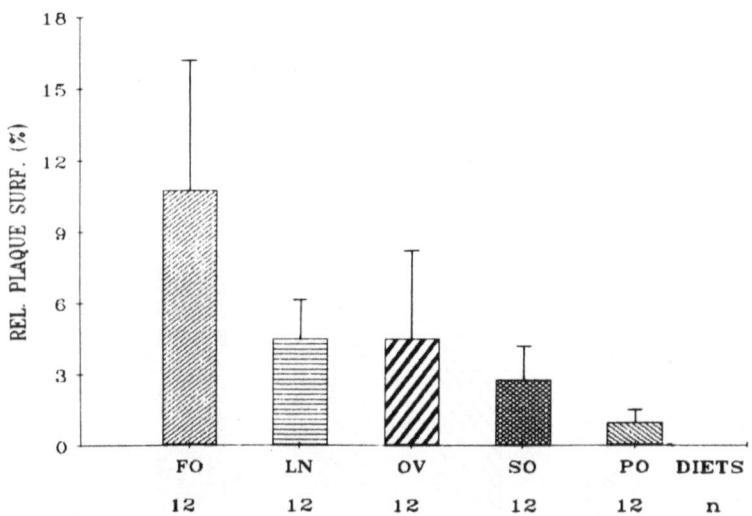

Fig.4. Effect of dietary palm oil on experimental atherosclerosis in rabbits
    FO : Fish oil   SO : Sunflowerseed oil
    LN : Linseed oil   PO : Palm oil
    OV : Olive oil
    a. (upper panel) : Number of plaques per 100 cm^2 aortic surface,
                       mean + s.e.m.
    b. (lower panel) : Relative plaque surface (%. mean + s.e.m.).

EFFECT OF PALM OIL ON SOME FACTORS CONTRIBUTING TO ATHEROSCLEROSIS

Blood lipid- and lipoprotein levels have been shown to be powerful risk indicators for atherosclerosis and for ischaemic heart disease in particular (21,21). Recent studies even indicate the cholesterol content of the blood to be a causal risk factor (22).

Increasing evidence also supports the concept that certain functions of blood platelets are implicated in atherogenesis (For a review: see ref. 23). Therefore, we measured blood lipids, lipoproteins and platelet aggregation in the rabbit atherosclerosis study.

Blood lipids and lipoproteins

After 1 year of feeding and at the end of the feeding period, blood was collected for lipid- and lipoprotein studies (Table 1). For triglyceride and VLDL (very low density lipoproteins) contents, no significant differences were observed between the groups. Total serum cholesterol content was reduced in the SO group; differences between the other groups were not statistically significant. Free cholesterol was highest, and esterified cholesterol was lowest in the fish-oil group. As compared with the fish oil- and the linseed oil groups, the LDL (low density lipoprotein) content of the serum was significantly reduced in the PO, SO and olive oil groups.

These results clearly demonstrate that palm oil, although it contains about 50% long chain saturated fatty acids, does not cause a rise in blood lipid levels upon prolonged feeding. Comparable results were described by Baudet et al.(24) for human volunteers. After feeding palm oil for a period of 5 months, they, too, found a low LDL content of the serum, and they, moreover, demonstrated that the ratio between HDL (high density lipoprotein)- and LDL-cholesterol was highest upon palm oil feeding (0.72) as compared with sunflowerseed oil (0.58), peanut oil (0.26) and milk fat (0.29).

A shorter feeding period has been associated with a cholesterol-increasing effect of palm oil in human volunteers (25). Therefore, the effect of the dietary fat type on blood lipid levels and their ratio's may be determined by additional factors than just the degree of (un)saturation of the dietary lipid,e.g. feeding period, treatment of the lipids before their usage, amount and composition of their non-triglyceride part, etc.. This aspect needs further clarification.

Platelet aggregation

Platelet aggregation in citrated platelet rich plasma was measured at the end of the feeding period (Fig.5). No significant differences between the groups were observed for aggregation induced by adenosine diphosphate, collagen and PAF-Acether. Thrombin-induced aggregation was increased upon fish-oilfeeding and was significantly reduced in the linseed oil group. Palm oil feeding did not cause significant effects on platelet aggregation measured in vitro.

Table 1. Effect of various dietary oils on serum lipid and lipoprotein contents of rabbits after 1 (n= 14-17) and 1.5 (n= 12) years of feeding

Parameter[2] (unit)	Period (years)	FO	LN	OV	SO	PO
Triglycerides (mM)	1	0.36 ± 0.047	0.30 ± 0.022	0.30 ± 0.024	0.33 ± 0.022	0.37 ± 0.040
Triglycerides (mM)	1.5	0.56 ± 0.070	0.47 ± 0.025	0.45 ± 0.029	0.45 ± 0.037	0.46 ± 0.031
VLDL (AU)	1	9 ± 0.6	9 ± 1.2	11 ± 1.3	9 ± 0.9	11 ± 0.6
Total Chol. (mM)	1	1.59 ± 0.199	1.64 ± 0.138	1.70 ± 0.099	1.44 ± 0.150	1.90 ± 0.170[3]
Total Chol. (mM)	1.5	1.64 ± 0.186	1.84 ± 0.147	1.85 ± 0.165	1.45 ± 0.197	2.13 ± 0.199
Chol. ester (mM)	1.5	1.03 ± 0.128	1.35 ± 0.114	1.39 ± 0.126	1.11 ± 0.156	1.59 ± 0.147
Free Chol. (mM)	1.5	0.61 ± 0.064	0.48 ± 0.039	0.46 ± 0.046	0.35 ± 0.041	0.54 ± 0.060
LDL (AU)	1	76 ± 16.6	77 ± 11.3	38 ± 4.9	36 ± 8.0	43 ± 7.6

1) For explanation of group code : see Fig. 4
2) AU = arbitrary units, measured by a turbidimetric technique
3) After omission of one significant outlyer

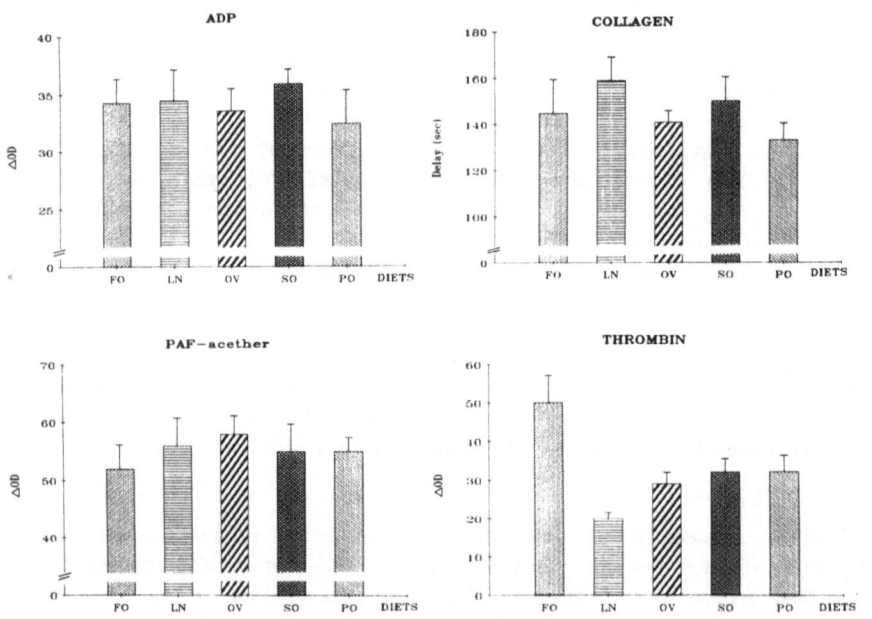

Fig.5. Effect of feeding various oils to rabbits on platelet aggregation
in citrated PRP (n= 6).
Upper left : ADP, final concentration (f.c.) 6.0 μg/ml
Upper right : Collagen, f.c. 12.5 μl/ml
Lower left : PAF-Acether, f.c. 1.0 ng/ml
Lower right : Thrombin, f.c. 0.5 U/ml
For explanation of group code : see Fig. 4 .

PROSTANOIDS, THROMBOSIS AND ATHEROSCLEROSIS

Prostaglandins and thromboxanes (prostanoids) are a series of modulating substances, derived from certain polyunsaturated fatty acids. Prostanoids serve a great number of regulatory functions in the body; they are formed enzymatically in almost every tissue and because of their extreme potency, in combination with a short half-life, they are considered local hormones.

It is now generally agreed that prostanoids formed by blood platelets and blood vessels, play an important role in haemostasis, thrombosis and related phenomena. Activated blood platelets produce ThromboxaneA$_2$ (TxA$_2$) which promotes platelet aggregation and, thereby, has a pro-thrombotic effect. Blood vessels produce prostacyclin (prostaglandin I$_2$, PGI$_2$) which de-activates blood platelets and breaks up platelet aggregates. Due to these contrasting activities it is thought that the balance between pro- and anti-thrombotic prostanoids is an important determinant of arterial thrombosis tendency (26). Earlier studies demonstrated that, indeed, arterial thrombosis tendency is higher, the higher the potency of blood platelets to produce prothrombotic TxA$_2$ (27). We, therefore, measured TxA$_2$ and PGI$_2$ production in the blood samples of rats, used before for aggregation measurements, applying a radioimmunoassay (RIA) for TxB$_2$, the stable hydration product of TxA$_2$, and a RIA for 6-keto-PGF$_{1\alpha}$,the major metabolite of PGI$_2$. Feeding of palm oil caused a significant reduction in the formation of prothrombotic TxA$_2$ upon activation of platelets with collagen. The PGI$_2$ production was no significantly different between the groups. Therefore, as compared with the SO and the low-fat control group, PO feeding significantly reduced the Tx/PGI ratio (Fig. 6a).

As compared with the sunflowerseed oil and olive oil groups, PO feeding to rabbits caused a lower TxB$_2$ formation by collagen-activated blood platelets (RIA) and a higher vascular PGI$_2$ production (platelet aggregation bioassay). Although these differences were not statistically significant, the resulting Tx/PGI ratio was significantly reduced in the PO group and comparable to the values obtained for the fish oil and linseed oil groups (Fig. 6b).

These effects of palm oil on the formation of pro- and anti-thrombotic prostanoids are likely to explain, at least in part, the anti-thrombotic and anti-atherosclerotic effects of palm oil.

ACKNOWLEDGEMENT

We thank Unilever Research, Vlaardingen, The Netherlands, for the excellent cooperation during the rabbit study, and for preparing the diets for the recent palm oil studies in rats.

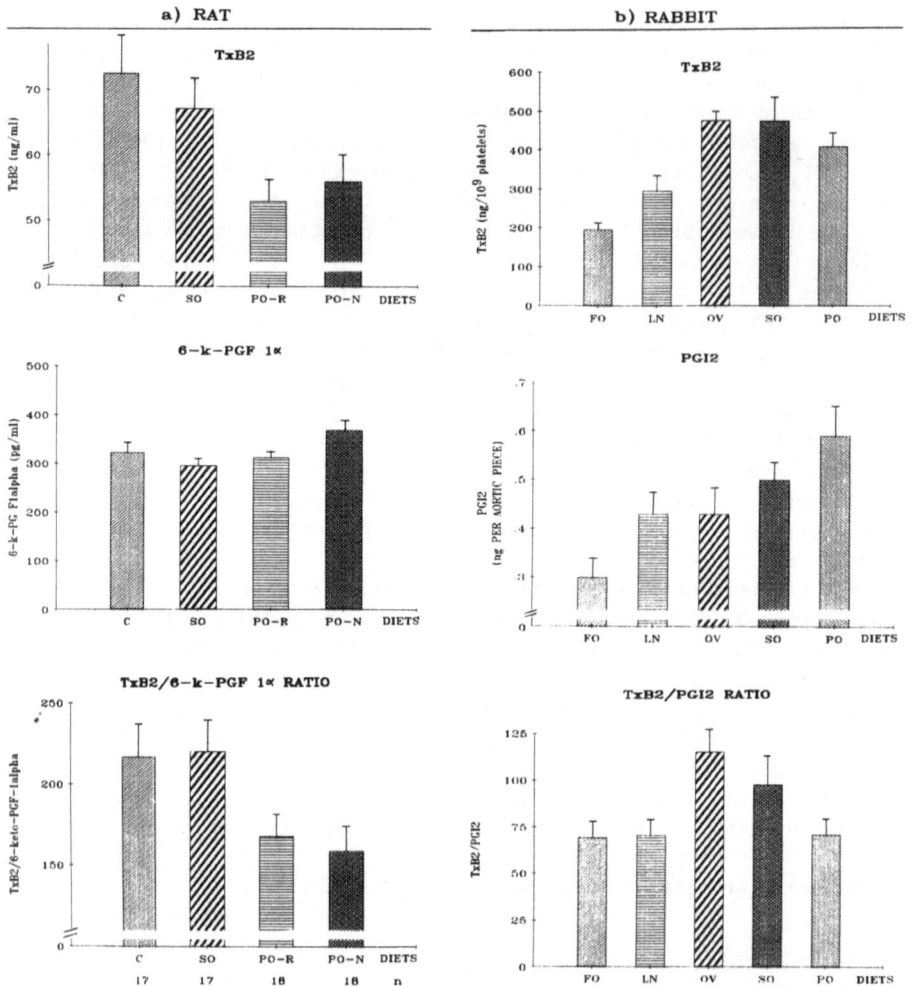

Fig. 6. Effect of feeding various oils to rats and rabbits on prostanoid production in whole blood, activated by collagen (rats), collagen-activated cPRP (rabbits, n= 6 pools of 2 animals), and mechanically stimulated aortic pieces (rabbit, n= 12).

For explanation of group code : see Figs. 2 (rat) and 4 (rabbit).

REFERENCES

1. R.O.Vles, J.Büller, J.J.Gottenbos and H.J.Thomasson, Influence of type of dietary fat on cholesterol-induced atherosclerosis in the rabbit, J.Atheroscl.Res. 4:170 (1964)

2. D.Kritchevsky, Experimental atherosclerosis in primates and other species, Ann.N.Y.Acad.Sci. 162:80 (1969)

3. H.Malmros, Dietary prevention of atherosclerosis, Lancet ii:479 (1969)

4. R.W.Wissler and D.Vesselinovitch, The effects of feeding various dietary fats on the development and regression of hypercholesterolemia and atherosclerosis, Adv.Exp.Med.Biol. 60:65 (1975)

5. G.Weber, Regression of arterial lesions: facts and problems, in: "International Conference on Atherosclerosis", L.A.Carlson,R.Paoletti,C.Sirtori and G.Weber eds., Raven Press, New York (1978)

6. C.F.Howard,Jr., The relationship of diet and atherosclerosis in diabetic Macace Nigra, Adv.Exp.Med.Biol. 60:13 (1975)

7. W.B.Kannel, Results of the epidemiologic investigation of ischaemic heart disease illustrated by the Framingham study, in "Ischaemic Heart Disease", de Haas,Hemker and Snellen eds.,University Press, Leiden (1970)

8. J.Stamler, Epidemiology of coronary heart disease, Med.Clin.North Am. 57:5 (1973)

9. S.Heyden, Epidemiological data on dietary fat intake and atherosclerosis with an appendix on possible side-effects, in :"The Role of Fats in Human Nutrition", A.J.Vergroesen ed.,Academic Press, London,New York (1975)

10. S.Renaud, R.Morazain, L.McGregor and F.Baudier, Dietary fats and platelet functions in relation to atherosclerosis and coronary heart disease, Haemostasis 8:234 (1979)

11. N.Kronmann and A.Green, Epidemiological studies in the Upernavik district, Greenland.Incidence of some chronic diseases, 1950-1974, Acta Med.Scand. 208:401 (1980)

12. G.Hornstra and A.Vendelmans-Starrenburg, Induction of experimental arterial occlusive thrombi in rats, Atherosclerosis 17:369 (1973)

13. G.Hornstra, Effect of type and amount of dietary fats on arterial thrombus formation (Chapter 4), in "Dietary Fats, Prostanoids and Arterial Thrombosis", G.Hornstra ed., Martinus Nijhoff Publisher, The Hague,Boston,London (1982)

14. G.Hornstra and R.N.Lussenburg, Relationship between the type of dietary fatty acids and arterial thrombosis tendency in rats, Atherosclerosis 22:499 (1975)

15. G.Hornstra, Location of dietary fat effect on arterial thrombus formation (Chapter 5), in "Dietary Fats, Prostanoids and Arterial Thrombosis", G.Hornstra ed., Martinus Nijhoff Publisher, The Hague, Boston,London (1982)

16. J.F.Mustard and M.A.Packham, Factors influencing platelet function: adhesion, release and aggregation, Pharmacol.Rev. 22:97 (1970)

17. C.M.Ingerman-Wojenski and M.J.Silver, A quick method for screening platelet dysfunctions using the whole blood lumi-aggregometer, Thromb.Haemost. 51:154 (1984)

18. G.Hornstra,Relationship between dietary fat type, platelet fatty acid composition and eicosanoid formation by activated platelets, (Chapter 8), in "Dietary Fats, Prostanoids and Arterial Thrombosis", G.Hornsta ed., Martinus Nijhoff Publisher, The Hague,Boston,London, (1982)

19. R.O.Vles, Effets des corps gras sur le myocarde de diverses especes animal: essais d'evaluation histometrique, Rev.Franc.Corps Gras 25:289 (1978)

20. B.Lewis, The LDL-theory and the HDL-hypothesis, in "Diets and Drug in the Atherosclerosis", G.Noseda,B.Lewis and R.Paoletti eds., Raven Press, New York (1980)

21. S.B.Hulley, Epidemiology as a guide to clinical decisions. The association between triglyceride and coronary heart disease, New Engl.J.Med. 302:1383 (1980)

22. Lipids Research Clinics Program. The Lipid Research Clinics Coronary Primary Prevention Trial Results. II.The relationship of reduction in incidence of coronary heart disease to cholesterol lowering, JAMA 251:365 (1984)

23. A.B.Chandler, Arterial thrombosis, platelets and atherogenesis, in "Dietary Fats, Prostanoids and Arterial Thrombosis", G.Hornstra ed.,Martinus Nijhoff Publisher, The Hague,Boston,London (1982)

24. M.F.Baudet, C.Dachet, M.Lasserre, O.Esteva and B.Jacotot, Modification in the composition and metabolic properties of human low density and high density lipoproteins by different dietary fats, J.Lipid Res. 25:456 (1984)

25. F.H.Mattson and S.M.Grundy, Comparison of effects of dietary saturated monounsaturated and polyunsaturated fatty acids on plasma lipids and lipoproteins in man, J.Lipid Res. 26:194 (1985)

26. S.Moncada and J.R.Vane, Unstable metabolites of arachidonic acid and their role in haemostasis and thrombosis, Brit.Med.Bull. 34:129 (1978)

27. G.Hornstra, The significance of prostanoids in the dietary fat effect on arterial thrombogenesis (Chapter 6), in "Dietary Fats, Prostanoids and Arterial Thrombosis", G.Hornstra ed., Martinus Nijhoff Publisher,The Hague,Boston,London (1982)

DIETARY FATTY ACIDS AND PLATELET COMPOSITION AND FUNCTION IN HUMAN

STUDIES

Serge Renaud

INSERM, Unit 63,
22 av. Doyen Lépine
69500 Bron, France

INTRODUCTION

The intake of saturated fatty acids is known to be the main environmental factor associated with coronary heart disease (1,2). The usual explanation for that association is that saturated fats increase serum lipids, especially cholesterol, which accumulate in arteries causing atherosclerosis. In the pathogenesis of coronary heart disease, in addition to atherosclerosis, thrombosis plays a leading role. Several investigators have shown that saturated fats, not only induce atherosclerosis, but also predispose to thrombosis in different experimental models (3,4).

Hornstra with the arterial loop technique has shown in rat that the severity of thrombosis was closely parallel to the level in the diet of fatty acids with more than 14 carbon atoms (5).

In our model of venous thrombosis in rat we found that the severity of lesions was markedly well related to the level of dietary saturated fatty acids, especially stearic acid (18:0) (6,7). Both the thrombotic tendency and the intake of saturated fatty acids were also closely related to thrombin induced aggregation (8,9) and to the clotting activity of the platelet phospholipids the most active on clotting i.e. phosphatidylserine and phosphatidylinositol. This platelet membrane phospholipid clotting activity has been frequently mentioned as platelet factor 3. The clotting test evaluating its activity is then the factor 3 clotting time or F_3-CT. Of interest is that the factor 3 activity is the only lipidic factor involved in coagulation and thus it is feasible that its activity might be changed by the type of dietary fat.

The modifications induced by saturated fats in the clotting activity of platelets and their response to thrombin were not related to serum cholesterol or other lipid fraction, but rather to changes in the fatty acid composition of platelet phospholipids (6) especially to the fatty acid 20:3(n-9) (10). That fatty acid derived from 18:0 by desaturation and elongation, is known to be synthesized in large amounts in essential fatty acid deficiency (11). However, in the absence of a deficiency in essential fatty acids, 20:3(n-9) seems to be in platelet or plasma phospholipids the most reliable marker of the saturated fat intake, at least in rat (10).

Platelet function tests and nutrients   To determine whether the
same relationship between the intake of saturated fatty acids and
platelet functions could be demonstrated in man, groups of farmers in
contrasted area concerning the mortality rate frome CHD, were compared
in France and Great Britain. For this purpose a mobile laboratory was
organized allowing the same instrumentation and the same team to perform
the studies in the different selected regions and countries. The
hematologic tests were performed immediately after the venesection,
while blood and food samples were frozen for determinations done in the
main laboratory. Dietary habits were evaluated by a 24 hour recall, the
weighing of all the food and drinks covering a second period of 24 hours
and chemical analysis of a duplicate sample of food also of a 24 hour
period. The pilot studies performed in France (12) and Great Britain
(13) have indicated that the clotting activity of platelets and their
response to thrombin-induced aggregation were more closely related to
the intake of saturated fatty acids than serum cholesterol. In the area
of the lower mortality rate from CHD, the intake of saturated fat was in
general lower as well as most of the platelet function tests. These
earlier results were confirmed subsequently in smokers from East and
West Scotland (14) and especially in more extensive studies in 260
farmers (40-45 years) from nine areas in France and Great Britain (15).
The results obtained concerning the clotting activity of platelets are
illustrated in figure 1.

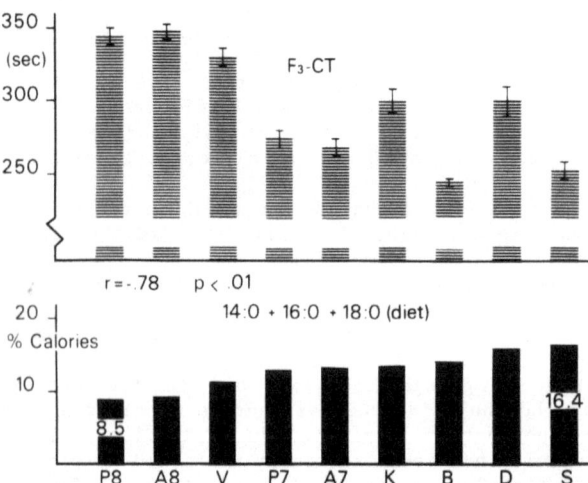

Fig. 1. Nine groups
of French and British
farmers studied over a
period of two years,
classified by increa-
sing order according
to their intake of
saturated fatty acids
(14:0 + 16:0 + 18:0)
(bottom). At the top
of the figure is shown
the corresponding
clotting activity of
of the whole platelets
(F_3-CT).

It can be noted that there is a significant inverse correlation
between the intake of saturated fatty acids and the F_3-CT, but certain
groups apparently do not follow the general trend. In figure 2 are
illustrated similar results, with dietary calcium being taken into
consideration in addition to saturated fatty acids, in multivariate
linear regression analysis. This time the correlation coefficient is
almost ideal (r = -.99).
    While in studies in animals only the type of fat can be different
from one group to another, populations differ rarely by only one factor
even if they are all farmers of the same age. Calcium (from food and
mostly water) was one of the dietary component being markedly different
in the various areas (range from 800 to 1400 mg/day).
    In man, calcium is known to bind to saturated fatty acids, forming
insoluble soaps, which are excreted in the feces (16). In animal
studies, we have also shown similar results with a resulting marked

improvement of platelet functions (clotting and thrombin aggregation) (17). The chief interest of the striking inhibitory effect of calcium on the intestinal absorption of saturated fatty acids resulting in beneficial effects on platelets, seems to be that calcium is the main factor of water hardness. Moreover, the regions with hard water have been shown in many countries to have a lower mortality rate from CHD (18) without much explanation for this preventive effect.

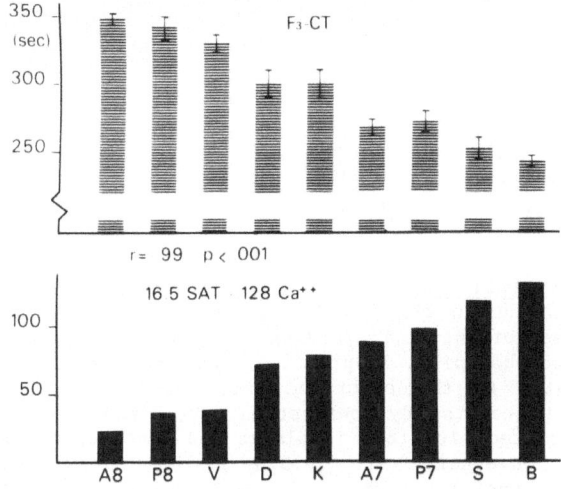

Fig. 2. The nine groups of French and British farmers from figure 1 classified according to their intake of saturated fat minus that of calcium per 3000 calories (multivariate analysis).

On thrombin-induced aggregation the same relationship with saturated fatty acids and calcium have been observed in the nine groups of farmers. Finally, also of interest is that those significant correlations were also observed on an individual basis in the 260 subjects as shown in Table 1.

Table 1. Multiple regression analysis in 260 French and British farmers between nutriments (chemical analysis) and platelet functions or serum cholesterol

Diet	F_2-CT	THR	Chol
SAT	*** .47	*** .28	.11
18:3	*** -.24	*** -.31	.0
18:2	-.08	-.04	*** -.20
Ca^{++}	** -.19	*** -.25	-.06

Standard partial regression coefficients. **P < .01   ***P < .001

By contrast, on an individual basis, cholesterol was not significantly correlated with the intake of either saturated fat or of calcium. The close relationship between dietary saturated fat and platelet aggregation to thrombin but not to ADP was also observed by other investigators in a pilot study comparing subjects from Italy (Canino), Finland (Nurmijari) and USA (Beltsville) (19). The results are illustrated in figure 3.

85

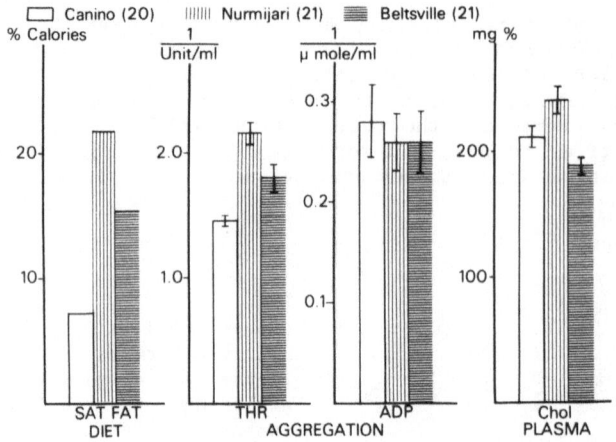

Fig. 3. Relationship between saturated fats, thrombin (THR) and ADP-induced aggregation in man. Chol: Cholesterol. Adapted from Agradi et al. (19), courtesy of the authors and Raven Press.

<u>Platelet functions and fatty acid composition</u>   As in animals, a higher intake of saturated fatty acids in the 260 French and British farmers did not result in an increase of the corresponding fatty acids in plasma total lipids or platelet phospholipids. In practice, the only fatty acid which was positively related to the saturated fatty acids was 20:3 (n-9), both in plasma and platelets. By contrast dietary linoleic acid (18:2) was positively correlated with 18:2 in plasma and platelets, but also with 22:4(n-6), and inversely with 20:3(n-9). Concerning the platelet function tests especially the F_3-CT and aggregation to thrombin, they were significantly correlated on a group and individual basis with 20:3(n-9). Therefore, the same close relationship between the intake of saturated fatty acids, the platelet function tests and 20:3(n-9) was also observed in man. This is of special interest since it has been shown by Kingsbury (20) that there was a 34 % increase in 20:3(n-9) in the cholesterol ester of patients with myocardial infarction. A similar result has been found in the platelet phospholipids (Riemersma, R.A., Wood, D., Oliver, M., Edinburgh, Scotland, personal communication). Therefore, 20:3(n-9) seems to be a marker of both the intake of saturated fat and the susceptibility to CHD. These results somehow substantiate the hypothesis of CHD being the consequence of a relative deficiency in EFA (21).

Finally, of interest concerning the possible role of 20:3(n-9) in CHD via an increase in the response of platelets, especially to thrombin, is its location in the phospholipids, mostly in PI (Table 2), the most active fraction in platelet metabolism.

Table 2.   Level of 20:3(n-9) in plasma and platelet phospholipids in five farmers from different regions

	PI	PS	PE	PC
Plasma	1.21 ± 0.13	0.41 ± .09	0.39 ± 0.3	0.30 ± .06
Plat.	0.59 ± .11	—	0.41 ± 0.3	0.13 ± .01

Mean ±S.E.

However, even if 20:3(n-9) is largely located in PI, it does not appear to be a good substrate for cyclooxygenase. By contrast it is one of the best substrate for lipoxygenase (22). Our group has been able to observe that 20:3(n-9) potentiates thrombin aggregation (23) by the production through the lipoxygenase pathway, of a 12-OH derivative exhibiting PGE_2-like activity (24).

III. DIET MODIFICATION AND PLATELET FUNCTIONS IN MAN

Studies in man have shown that platelet functions can be lowered by diet modification. Hornstra et al. (25) as well as Fleischman et al. (26) have observed a decreased tendency for platelets to form aggregates in the filtragometer after an increased intake of linoleic acid. Under similar circumstances, O'Brien et al. (27), Jakubowski and Ardlie (28) noted a decrease in different tests related to platelet functions. On platelet aggregation Iacono et al. (29) observed that a diet with a P/S ratio of one induced, in six weeks, a significant reduction in the aggregation to thrombin and collagen, but not to ADP and epinephrine. Our earlier studies in Moselle (East of France) have indicated that the intake of saturated fat was higher (15% calories) than in farmers from Var (11% calories) in relation to a lower response of all the platelet function tests in Var (South of France) (12,15). To determine whether platelet functions in Moselle depended mainly on the dietary habits, four groups of 25 farmers were studied each year for three years, before and after their intake of saturated fat was reduced to approximately 10% calories (30). For the first year, only two groups had their diet modified, the two other groups serving as controls. Finally, after two years, the diet was also changed in the control groups to compare the respective beneficial effects of 18:2 and 18:3 since results shown earlier in the nine groups of farmers indicated that 18:3 might improve more platelet functions than 18:2.

After one year of diet modification the P/S passing from 0.32 to approximatively one in the 50 farmers having modified their diet, the clotting activity of platelets (F_3-CT) and the response of platelets to thrombin-aggregation were drastically reduced to levels even lower than those observed in Var. By contrast, the response to ADP-induced aggregation and to adrenaline were increased as compared to the previous year. Serum cholesterol was decreased by approximately 10%. In the 50 farmers having not modified their diet, no significant changes were observed in any of the platelet or lipid parameters as compared to the values of the previous year.

We had already observed in rat (9) that platelets of animals on a saturated fat diet were more susceptible to thrombin and less to ADP, while animals fed a diet rich in polyunsaturated fat exhibited a high response of their platelets to ADP and a low response to thrombin. Thus a high response of platelets to ADP could be the normal picture of populations with a low-mortality rate from CHD. Nevertheless, in areas we examined so far with a lower CHD incidence (South France, East Scotland), the most significant result was a low response of platelets to clotting and to thrombin, but the response to ADP was also lower than in areas with a higher CHD incidence (East France, West Scotland). Thus the aim was to improve the diet modification such that all the platelet function tests will be decreased by the diet. The fact that in Var the P/S ratio was only 0.5 - 0.6 suggested, in subsequent studies, to decrease the P/S ratio to lower values. In 1978, the two control groups had also their diet modified, one group (P/S = 0.8) with oil and margarine (38% polyunsaturated) from sunflower seed supplying mostly 18:2 and the other (P/S = 0.6) with oil and margarine (26% polyunsaturated) from rapeseed (without erucic acid) supplying 18:2 and 18:3. One year later, all the platelet function tests examined (F_3-CT, aggregation to thrombin, ADP, collagen) were considerably decreased in the two groups (Figure 4), cholesterol being significantly decreased only in the group given fats from sunflower seed. It has to be emphasized that the saturated fat intake was decreased from 15.4 to 11.4% of calories in the sunflower group and from 14.4 to 10.1%, in the rapeseed group (30).

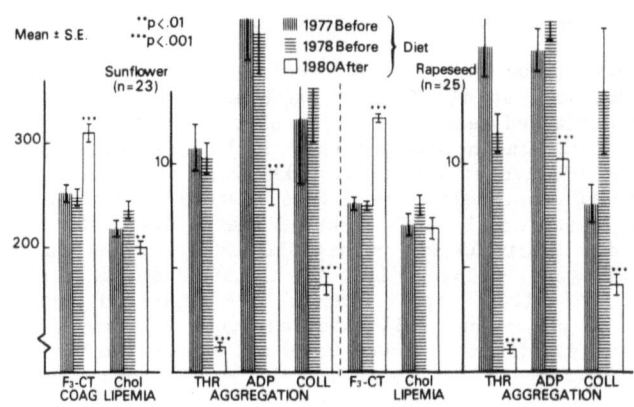

Fig. 4. Serum cholesterol and platelet functions in Moselle farmers two years before and one year after diet modification, the main fats (oil and margarine) being either from sunflower seed or rapeseed. Aggregation to thrombin (THR), ADP, collagen (COLL). Chol: serum cholesterol. Coag: coagulation. (A prolongation of the clotting time indicates a decrease in the clotting activity of platelets). Adapted from Renaud et al. (30), courtesy of Amer. J. Clin. Nutr.

The dietary habit modifications were reflected in the fatty acid composition of plasma and platelet lipids. In figure 5 are shown the main changes in the fatty acid platelet phospholipids, in the two groups of farmers studied in figure 4. While with sunflower seed oil, there was an increase in 18:2 and a decrease in 20:3(n-9) and 20:4(n-6), with rapeseed oil, the increase was in 18:1 and 20:5(n-3), and the decrease mostly in 20:4(n-6).

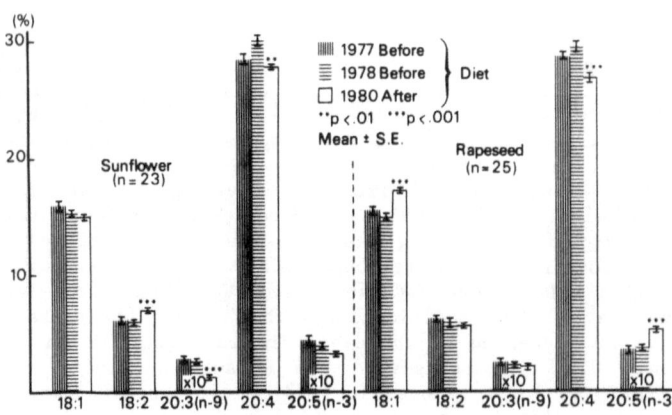

Fig. 5. Platelet phospholipid fatty acids in the subjects of figure 7. Adapted from Renaud et al. (30), courtesy of Amer. J. Clin. Nutr.

Since the results of the platelet function tests (Figure 4) were somewhat similar in the groups with sunflower and rapeseed oil, comparison with the platelet fatty acid composition does not give obvious clues as to the main fatty acid changes responsible for the observed amelioration in platelet functions. It can be emphasized that a sizeable increase in the level of 18:2 is not essential for drastically improving the platelet functions, as shown in the group with rapeseed. The only modification identical for the 2 groups was the decreased in 20:4(n-6). Nevertheless further statistical analysis of the results which comprises 14 groups of subjects studied over a period of four years (a total of 340 sets of determinations) show that, as in previous studies dietary saturated fatty acids were mostly correlated with F_3-CT and thrombin aggregation (r =.92 and .94 respectively, $p<.001$) ; 18:3, inversely related to F_3-CT and thrombin aggregation (r =.75 and .66 respectively, $p<.01$), and positively related to 20:5(n-3) (r =.73 $p<.01$). Finally the platelet function F_3-CT and thrombin aggregation were mostly related, in the platelet phospholipids, to 20:3(n-9) (r = .72, $p<.01$ and .79, $p<.001$ respectively), but not to 20:4(n-6) or any other fatty acid determined in the study (30).

However another platelet parameter of interest seems to be cholesterol which was determined in carefully washed platelets. Platelet cholesterol appears to be inversely related to serum cholesterol and positively related to the intake of polyunsaturated fatty acids, especially 18:3(n-3) (30). It seems also to be positively correlated with ADP induced aggregation a result concordant with those obtained by enriching platelets with cholesterol 'in vitro' (31).

IV. CONCLUSIONS

In addition to the well-known effect on serum cholesterol, saturated fats, both in animals and in free-living populations, appear to have a striking effect to stimulate certain platelet functions, mainly their clotting activity and response to thrombin. These platelet functions appear to be related to the level in plasma and platelet phospholipids of 20:3(n-9), a fatty acid derived from saturated fats, and apparently, the most reliable marker of their intake.

Of interest is that saturated fats are known from sound studies to be the main environmental factor associated with CHD.

By contrast the beneficial effect of a high intake polyunsaturated fatty acids on CHD except perhaps for the (n-3) family fatty acids (32), has not been much observed in prospective studies. On platelets, at least in the studies reported here, a marked increase in the polyunsaturated fatty acids in addition to the decrease of saturated fat to 10% of calories do not seem to improve much their functions. When given in large amounts both in animals and in man, polyunsaturated fats even induce a hyperaggregability to ADP. In our subjects, this result was observed mostly when the P/S ratio was above one. By contrast, long-term consistent improvements of all the platelet functions could be obtained with P/S ratio of 0.6 to 0.8, in other terms with intake of polyunsaturated fatty acids close to those observed in mediterranean countries, with a much lower mortality rate from CHD. By contrast in prevention trials with P/S ratio higher than one (33) or even than two (34), the results obtained by the diet modification were rather inconclusive while a P/S ratio of less than one (35) (apparently 0.7, Hjerman, I. ; personal communication) has been associated in a prospective study with a 47% reduction in CHD. The untoward effects on platelet aggregation to ADP could eventually explain the disappointing results observed with P/S ratio higher than one.

ACKNOWLEDGMENTS

These studies were supported in part by Grants from CETIOM and La Fondation pour la Recherche Médicale.

REFERENCES

1. A. Keys, Coronary heart disease in seven Countries, <u>Circulation</u> 41: (Suppl 1) (1970).
2. A. Keys, A., "Seven Countries", Harvard University Press, Cambridge and London (1980).
3. A. Nordøy, J.T. Hamlin, A.B. Chandler, and H. Newland, The influence of dietary fats on plasma and platelet lipids and ADP induced platelet thrombosis in the rat, <u>Scand. J. Haemat.</u> 5:458 (1968).
4. J.F.Mustard, H.C. Rowsell, E.A. Murphy, and H.G. Downie, Diet and thrombus formation: Quantitative studies using an extracorporel circulation in pigs, <u>J. Clin. Invest.</u> 42,1783 (1963).
5. G. Hornstra, Dietary fats and arterial thrombosis, <u>Haemostasis</u> 2:21 (1973/74).
6. P. Gautheron, and S. Renaud, Hyperlipemia induced hypercoagulable state in rat. Role of an increased activity of platelet phosphatidyl-serine in response to certain dietary fatty acids, <u>Thromb. Res.</u> 1:353 (1972).
7. S. Renaud, Thrombotic, atherosclerotic and lipemic effects of dietary fats in the rat, <u>Angiology</u> 20:657 (1969).
8. S. Renaud, and J. Godu J, Induction of large thrombi in hyperlipemic rats by epinephrine and endotoxin, <u>Lab. Invest.</u> 21:512 (1969).
9. S. Renaud, R. Kinlough,and J.F. Mustard, Relationship between platelet aggregation and the thrombotic tendency in rats fed hyperlipemic diets, <u>Lab. Invest.</u> 22:339(1970).
10. L. McGregor, R. Morazain, and S. Renaud, A comparison of the effects of dietary short and long chain saturated fatty acids on platelet functions, platelet phospholipids, and blood coagulation in rats, <u>Lab. Invest.</u>, 43:438 (1980).
11. R.T. Holman, The ratio of trienoic-tetraenoic acids in tissue lipids as a measure of essential fatty acid requirement, <u>J. Nutr.</u> 70:405 (1960).
12. S. Renaud, E. Dumont, F. Godsey, A. Suplisson, and C. Thevenon, Platelet functions in relation to dietary fats in farmers from two regions of France, <u>Thromb. Haemost.</u> 40:518 (1979).
13. S. Renaud, R. Morazain, F.Godsey, E. Dumont, I.S. Symington, E.M. Gillanders, and J.R. O'Brien, Platelet functions in relation to diet and serum lipids in British farmers, <u>Br. Heart J.</u> 46:562 (1981).
14. S. Renaud, E. Dumont, F. Baudier, and I.S. Symington, Effect of smoking and dietary saturated fats on platelet functions in Scottish farmers, <u>Cardiovasc. Res.</u> 19:155 (1985).
15. S. Renaud, R. Morazain, F. Godsey, E. Dumont, C. Thevenon, J.L. Martin, and F. Mendy, Nutrients, platelet function and composition in nine groups of French and British farmers, <u>Atherosclerosis</u> 59:(April) (1986).
16. A.K. Bhattacharyya, C. Thera, J.T. Anderson, F. Grande, and A. Keys, Dietary calcium and fat, effect on serum lipids and fecal excretion of cholesterol and its degradation products in man, <u>Amer. J. Clin. Nutr.</u> 22:1161 (1969).
17. S. Renaud, M. Ciavatti, C. Thevenon, and J. P. Ripoll, Protective effects of dietary calcium and magnesium on platelet function and atherosclerosis in rabbits fed saturated fat, <u>Atherosclerosis</u> 47: 187 (1983).
18. G.W. Comstock, The epidemiologic perspective - Water hardness and cardiovascular disease, <u>J. Environm. Path. Toxicol.</u> 4:9. (1980).

19. E. Agradi, A. Carvalho, R. Dougherty, A. Ferro-Luzzi, A. Galli, C. Galli, G. Gianfranceshi, J. Iacono, R. Paoletti, and L. Sautebin, Epidemiological studies on dietary lipids, human plasma lipids and platelet lipids and function, in: "International Conference on Atherosclerosis," L.A. Carlson, R. Paoletti, C.R., Sirtori, and G. Weber, ed.,Raven Press, New York (1978).

20. K.J. Kingsbury, C. Brett, and R. Stovold, Abnormal fatty acid composition and human atherosclerosis, Postgrad. Med. J. 50:425 (1974).

21. M.F. Oliver, Fats and atheroma (letter), Br. Med. J. 1:889 (1979). Brit. Med. J. 1:889.(1979).

22. P. Needleman, A. Wyche, L. Leduc, S.K. Sankarappe, B.A. Jakschik, and H. Sprecher, Fatty acids as sources of potential "magic bullets" for the modification of platelet and vascular function. Proc. Natl. Acad. Sci. USA 76:944 (1979).

23. M. Lagarde, M. Burtin, H. Sprecher, M. Dechavanne, and S. Renaud, Potentiating effect of 5,8,11-eicoatrienoic acid on human platelet aggregation, Lipids 18:291 (1983).

24. M. Lagarde, M. Burtin, M. Rigaud, H. Sprecher, M. Dechavanne, and S. Renaud, Prostaglandin E_2-like activity of 20:3 n-9 platelet lipoxygenase end-product, FEBS Letters 181:53.(1985).

25. G. Hornstra, B. Lewis, A. Chait, O. Turpeinen, M.J. Karvonen, and A.J. Vergroesen, Influence of dietary fat on platelet function in men, Lancet 1:1155 (1973).

26. A.I. Fleischmann, D. Justice, M.L. Bierenbaum, A. Stier, and A. Sullivan, Beneficial effect of increased dietary linoleate upon in vivo platelet function in man. J. Nutr. 105:1286 (1975).

27. J.R. O'Brien, M.D. Etherington, S. Jamieson, A.J. Vergroesen, A.J. and F. Ten Hoor, Effect of a diet of polyunsaturated fats on some platelet-function tests, Lancet 2:995 (1976).

28. J.A. Jakubowski and N.G. Ardlie, Modification of human platelet function by a diet enriched in saturated or polyunsaturated fat, Atherosclerosis 31:335 (1978).

29. J.M. Iacono, R.A. Binder, M.W. Marshall, N.W. Schoerne, J.A. Jencks, and J.F. Mackin, Decreased susceptibility to thrombin and collagen platelet aggregation in man fed low fat diet, Haemostasis 3:306 (1975).

30. S. Renaud, F. Godsey, E. Dumont, C. Thevenon, E. Ortchanian, and J.L. Martin, Influence of long-term diet modification on platelet function and composition in Moselle farmers, Amer. J. Clin. Nutr. 43:136 (1986).

31. S.J. Shattil, R. Anayagalindo, J. Bennett, R.W. Colman and R.A. Cooper, Platelet hypersensitivity induced by cholesterol incorporation, J. Clin. Invest. 55:636 (1975).

32. S.H. Jr. Goodnight, W.S. Harris, W.E. Connor, and D.R. Illingworth, Polyunsaturated fatty acids hyperlipemia and thrombosis, Arteriosclerosis 2:87 (1982).

33. J.N. Morris and K.P. Ball, Controlled trial of soya-bean oil in myocardial infarction. Report of a research committee to the Medical Research Council, Lancet 2:693 (1968).

34. S. Dayton, M.L. Pearce, S. Hashimoto, W.J. Dixon, and U. Tomyasu, A controlled clinical trial of a diet high in unsaturated fat in preventing complication of atherosclerosis, Circulation 40(Suppl 2):1 (1969).

35. I. Hjerman, K. Velve Byre, I. Holme,and P. Leren, Effect of diet and smoking intervention on the incidence of coronary heart disease: report from the Oslo Study Groups of a randomized trial in healthy men, Lancet 2:1303 (1981).

COMPARATIVE EVALUATION OF OLIVE OIL AND OF CORN OIL IN THE PREVENTION OF ATHEROSCLEROSIS

Cesare R. Sirtori and Cristina Manzoni

Institute of Pharmacological Sciences and Center E.
Grossi Paoletti
University of Milano, 20129 Milan, Italy

INTRODUCTION

Regional cardiovascular morbidity and mortality data in the Western hemisphere shows a particularly low prevalence in the Mediterranean region[1]. In spite of the recent fall of cardiovascular disease in the US, mortality data are still about twice higher than those occurring in the Mediterranean nations[2]. The currently growing popularity of the so-called "Mediterranean diet"[3] (characterized, among other things, by a moderately low intake of polyunsaturated fatty acids and by a high intake of monounsaturates from olive oil) is demonstrated by articles in a variety of lay publications[4]. There is, however, still no adequate basis to explain a direct influence of any nutrient intake on the low cardiovascular mortality in this region. Among the different hypotheses, some have been stressed: low salt intake[5], higher consumption of fiber-rich food[6] or of red wine[7].

The beneficial activity of polyunsaturated fatty acid (PUFA) rich diets on plasma lipids and platelet function has been repeatedly described[8,9], whereas the activity of olive oil has never been adequately tested in patients; at most, olive oil has been considered to have a "neutral" effect on plasma lipids[10]. Recent clinical reports show, however, that an elevated oleic acid intake may beneficially influence plasma lipids[11,12]. When given to normolipidemic and hypertriglyceridemic individuals, a formula diet rich in oleic acid may prove to be as hypolipidemic as a similar diet rich in linoleic acid. The object of this presentation will be to examine results from the most significant clinical trials, both from ourselves and from other research groups, investigating on the relative merits of diets rich in monounsaturates (prevalently oleic acid) and polyunsaturates (mostly linoleic acid in corn oil), in terms of the reduction of plasma lipids and eventual changes in platelet aggregability and eicosanoid metabolism.

93

CURRENT EVIDENCE FOR THE ACTIVITY OF POLYUNSATURATES ON LIPID METABOLISM

Studies on the lipid-lowering activity of different dietary fatty acids have generally indicated (in normolipidemic individuals) that the administration of PUFA leads to a reduction of cholesterolemia, predictable from standard formulas[10]. Studies in patients with severe hyperlipidemia are also consistent with the validity of the suggested mathematical formulas[13]. More recently, however, caution has been suggested on the use of the formulas when total dietary fat is reduced; apparently, upon reducing the proportion of dietary energy from about 39 to 23%, there is little difference in the final cholesterolemia, if the P/S ratio is changed from 0.4 to 0.9[14].

PUFA, when compared to a highly saturated diet, also generally lead to a reduced platelet reactivity with a lower formation of pro-aggregatory prostaglandin metabolites[15]. Very recent findings indicate that oleic and linoleic acids, when added in vitro to human platelets, determine similar changes in the membrane physico-chemical properties, while reducing the reactivity to standard stimuli in an analogous manner[16]. In a parallel study in rabbits, we could note that the arterial wall production of 6-keto-PGF$_{1\alpha}$, the major metabolite of the anti-aggregatory prostacyclin, is higher after a diet rich in olive, rather than in corn oil[17].

Oleic vs Linoleic Acid Activity when Given in Formula Diets

As above indicated, few clinical data are available on the effects of a diet rich in olive oil on metabolic and thrombotic indexes. Classically, according to the Keys' formula[10], oleic acid intake should not lead to any significant change in cholesterolemia. Recently, however, Mattson and Grundy[11], in normolipidemic subjects and in hypertriglyceridemic patients, compared three liquid formula diets, all containing 40% of total calories as fat, of which: Control diet: 49.7% saturates (SFA), 40.0% monounsaturates (MUFA), 10.1% PUFA; Mono diet: 8.6% SFA, 73.4% MUFA, 17.9% PUFA; Poly diet: 11.2% SFA, 14.7% MUFA and 73.3% PUFA. A total of 20 volunteers were treated for 4 weeks on each diet. In the 12 normolipidemic subjects (Table 1), both the Mono and the Poly diets reduced cholesterolemia to a similar degree, compared to the Control diet (respectively -12.5% and -15.6%). However, the Poly diet also reduced HDL-cholesterol by 9.8%, vs essentialy no change with the Mono diet. In the 8 type IV patients, following the same protocol, similar changes of total cholesterolemia were recorded (Table 1). In these patients, changes of both plasma triglycerides and HDL-Cholesterol were minimal and non statistically significant.

The hypothesis that a diet rich in oleic acid may be better suited for the preventive management of patients with a high cardiovascular risk, was more recently re-evaluated by Grundy[12]. This Author again tested three liquid formula diets. Two of the diets had a relatively high fat content. The High-Sat and the High-Mono diets contained 40% of total calories as fat, 43% as carbohydrates and 17% as milk proteins. The third diet was Low-Fat (20% of total calories), containing equal quantities of SFA, MUFA and PUFA. The 11 selected normal volunteers participated in two different studies, both of which showed a definite plasma cholesterol lowering activity of the High-Mono and of the Low-Fat diets. In both studies, a clear depressing potential on HDL-cholesterol by the Low-Fat diet could be demonstrated. The final levels (pool of two studies) of plasma total cholesterol, triglycerides, low density and high density lipoprotein (LDL and HDL) cholesterol for the High-Mono and the Low-Fat diets are reported in **Table 2**. The reduction of total cholesterol vs the High-Sat diet was more dramatic with the High-Mono (13% vs 8% for the Low-Fat); similar differences were noted for LDL cholesterol (21% and 15% reductions). The Low-Fat diet also raised triglyceridemia.

Table 1. Plama Lipid and Lipoprotein Changes at the End of a Saturated (SAT), Monounsaturated (MONO) and Polyunsaturated (POLY) Liquid Formula Diet, in Normolipidemic (NL) and Type IV Subjects

	SAT	MONO	POLY
NL (n=12)			
Tot. Chol.	227.8 + 15.6	196.7 + 9.3*	190.8 + 12.6**
Triglycerides	143.0 + 12.6	134.4 + 12.1	133.0 + 14.2
HDL-Chol.	41.9 + 3.6	41.0 + 3.4	36.8 + 2.6**
Type IV (n=8)			
Tot. Chol.	219.1 + 15.1	196.8 + 9.2	191.5 + 8.2*
Triglycerides	433.8 + 57.9	420.8 + 53.4	378.5 + 61.9
HDL-Chol.	34.1 + 1.6	33.5 + 1.5	32.9 + 14.1

Data are expressed as mg/dl, means + SEM

*p < 0.05, **p < 0.01 vs SAT   (from MATTSON and GRUNDY, 1985)

TABLE 2. Effects of the Change from a High-Mono[a] to a Low-Fat[b] Diet on Plasma Lipids and Lipoproteins in Normal Volunteers (n= 11)

	Total Cholesterol	Total Triglycerides	LDL Cholesterol	HDL Cholesterol	LDL/HDL Cholesterol Ratio
High-Mono	208 ± 7	178 ± 12	136 ± 7	39 ± 2	3.5 ± 8
Low-Fat	222 ± 9*	235 ± 29	147 ± 10	32 ± 2**	4.7 ± 1.3**

Data are expressed as mg/dl; means ± SEM

[a] 40% of calories from fat with 28% monounsaturates

[b] 20% of calories from fat with P/S 1.0

*   p 0.05,   p 0.01 vs High-Mono
**

(from GRUNDY, 1986)

These studies have stimulated considerable interest in a possible change of nutritional policies for the prevention of heart disease, also in view of the newer guidelines, indicating "ideal" levels of plasma cholesterol[18]. However, these studies made use of liquid formula diets, which may not provide a physiological type of therapeutic diet. We have, therefore, evaluated a similar dietary approach using, however, a normal everyday Italian type of diet, i.e. with a relatively low fat content, compared to an average American diet. Our study also examined, in addition to plasma lipid/lipoprotein levels, platelet reactivity and eicosanoid metabolism after the different diets.

Personal Experience: Dietary Olive Oil vs Corn Oil

A group of patients with a high atherosclerosis risk was offered randomly, two 8-week dietary periods, one with corn oil and one with olive oil as the major dietary fat. This study, designed for a free-living population, evaluated both the lipid-lipoprotein changes as well as the alterations in the platelet aggregability and eicosanoid metabolism, also believed to be beneficially affected by diets with an elevated P/S fatty acid ratio[9,15]. Participating subjects were selected from those referred to our Lipid Clinic because of clinical hyperlipidemia with a high atherosclerosis risk and/or evidence of clinical disease (hypertension, coronary heart disease, peripheral vascular disease. A total of 84 patients were screened for potential selection and, following rigid criteria, 26 of these, 13 males and 13 females (age range 39-65 y), entered the protocol. After a 4-week pretrial period with the major objective of evaluating the daily caloric intake, in order to maintain the actual body weight of each subject, and to standardize fat consumption (saturated fat at about 10% of total daily calories), the patients were started, according to a table of random numbers, on the following sequences (8 weeks each) of dietary fats: olive oil-corn oil (OLIVE-CORN) and corn oil-olive (CORN-OLIVE)[19].

The experimental diets were designed to have the same nutrient composition and cholesterol content, i.e. protein 18-21%, carbohydrate 49-52%, total lipids 30%, and cholesterol 100 mg/1,000 KCal. The daily caloric intake, as assessed in the pretrial period, was maintained with the experimental diets. Patients were seen at intervals, the biochemical evaluations including plasma lipids (total cholesterol and ultracentrifugal fractions, apolipoproteins AI and B, triglycerides), whereas platelet studies evaluated the sensitivity to major aggregants (adrenaline, collagen and arachidonic acid), and the levels of products of the eicosanoid pathway, i.e. malondialdehyde (MDA) and thromboxane B_2 (TXB_2).

Twenty-three of the 26 patients completed the protocol: 13 (6 normolipidemic and 7 hyperlipidemic) followed the OLIVE-CORN sequence; the other 10 (7 hyperlipidemic and 3 normolipidemic) the opposite sequence. No significant weight changes were recorded during the two diets, and blood pressure variations were also minimal.

Minimal changes in plasma lipid-lipoprotein levels and in the platelet parameters were noted during the 4 week run-in period (**Tables 3 and 4**). During the OLIVE-CORN sequence, OLIVE caused a small total cholesterol reduction; a significant drop was noted with CORN (-7.7% vs baseline, $p < 0.01$) (**Table 3**). In the 10 patients completing the CORN-OLIVE sequence, the effect of the CORN diet on plasma cholesterol was statistically significant (-6.8%, p 0.05) and was not followed by any further changes during OLIVE. Comparison of the 13 patients with elevated cholesterol (either IIA and IIB) with the 10 normo-cholesterolemics (9 normolipidemics and 1 type IV), irrespective of the sequence of treatments, failed to show any difference in the absolute cholesterol responses. The mean reduction of total cholesterol after corn oil in the 13 hypercholesterolemic patients was of 19.5 mg/dl (-6.6% vs base) and of 10.8% mg/dl with olive (-3.7%); in the 10 patients with normal total and LDL cholesterol, there was a reduction of 18.9 mg/dl with CORN (-10.5%) and of 8.5 mg/dl with OLIVE (-4.8%).

LDL cholesterol levels reflected, to a large extent, the described variations of total cholesterolemia. In the OLIVE-CORN sequence, LDL cholesterol was only slightly reduced by OLIVE, and significantly (to -9.2%, $p < 0.05$) by CORN. In the other sequence, CORN caused an initial 6.9% fall of LDL cholesterol, followed by stabilization with OLIVE. Changes in VLDL cholesterol and in total and VLDL triglycerides were mostly of a small degree and non statistically significant. HDL cholesterol fell with CORN, especially (-4.6%) when given first, returning to baseline with OLIVE. When OLIVE was given first, HDL cholesterol tended to rise.

The LDL/HDL cholesterol ratios, rated by many as a reliable index of the atherogenic risk, were improved by both diets. In the OLIVE-CORN sequence, from a base of 4.59, there was a reduction to 4.35 with OLIVE and to 4.19 with CORN. In the opposite sequence, from a last baseline value of 4.44, there was a fall to 4.32 with CORN and to 4.16 with OLIVE. When pooling all data from the two sequence, the HDL/LDL cholesterol values were: base 4.51; OLIVE 4.25; CORN 4.25.

Both diets affected the apoprotein AI and B levels. In the OLIVE-CORN sequence, a 5.2% ($p < 0.05$) rise of apo AI was detected after OLIVE, no further change occuring with CORN. In the opposite sequence, apo AI was significantly increased at the end of OLIVE (+4.3%, $p < 0.05$). Apoprotein B levels were moderately reduced by OLIVE when given first, with no further change with CORN, in spite of the more significant hypocholesterolemic activity. A dramatic apo B reduction occurred with CORN when given first (-9.5%, $p < 0.01$), with a further slight fall with OLIVE. The AI/B ratio, also rated as an important risk

TABLE 3. Plasma Lipid, Lipoprotein, Apolipoprotein, Glucose and Uric Acid Levels during a Comparative Study between Low Lipid Diets Rich in Olive or Corn Oil

| | OLIVE CORN (n=13) | | | | CORN OLIVE (n=10) | | | |
	Pre-trial (-4 weeks)	Base (0)	End of OLIVE	End of CORN	Pre-trial (-4 weeks)	Base (0)	End of CORN	End of OLIVE
Cholesterol (mg/dl)								
Total	246.4±13.2	240.7±11.0	237.4±15.1	222.2±14.9*	276.1±18.9	273.1±23.0	254.6±19.7*	254.3±17.2*
VLDL	30.6±4.4	27.8±4.1	27.8±3.4	24.5 3.1	23.3±4.4	22.5±3.9	20.3±4.1	19.0±3.2
LDL	182.1±10.9	174.9±9.4	170.5±11.6	158.8±12.9*	205.7±17.4	204.5±19.8	190.3±16.8*	189.7±15.6*
HDL	39.2±3.1	38.1±3.0	39.2±2.8	37.9±2.5	46.8±2.4	46.1±1.9	44.0±1.8*	45.6±2.0
Triglycerides	181.4±23.9	156.9±17.5	171.4±15.5	162.2±17.4	154.9±30.6	129.8±20.7	129.8±20.7	132.1±16.7
Apolipoproteins								
AI	135.5±4.3	127.0±3.2	135.5±3.1*	132.8±3.4	134.3±4.0	135.5±4.3	134.4±3.4	141.4±2.7*
B	100.0±3.3	99.6±3.2	96.2±3.2	96.5±3.3	107.1±8.7	108.0±6.1	97.7±4.4*	95.5±4.6
Glucose	104.5±3.9	102.1±5.1	95.5±4.3*	100.6±5.5	100.4±4.5	98.1±2.5	98.1±2.5	92.8±3.4**
Uric acid	6.0±0.4	5.0±0.4	5.5±0.4	5.6±0.4	5.2±0.6	5.7±0.5	5.5±0.6	4.9±0.5

Data are expresssed as means ± SEM

* p <0.05; ** p <0.01 vs Base; p <0.05; p <0.01 vs other diet

(from SIRTORI et al, 1986)

Table 4. Platelet threshold aggregatory concentrations (TACs), malondialdehyde and thromboxane B_2 (TXB_2) formation during the corn vs olive dietary study

		OLIVE	CORN (n=12)			CORN	OLIVE (n=9)	
	Pre-trial (-4 weeks)	Base (0)	End of OLIVE	End of Corn	Pre-trial (-4 weeks)	Base (0)	End of CORN	End of OLIVE
TACs								
Adrenaline (µg/ml)	0.16±0.05	0.15±0.05	0.22±0.05*	0.25±0.10*	0.09±0.05	0.06±0.03	0.09±0.02	0.08±0.02
Collagen (µg/ml)	0.44±0.08	0.33±0.07	0.41±0.07*°	0.33±0.07	0.27±0.05	0.33±0.06	0.29±0.04	0.42±0.08*°
Arachidonic acid (mM)	0.38±0.04	0.32±0.04	0.29±0.03	0.41±0.04*°	0.31±0.01	0.30±0.05	0.35±0.07*	0.38±0.05*
MDA								
(nmoles/3x10^8 platelets)	0.45±0.05	0.43±0.05	0.42±0.06	0.47±0.05	0.35±0.06	0.33±0.05	0.35±0.04	0.38±0.07
TBX$_2$								
(ng/ml PRP)	141.4±21.2	186.7±19.9	146.2±16.8*°	172.7±22.7	163.5±26.2	164.7±21.4	140.7±21.4	177.4±23.5

Data are expressed as means ± SEM

* p < 0.05 vs Base; ° p < 0.05 vs other diet

(from Sirtori et al, 1986)

index, improved with both diets, but more markedly with OLIVE (Figure 1). Uric acid levels did not change significantly with either dietary treatment, whereas plasma glucose was significantly lower in the OLIVE periods (**Table 3**), with an overall mean reduction of -6.2% vs CORN administration.

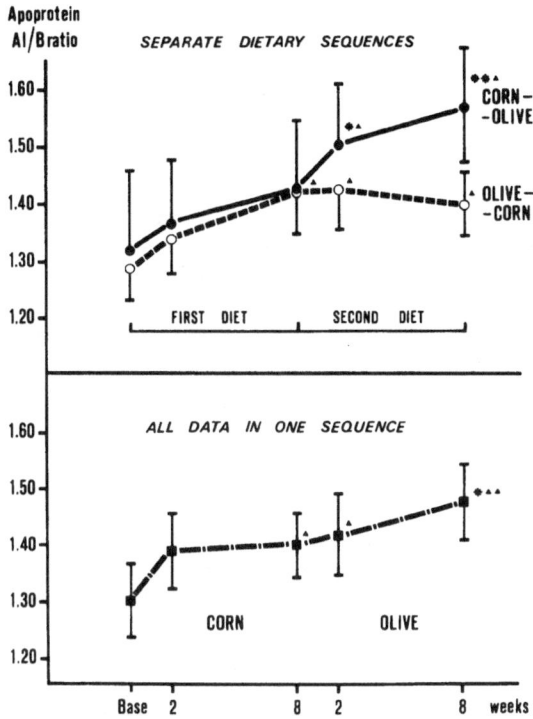

Fig. 1. Apoprotein AI/B ratios in the patients during the experimental diets (CORN—OLIVE, n=10; OLIVE—CORN, n=13). Data are expressed as X+SEM for each separate dietary sequence and pooling all data into a single sequence. Normal apo AI/B ratios for healthy people in this age range are: males 1.55+5.6; females 1.67+8.4. ▲ p <0.05, ▲▲ p <0.01 vs Base; * p <0.05, ** p <0.01 vs diet given in the first sequence.

Platelet data from only 21 patients were available; one patient, in fact, presented with a spontaneous platelet aggregation, not allowing quantitative analyses of the dietary effects on aggregability and/or release of prostaglandin metabolites; the other was an occasional aspirin consumer. In the OLIVE-CORN sequence (n=12), a significant increase of the threshold aggregatory concentration (TAC) to adrenaline occurred with both diets, more marked with CORN. CORN also determined a statistically significant increase of the TAC to arachidonic acid, whereas OLIVE, in the last sampling, was associated with a significant reduction of the TXB_2 release (**Table 4**). In the CORN-OLIVE (n=9) sequence, only a significant increase of the TAC for collagen, both vs base and vs CORN, may be noted at the end of the OLIVE treatment. No significant variations occurred in the release of MDA or of TXB_2.

CONCLUSIONS

In the last two years, clinical and experimental[17] studies have re-evaluated the comparative activities of dietary monounsaturated and polyunsaturated fatty acids. Although the beneficial activity of PUFA-rich diets on plasma lipids, in particular on total cholesterol, has never been disputed, some more recent related and unrelated findings have cast same doubts on the real advantages for human health of diets with a marked enrichment with PUFA. In particular, lipoprotein changes after PUFA-rich diets have generally been caracterized by reduced HDL-cholesterol[20], as confirmed by the more recent clinical studies. This finding may carry an unquantifiable disadvantage in terms of cardiovascular prevention[21]. Moreover, polyunsaturated diets may lead to an increased formation of peroxide radicals, possibly harmful, both for an increased carcinogenic risk and reduced life span expectancy[22].

The above described studies have re-examined the relative activity of diets rich in polyunsaturated and monounsaturated fatty acids on lipid metabolism, as well as on platelet function and eicosanoid metabolism. Classical studies by Keys et al[10] had shown a significant activity of PUFA-rich diets on plasma cholesterol, predictable from standard formulas, vs a negligible activity of MUFA-rich diets. These studies were carried out with mixed oil formulas and with variable caloric intake in healthy outpatient volunteers. The use of liquid formula diets by the Grundy group[11,12], suggested that MUFA-rich diets may be as effective and possibly more desirable than PUFA diets. In particular, there is clear evidence that a MUFA-rich diet may be preferable to a very low fat diet with PUFA[12].

Our studies have been carried out with standard solid diets, only varied in the fat content. Lower percentages of fat calories were present in the basal diet (30% vs 40% in the American studies) and we used a physiological, non formulated diet. Olive oil did not appear to reduce total cholesterolemia per se, but it did maintain the reduction achieved with the CORN diet. The OLIVE diet induced, moreover, significant lipoprotein and particulary apolipoprotein changes. Although the predictive value of apoprotein changes, in terms of cardiovascular morbidity and mortality is disputed, most studies agree on the validity of their determination[23]. Compositional studies by Vega et al[24] described a parallel change of apo B and LDL cholesterol levels in hyperlipidemic patients after PUFA, with a concomitant reduction of HDL-cholesterol and of apo AI, indicating that the chemical composition of the particles is not altered by this diet. Similar, although less dramatic, findings were collected in our study. With OLIVE, some discrepancies were noted between the reduction of LDL cholesterol and of apo B (this latter generally more marked). Furthermore, the AI/B ratios were significantly elevated by the OLIVE diet.

Platelet studies, showing less dramatic variations, compared to elsewhere reported findings[9,15] provided, however, evidence that the two diets exert at least similar effects. In vitro data on endothelial cells has shown an inibitory activity of 18:2 on prostacyclin production, vs no effect of 18:1[25]. There was no specific antagonism to any of three aggregants used in the study by either diet, thus indicating that these may act similarly, somewhat reducing platelet sensitivity. An unexpected finding from our study was the significant reduction of plasma glucose levels induced by the OLIVE diet in both dietary sequences. Somewhat similar observations had been previously reported by Katsilambros et al[26], who examined diabetics receiving a standard test meal, and a similar meal containing vegetables and olive oil (15 ml). There was a clear reduction of the plasma glucose responses after the latter meal, attributed by the Authors to a delayed gastric emptying induced by the oil. These definitely interesting findings should be re-examined in properly designed clinical trials.

The demonstration that a low lipid diet rich in olive oil, with a relatively low P/S ratio, exerts a favorable effect on plasma lipid and apolipoprotein levels as well as on platelet function, compared to a reference diet rich in PUFA, is new. These observations suggest that a dietary replacement may not always be necessary in high-risk individuals consuming a traditional Mediterranean diet, low in total and saturated fat. These findings also point out that part of the apparent protection against cardiovascular disease noted in this geographical region may be related to the intake of olive oil as the major dietary fat.

ACKNOWLEDGEMENT

Supported in part by a Grant from the European Economic Community (Direction Génerale de l'Agriculture, Direction C. Division 3).

REFERENCES

1. V. K. Ovacov and V.A. Bystrova, Present trends in mortality in the age group 35-64 in selected developed countries between 1950-73, Wld. Hlth. Stat. Quart. 31: 208-26 (1978)
2. W. B. Kannel, J. T. Doyle, C.D. Jenkins, L. Kuller, R.D. Podell, and J. Stamler, Optimal resources for primary prevention of athero-sclerotic diseases, Circulation 70: 157-205A (1984)
3. A. Ferro-Luzzi, P. Strazzullo, P. Scaccini, A. Siani, S. Sette, M. A. Mariani, T. Mastranzo, R. M. Dougherty, and J. M. Jacono, Changing the Mediterranean diet: effects on blood lipids, Am. J. Clin. Nutr. 40: 1927-37 (1984)

4. R. Salvadori, La Dieta Mediterranea, Rome, Italy: Idee Libro Publ. (1984)

5. J. D. Swales, Dietary salt and hypertension, Lancet i: 1177-9 (1980)

6. A. Rivellese, G. Riccardi, A. Giacco, D. Pacioni, S. Genovese, P. L. Mattioli, and M. Mancini, Effect of dietary fiber on glucose control and serum lipoproteins in diabetic patients, Lancet ii: 447-50 (1980)

7. A. S. St Leger, A. L. Cochrane, and F. Moore, Factors associated with cardiac mortality in developed countries with particular reference to the consumption of wine, Lancet i: 1017-20 (1979)

8. R. L. Jackson, O. D. Taunton, J. D. Morrissett, and A. M. Gotto Jr., The role of dietary polyunsaturated fat in lowering plasma cholesterol in man, Circ. Res. 42: 447-53 (1978)

9. J. R. O'Brien, M. D. Etherington, S. Jamieson, A. J. Vergroesen, and F. Ten Hoor, Effect of dietary polyunsaturated fats on some platelet-function tests, Lancet ii: 995-7 (1976)

10. A. Keys, J. T. Anderson, and F. Grande, Serum cholesterol response to changes in the diet. I. Iodine value of dietary fat versus 2S-P, Metabolism 14: 747-58 (1965)

11. F. M. Mattson and S. M. Grundy, Comparison of effects of dietary saturated, monounsaturated and polyunsaturated fatty acids on plasma lipids and lipoproteins in man, J. Lipid Res. 26: 194-202 (1985)

12. S. M. Grundy, Comparison of monounsaturated fatty acids and carbohydrates for lowering plasma cholesterol, N. Engl. J. Med. 314: 745-8 (1986)

13. E. J. Schaefer, R. I. Levy, N. D. Ernst, F. Van Sant, and H. B. Brewer Jr., The effects of low-cholesterol, high polyunsaturated fat, and low-fat diets on plasma lipid and lipoprotein cholesterol levels in normal and hypercholesterolemic subjects, Am. J. Clin. Nutr. 34: 1758-63 (1981)

14. T. Kuusi, C. Ehnholm, J. K. Huttunen, E. Kostiainen, P. Pietinen, U. Leino, U. Uusitalo, T. Nikkari, J. M. Jacono, and P. Puska, Concentration and composition of serum lipoproteins during a low-fat diet at two levels of polyunsaturated fat, J. Lipid Res. 26: 360-7 (1985)

15. J. H. Brox, J. E. Killie, S. Gunnes, and A. Nordoy, The effect of cod liver oil and corn oil on platelets and vessel wall in man, Thromb. Haemost. 46: 604-11 (1981)

16. E. Mc Intyre, R. L. Hoover, M. Smith, M. Steer, C. Lynch, M. J. Karnovsky, and E. W. Salzman, Inhibition of platelet function by cis-unsaturated fatty acidis, Blood 63: 848-57 (1984)

17. I. Masi, E. Giani, C. Galli, E. Tremoli, and C. R. Sirtori, Diets rich in saturated, monounsaturated, and polyunsaturated fatty acids differently affect plasma lipids, platelets, and arterial wall eicosanoids in rabbits, Ann. Nutr. Metab. 30: 66-72 (1986)

18. Lowering Blood Cholesterol to Prevent Heart Disease, J. Am. Med. Ass. 253: 2080-86 (1985)

19. C. R. Sirtori, E. Tremoli, E. Gatti, G. Montanari, M. Sirtori, S. Colli, G. Gianfranceschi, P. Maderna, C. Zucchi Dentone, G. Testolin, and C. Galli, Controlled evaluation of fat intake in the Mediterranean diet: Comparative activities of olive oil and corn oil on plasma lipids and platelets in high risk patients, Am. J. Clin. Nutr. in press

20. J. Shepherd, C. J. Packard, J. R. Patsch, A. M. Gotto Jr., and O. D. Taunton, Effects of dietary polyunsaturated and saturated fats on the properties of high-density lipoproteins and the metabolism of apolipoprotein A-I, J. Clin. Invest. 61: 1582-92 (1978)

21. G. Heiss, N. J. Johnson, S. Reiland, C. E. Davis, and H. A. Tyroler, The epidemiology of plasma high-density lipoprotein cholesterol levels. The Lipid Research Clinics Program Prevalence Study, Circulation 62 (Suppl. IV): 116-36 (1980)

22. D. Harman, The aging process, Proc. Natl. Acad. Sci. USA 78: 7124-28 (1981)

23. J. D. Brunzell, A. D. Sniderman, J. J. Albers, P. O. Kwiterovich Jr., Apoproteins B and A-I and coronary artery disease in humans, Arteriosclerosis 4: 79-83 (1984)

24. G. Lena Vega, E. Groszek, R. Wolf, and S. M. Grundy, Influence of polyunsaturated fats on composition of plasma lipoproteins and apoproteins, J. Lipid Res. 23: 811-22, (1982)

25. A. A. Spector, J. C. Hoak, G. L. Fry, G. M. Denning, L. L. Stoll, and J. B. Smith, Effect of fatty acid modification on prostacyclin production by cultured human endothelial cells, J. Clin. Invest. 65: 1003-12 (1980)

26. N. Katsilambros, Ph. Phipippides, and J. Boletis, Blood sugar changes in diabetics after a test meal containing vegetables and olive oil, Acta Diabet. Lat. 19: 185-86 (1982)

MINOR FATTY ACIDS OF THE N-6 SERIES AND PLATELETS

M. Lagarde, E. Vericel, M. Guichardant, S. Durbin and
M. Dechavanne

INSERM U.63, Institut Pasteur, Laboratoire d'Hémobiologie
Faculté de Médecine Alexis Carrel, 69372 Lyon Cedex 08
France

INTRODUCTION

Fatty acids of the n-6 family in mammalian cells are all polyunsaturated since they derive from linoleic acid (18:2n-6) which is essential (figure 1).

Figure 1. Scheme of polyunsaturated fatty acids biosynthesis in mammalians. This concerns the n-6 series. E : elongase

$$\Delta6\text{-desaturase} \quad E \qquad \Delta5\text{-desaturase} \qquad E \qquad \Delta4\text{-desaturase}$$
$$18:2n\text{-}6 \longrightarrow 18:3n\text{-}6 \longrightarrow 20:3n\text{-}6 \longrightarrow 20:4n\text{-}6 \longrightarrow 22:4n\text{-}6 \longrightarrow 22:5n\text{-}6$$

$$E$$
$$18:2n\text{-}6 \longrightarrow 20:2n\text{-}6$$

In numerous cells, including blood and vascular cells, these acids are quantitatively most important. But, except for arachidonic acid (20:4n-6) and in a lesser extent linoleic acid (18:2n-6) which are major components of the polyunsaturated pool, the other members of the family are generaly quantitatively minor. They may have however important biological activities, as far as platelet functions are concerned for instance, and their activities might be relevant under special nutritional conditions. This paper will briefly review data concerning these fatty acids and will also bring recent results about in vitro metabolism of dihomogamma-linolenic acid (20:3n-6) and in vitro metabolism of its precursor, gamma-linolenic acid (18:3n-6).

EFFECT AND METABOLISM OF N-6 FAMILY FATTY ACIDS

18:2n-6 is known for years as a potential depressor of platelet functions as shown by decreasing the occurence of the arterial thrombosis in rat (1) or platelet aggregation in diabetics (2). However, it might inhibit the generation of prostacyclin by endothelial cells in culture (3). It is a poor substrate of platelet lipoxygenase but seems to be well oxygenated by 15- or n-6-lipoxygenases. Endothelial cells, for instance, produce substantial amounts of 13-OH-18:2 from their endogenous 18:2n-6. This lipoxygenase product inhibits adhesion to the endothelium (4).

Gamma-linolenic acid (18:3n-6) is found in significant quantities around 10% of total acids) in evening primrose, gooseberry and blackcurrant seeds (5). Feeding 18:3n-6 would bypass the Δ6-desaturase step, the limiting enzyme for producing dihomo-gamma-linolenic acid (20:3n-6). Intake of 18:3n-6 would then be of interest under situations where Δ6-desaturase is depressed (6). Our personal experience concerning 18:3n-6 intake will be reported in the next paragraph of this paper.

In contrast to 18:2n-6, 18:3n-6 is oxygenated by both platelet cyclo-oxygenase and lipoxygenase (7). The former enzyme produces 10-OH-6,8-15:2 instead of 12-OH-5,8,10-17:3 (HHT) from 20:4n-6. As expected, the latter enzyme provides 10-OH-6,8,12-18:3 but also substantial quantities of 13-OH-6,9,11-18:3 (7), indicatihg that human platelets exhibit a 15-or n-6-lipoxygenase activity.

20:3n-6 has been extensively studied, being the precursor of PGE1, a potent inhibitor of platelet aggregation (8). In dietary supplement, 20:3n-6 inhibits platelet functions (9,10) although recent data disagree this assumption (11)

20:3n-6 and 20:4n-6 are almost equivalent substrates for platelet cyclooxygenase but PGH2 is much more easily converted by thromboxane synthase than PGH1. Subsequently, more PGE1, D1 and F1α are produced than PGE2, D2 and F2α (12,13). 20:3n-6 is also substrate of platelet lipoxygenase, leading to the formation of 12-OH-8,10,14-20:3. This pathway also leads to the formation of small amounts of di- and tri-hydroxy derivatives (12). These amounts seem however slightly higher than those obtained from 20:4n-6 through the same pathway (14). We have recently found that human platelets convert 20:3n-6 into a cyclooxygenase dependent monohydroxy derivative, 15-OH-8,11,13-20:3. This result differs from that of hamberg (8) who reported that the corresponding molecules from 18:3n-6, 13-OH-6,9,11-18:3, was not cyclooxygenase dependent. According to our result, 15-OH-8,11, 13-20:3 would rather be a by-product of an aborted cyclooxygenase activity as it has been described previously in smooth muscle cells with the formatior of 15-HETE from 20:4n-6 (15). The rate production of 15-OH from 20:3n-6 is higher than that of 12-OH. Its relative amount is highest at low concentrations of 20:3n-6 suggesting it might be of physiological relevance. Wether it is biologically active is not know yet.

Adrenic acid (22:4n-6) seems to be produced from 20:4n-6 in platelets, at least under certain conditions (16). 22:4n-6 is a strong inhibitor of TXA2 formation, while it is converted into dihomo TXA2 (17), which is biologically inactive (18). This differs from its oxygenation into dihomo PGI2 by endothelial cells (19), this derivative sharing the antiaggregating effect of PGI2, although in a lesser extent (18). This makes 22:4n-6 a potential inhibitor of platelet-endothelial cell interactions.

22:5n-6 is the end terminal fatty acid of the family. In linolenic acid (18:3n-3) deficiency it may increase, especially in tissues normally rich in 22:6n-3. In these cases it might replace 22:6n-3 in cellular structures (20). The effect of 22:5n-6 upon platelet behaviour is not known. Platelet cyclooxygenase and thromboxane synthase convert it into Δ4-dihomo thromboxane and 14-OH-4,7,10,12-19:4 as 22:4n-6 is converted into dihomo thromboxane and 14-OH-7,10,12-19:3. Similarly 22:5n-6 and 22:4n-6 are oxygenated by platelet lipoxygenase into 14-OH derivatives. This differs from 22 carbon fatty acids of the n-3 family, 22:5n-3 and 22:6n-3, which are oxygenated into both 11- and 14-OH compounds (21,22). The biological activity of these products remains to be determined.

Finally, a certain amount of 20:2n-6 is found in platelets. To our knowledge, the metabolism and the effect of this fatty acid has never been studied.

Table 1 summarizes some data concerning the oxygenation of minor fatty acids of the n-6 family.

Table 1. Summary of the oxygenation of n-6 family
fatty acids via cyclooxygenase (CO) and
lipoxygenase (LO). E.C. endothelial cells

	CO	LO
18:2n-6	-	13-OH-18:2 (E.C.)
18:3n-6	10-OH-15:2	10- and 13-OH-18:3
20:3n-6	12-OH-17:2,TXB1, PGE1,D1,F1α, 15-OH-20:3	12-OH-20:3
20:4n-6	12-OH-17:3,TXB2, PGE2,D2,F2α	12-OH-20:4
22:4n-6	14-OH-19:3,dihomo TXB2,dihomo PGI2 (E.C.)	14-OH-22:4
22:5n-6	14-OH-19:4,Δ4- dihomo TXB2	14-OH-22:5

INTAKE OF 18:3N-6

In recent experiments, 1g/day of 18:3n-6 from evening primerose oil
was ingested by elderly people (74-95 years) taking no anti-platelet drug
and having a constant diet throughout the study. After two months of 18:3n-6
supplement, subjects stayed without supplementation for two months and then
were fed equal amount of 18:2n-6 from sunflower oil. Another group of subjects were given 18:2n-6 first. In this cross over study (16 patients),
platelet functions, plasma and platelet fatty acid compositions were investigated (23).

After either supplement bleeding time was not altered. The level of
plasma β-thromboglobulin and platelet factor 4, an index of in vivo platelet activation, was not modified. Platelet aggregation induced by various
agents, in the presence or absence of plasma, was not significantly altered, except for a decreased response to collagen (platelet-rich plasma)
after 18:2n-6. The oxygenation of exogenous arachidonic acid via platelet
cyclooxygenase and lipoxygenase was not modified. Similarly, when stimulated with thrombin, endogenous formation of thromboxane B2 was not significantly decreased, except when subjects after 18:3n-6 were compared to
those after 18:2n-6. Comparing these two groups, platelet vitamin E was
also significantly decreased.

Although such intakes did not succeed for modifying platelet functions,
significant enhancements of the dietary fatty acids and/or some of their
polyunsaturated fatty acid metabolites could be observed in plasma and/or
platelet lipids. However, changes in fatty acid profiles in plasma lipids
were more pronounced than in platelets. Considering plasma phospholipids
and sterylesters, the most stable pools, the 18:2n-6 intake induced significant decrease of n-9 family fatty acids like 18:1n-9 and 20:3n-9, and n-3
ones like 20:5n-3 and 22:5n-3. In contrast, 18:2n-6 was more elevated. After
18:3n-6 intake, the most drastic changes concerned the increased 20:3n-6
and the decreased 18:1n-9. But the potential interest of enhancing the
20:3n-6 proportions could be attenuated by the significant increase of
20:4n-6 and decrease of 20:5n-3 and 22:4n-6.

In platelets, the 18:2n-6 supplement did not induce much alterations
of fatty acid distribution in either phosphatidyl-choline (PC), -ethanola-
mine (PE) or-inositol (PI), except for a significant decreased 18:1n-9
accompanied with a tendency of increased 20:4n-6 in PI.

More drastic changes could be observed after 18:3n-6. They were significant depletions of 16:1n-7 and 20:1n-9 and enhancement of 20:3n-6 in PC.
In PE the intake significantly decreased 16 DMA and 20:5n-3. In PI, 18:3n-3

was decreased and 20:4n-6 increased. Also 16:0 was depleted and 18:0 reciprocally enhanced.

In summary we can notice that several changes could be observed after fatty acid supplements, especially after 18:3n-6. However, in most cases, variations of fatty acids believed to be beneficial in respect to platelet functions, would be counterbalance by enhancements of pro-aggregatory fatty acids e.g. 20:4n-6 and depletions of anti-aggregatory ones e.g. 20:5n-3. This might explain why intakes of either 18:2n-6 or 18:3n-6 did not significantly alter platelet functions.

In the same experiments with diabetes mellitus, similar results were obtained although the variations in fatty acid compositions were somewhat different (unpublished observations).

CONCLUSION

In spite of the potential interest of certain minor fatty acids of the n-6 series upon platelet functions, the beneficial effect of dietary manipulations with these acids is not obvious. As compared to this approach, that which uses n-3 family fatty acids in much more promissing. In this respect, the benefit brought by fish fats intake is now well documented (24).

REFERENCES

1. Hornstra, G. 1974, Dietary fats and arterial thrombosis, Haemostatis, 2:21.

2. Houtsmuller, H. E., Van Hal-Ferwerde, J., Zahn, K. J. and Henkes, H. E., 1981, Favorable influences of linoleic acid on the progression of diabetic micro- and macro-angiopathy in adult onset diabetes mellitus, Progr. Lipid Res., 20:337.

3. Spector, A. A., Hoak, J. C., Fry, G. L., Denning, G. M., Stoll, L. L. and Smith, J. B., 1980, Effect of fatty acid modification on prostacyclin production by cultured human endothelial cells, J. Clin. Invest., 65:1001.

4. Buchanan, M. R., Haas, T. A., Lagarde, M. and Guichardant, M., 1985, 13-hydroxyoctadecadienoic acid is the vessel wall chemorepellant factor, LOX, J. Biol. Chem., 260:16056.

5. Traitler, H., Winter, H., Richli, U. and Ingenbleek, Y., 1984, Characterization of gamma-linolenic acid in Ribes seed, Lipids, 19:923.

6. Brenner, R. P., 1981, Nutritional and hormonal factors influencing desaturation of essential fatty acids, Progr. Lipid Res., 20:41.

7. Hamberg, M., 1983, W6-oxygenation of 6,9,12-octadecatrienoic acid in human platelets, Biochem. Biophys. Res. Commun., 117:593.

8. Kloeze, J., 1969, Relationship between chemical structure and platelet aggregation activity of prostaglandins, Biochim. Biophys. Acta, 187:285.

9. Willis, A. L., Comai, K., Kuhn, D. C. and Paulsrud, J., 1974, Dihomo-gamma-linolenate suppresses platelet aggregation when administered in vitro or in vivo, Prostaglandins, 8:509.

10. Kernoff, P. B. A., Willis, A. L., Stone, K. J., Davies, J. A. and McNicol, G.P., 1977, Antithrombotic potential of dihomo-gamma-linolenic acid in man, Brit. Med. J., 2:1441.

11. Szczeklik, A., Gryglewski, R. J., Sladek, K., Kosta-Trabka, E. and Zmuda, A., 1984, Dihomo-gamma-linolenic acid in patients with athe- rosclerosis : effects on platelet aggregation, plasma lipids and low-density lipoprotein-induced inhibition of prostacyclin generation, Thrombos. Haemostas. 48:344.

12. Falardeau, P., Hamberg, M. and Samuelsson, B., 1976, Metabolism of 8,11,14-eicosatrienoic acid in human platelets. Biochim. Biophys. Acta 441:193.

13. Lagarde, M., Gharib, A. and Dechavanne, M., 1977, Different utilization of arachidonic and dihomo-gamma-linolenic acids by human platelet pros- taglandin synthetase, Biochimie, 59:935.

14. Jones, R. L., Kerry, P. J., Poyser, N. L., Walker, I. C. and Wilson, N. H., 1978, The identification of trihydroxy eicosatrienoic acids as products from the incubation of arachidonic acid with washed human platelets, Prostaglandins, 16:583.

15. Bailey, J. M., Bryant, R. W., Whiting, J. and Salata, K., 1983, Cha- racterization of 11-HETE and 15-HETE, together with prostacyclin, as major products of the cyclooxygenase pathway in cultured rat aorta smooth muscle cells, J. Lipid. Res., 24:1419.

16. Davenas, E., Ciavatti, M., Nordoy, A. and Renaud, S., 1984, Effects of dietary lipids on behaviour lipid biosynthesis and lipid composition in rat platelets, Biochim. Biophys. Acta 793:278.

17. Van Rollins, M., Horrocks, L. and Sprecehr, H., 1985, Metabolism of 7,10,13,16-docosatetraenoic acid to dihomo-thromboxane, 14-hydroxy- 7,10,12-nonadecatrienoic acid and hydroxy fatty acids by human pla- telets, Biochim. Biophys. Acta, 833:272.

18. Sprecher, H., Van Rollins, M., Sun, F., Wyche, A. and Needleman, Ph., 1982, Dihomo prostaglandins and thromboxane. A prostaglandin family from adrenic acid that may be preferentially synthesized in the kidney, J. Biol. Chem., 257:3912.

19. Campbell, W. B., Falck, J. R., Okita, J. R., Johnson, A. R. and Callhan, K. S., 1985, Synthesis of dihomo prostaglandins from adrenic acid (7,10,13,16-docosatetraenoic acid) by human endothelial cells, Biochim. Biophys. Acta, 837:67.

20. Tinoco, J., 1982, Dietary requirements and functions of α-linolenic acid in animals, Progr. Lipid. Res., 21:1.

21. Alvedano, M. I. and Sprecher, H., 1983, Synthesis of hydroxy fatty acids from 4,7,10,13,16,19-[1-^{14}C]-docosahexaenoic acid by human pla- telets, J. Biol. Chem., 258:9339.

22. Guichardant, M. and Lagarde, M., 1985, Studies on platelet lipoxygenase specificity towards icosapolyenoic and docosapolyenoic acids, Biochim. Biophys. Acta, 836:210.

23. Vericel, E., Lagarde, M., Mendy, F., Courpron, Ph. and Dechavanne, M., 1986, Effect of linoleic acid and gamma-linolenic acid intake on pla- telet functions in elderly people, Thrombos. Res., In press.

24. Goodnight, S. H., Harris, W. S., Connor, W. F. and Illingworth, D. R., 1982, Polyunsaturated fatty acids, hyperlipidemia and thrombosis, Arteriosclerosis, 2:87.

FATS FROM MARINE ANIMALS IN HUMAN NUTRITION

J. Dyerberg

Head, Department of Clinical Chemistry
Aalborg sygehus, section North
DK-9100 Aalborg

Introduction

In dietary recommandations e.g.[1,2], aiming at preventing the increase in morbidity and mortality from atherosclerotic disorders afflicting many western societies, a common nominator has been a reduction in the overall fat consumption to 30-35% of energy intake, with one third coming from saturated fat, balanced with monounsaturated and polyunsaturated fats, which should each account for about 10% of energy intake.

The reason to the recommandation of polyunsaturated fats has been their ability to lower blood cholesterol levels[3] and in the implementation of the recommandations one will find that the term polyunsaturated fats has been used nearly synonymous with n-6 polyunsaturated fatty acids (n-6 PUFAs). The reason to this is not clear, but may steem from the concept of essentiallity introduced for n-6 PUFAs[4]. n-6 PUFAs are essential for normal growth and homeostasis of mamals, but the concept of essentiality was not introduced to characterize these fatty acids as preventing athero-thrombotic disorders when abundantly supplied. Another cause to the exclusivity in PUFA promotion found in several dietary guidelines is of course that n-6 PUFAs are easily obtainable, being major constituents in vegetable oils as corn oil, sunflower-seed oil and rape-seed oil[5] and consequently in food items processed from these products e.g. the soft margarines. The introduction of n-3 PUFAs as an essential fatty acid family both for humans and for other mamals[6,7]emphasizes the basic needs also for these food items. It is, however, their modulation of metabolic processes

113

and the implications of that for morbidity pattern and expression, when supplied in amounts far above the minimal requirements, as it is for the n-6 family, that derserves our interest, when we focuse on PUFAs and their role in dietary advices. It is my opinion that we in that respect have neglected the n-3 PUFA family, and it is about time to ask how much n-6 PUFA, and how much n-3 PUFA should we recommend to be included in the diet?

The Source of n-3 and n-6 PUFAs

Polyunsaturated fatty acids of marine origin are formed in the uni- and polycellular plants (phytoplankters and algae) of the sea, pass up the food chain and are incorporated in the lipids of sea-living animals[8]. The PUFAs will nearly exclusively belong to the n-3 20- and 22-carbon fatty acids, 20:5 n-3 eicosapentaenoic acid, 22:5 n-3 docosapentaenoic acid, and 22:6 n-3 docosahexaenoic acid, probably due to the low environmental temperature requiring highly unsaturated molecules, the n-3 structure allowing one more double bond that the n-6 structure to be introduced in the molecule.

The terrestial food chains will as the marine, contain both n-3 and n-6 PUFAs, but will in contrast to the marine organisms be dominated by the n-6 family. Furthermore the chain length will generally not exceed 18-carbon atoms. This means that our edible plant oils and the PUFAs in domestic fed animals will be dominated by linoleic acid (18:2 n-3) and only to a small extent contain linolenic acid (18:3 n-3). Arachidonic acid (20:4 n-6) is not a major nutritional component in avarage food. 18-carbon PUFAs can be chain-elongated and desaturated in mamals including man, but linoleic acid and linolenic acid will compete for the same enzymes, favouring the n-6 family, giving rise to the formation of arachidonic acid. Even when given at a high dose, linolenic acid is not converted to any greater extent to 20-carbon fatty acids in humans[9,10]. We gave linseed oil equivalent to 23.5 g linolenic acid per day and subsequently cod liver oil with 4.3 g eicosapentaenoic acid per day each for 8 days to a healthy volunteer. During the linseed oil period 18:3 n-3 rose markedly in the platelet lipids, whereas 20:5 n-3 only increased very slightly. In contrast to that cod liver oil, at a much lower n-3 PUFA dose, increased the 20:5 n-3 content markedly (table I).

The practical consequenses of experiments like that is that if it is considered advantageous to increase the levels of n-3 PUFAs of carbon 20 and 22 chain length in organ lipids, these fatty acids have to be supplied as such. The use of 18:3 n-3 will in that respect be of little value. The sea foods are excellent scources for long chained n-3 PUFAs. Much research

114

TABLE I. Percent composition of linolenic (18:3)- and eicosapentaenoic (20:5) acid in plasma lipids after feeding codliver oil and linseed oil.

		Codliver oil[a]		Linseed oil[b]	
		Before	After	Before	After
18:3	CE	tr	tr	tr	3.0
	TG	tr	tr	0.5	11.9
	PL	tr	tr	tr	2.2
20:5	CE	0.7	7.1	0.9	1.4
	TG	tr	5.3	0.5	0.6
	PL	0.7	7.1	.0.8	1.5

CE: cholesterol ester.
TG: triglycerides.
PL: phospholipids.
tr: traces.
a.   Equivalent to 4.3 g 20:5/day for 8 days.
b.   Equivalent to 23.5 g 18:3/day for 8 days.

is, however, needed to enlargen our knowledge of the selectivity of incorporation of different fatty acids into organ lipids, their biological role and in mastering the technique of introducing these fatty acids in processed food items.

Health aspects of n-3 PUFAs

The basic question before introducing n-3 PUFAs in food technology and dietary advices is of course, whether we have enough evidence to support the recommandation of their use. Not anticipating any conclusion, it may be pertinent to stress the author's own opinion, which is that the evidence is at least as convincing as what supports the use of n-6 PUFAs!

The interest in health aspects of n-3 PUFAs has mainly focused on cardiovascular disorders, but other health aspects have recently caugth attention. The following will be a brief survey of the present status of research in these areas.

The first report of a favourable influence of a high sea-food intake on morbidity and mortality from cardiovascular disorders was from Norway[11]. It was, however, considered a result of a concomitant fall in dairy fat intake, even if the speed of the fall in mortality from circulatory diseases in Norway during the second world war was not easily understood as a consequence of a fall i serum cholesterol. A clinical study giving patients with cardiovascular diseases a sea-food diet was neglected, mainly due to its design which does not allow solid conclusions. The outcome was, however, very positive in favour of a prophylactic effect of sea food[12]. It was not untill our studies in Greenland Eskimos[13,14] givning evidence for a causal relationship between a high n-3 PUFA intake, a favourable shift in well established atherothrombotic risk factors and a low mortality from AMI, that attention was drawn towards a possible beneficial effect of n-3 PUFAs on thrombotic disorders.

The same findings as in Greenlanders were recorded in Japanese [15] equally characterized by a high fish consume and a low AMI morbidity.

In persuing the mechanism by which n-3 PUFAs may influence cardiovascular morbidity a spectrum of favourable biological effects have been recorded. n-3 PUFAs have, compared to n-6 PUFAs, a more profound hypolipidemic effect especially on triglycerides and VLDL, but also on cholesterol and LDL both in normals and i patients with hyperlipidemia[16,17] due to suppression of VLDL production[18]. Whereas cholesterol rich LDL-particles are well established risk factors for ischemic heart disease in men, triglyceride rich lipoprotein particles have a strong relation to the severity of coronary atherosclerotic lesions in women![19]

Eicosanoid biology is influenced by substrate competition of 20:5 n-3 versus 20:4 n-6, shifting the hemostatic balance in an antithrombotic direction, as first described by us in our Eskimo examinations[14]. The mechanism seems to be an inhibition of thromboxane A2 (TXA2) formation, a generation in small amounts of biologically inactive TXA3 and a biologically important production of prostacyclin I3 (PGI3). PGI3 is as active as PGI2 which, in contrast to TXA2, will not be suppressed by n-3 PUFAs.[20,21] The result of the shift in hemostasis will be a lesser platelet aggregability and a conservation of the antithrombotic properties of the vasculature.

The relative importance of the different biological effects of n-3 PUFAs is difficult to assess and when it comes to more subtle effects these difficulties enlargens. n-3 PUFAs influence circulating anticoagulants[22], have a sligth hypotensive effect [22,23,24] and may lower blood viscosity[25] all of which may beneficially influence athero-thrombotic development. The finding of Kromhout et al.[26] of an inverse relationship between fish consumption and 20 year mortality from coronary heart disease on a very moderate fish consume, stress the lack of sufficient insight into the mechanism by which seafood diets may operate. On the other hand it makes nutritional advices more easy to handle, than when based on epidemiological data from heavy fish consuming societies, maybe reflecting a "supersaturated" situation.

n-3 PUFAs in addition to their role in platelet aggregation, thrombosis and atherosclerosis may alter host responses to pro-inflammatory stimuli by modulating 5-lipoxygenase pathway products[27]. Arachidonic acid (20:4 n-6) results in leucotrienes of the 4-series, whereas eicosapentaenoic acid (20:5 n-3) yields leucotrienes of the 5-series with different activities from those of the 4-series. It has been shown in animal studies that autoimmune disease may be diminished by feeding fish oil[28]. The implications for these disorders in man is unknown but inspiring research is being conducted in psoriasis[29] in rheumatoid arthritis[30] and in malignant disorders.

The present survey of health aspects of n-3 PUFAs has been deliberately general not including the data of the works referred to. Let me round this survey off by showing (Table II) first a spectrum of diseases in a fish eating society with frequencies markedly differant from that of a non-fish eating community. Hopefully to sharpen the interest for research into possible mechanisms to these differences. Secondly by taking some of the morbidity data surveyed here and putting them into a framework, including biochemical and biological findings to illustrate yet unproven but possible interactive relations (fig.1). Anyone is allowed to make his own correlations and to add new evidence as it appears. To day the relation between a high seafood intake and a low myocardial infarction rate has been strengthened by epidemiological, demographic and experimental evidence.

TABLE II.   Age-adjusted differences in morbidity from chronic diseases between Greenland Eskimos and Danes.

	Eskimos/Danes
Acute Myocardial Infarction	1/10
Apoplexy	2/1
Psoriasis	1/20
Diabetes	rare
Bronchial Asthma	1/25
Malignant disorders	1/1
Thyreotoxicosis	rare
Multiple sclerosis	0
Polyarthritis chronica	low

Data from Kromann and Green [31].

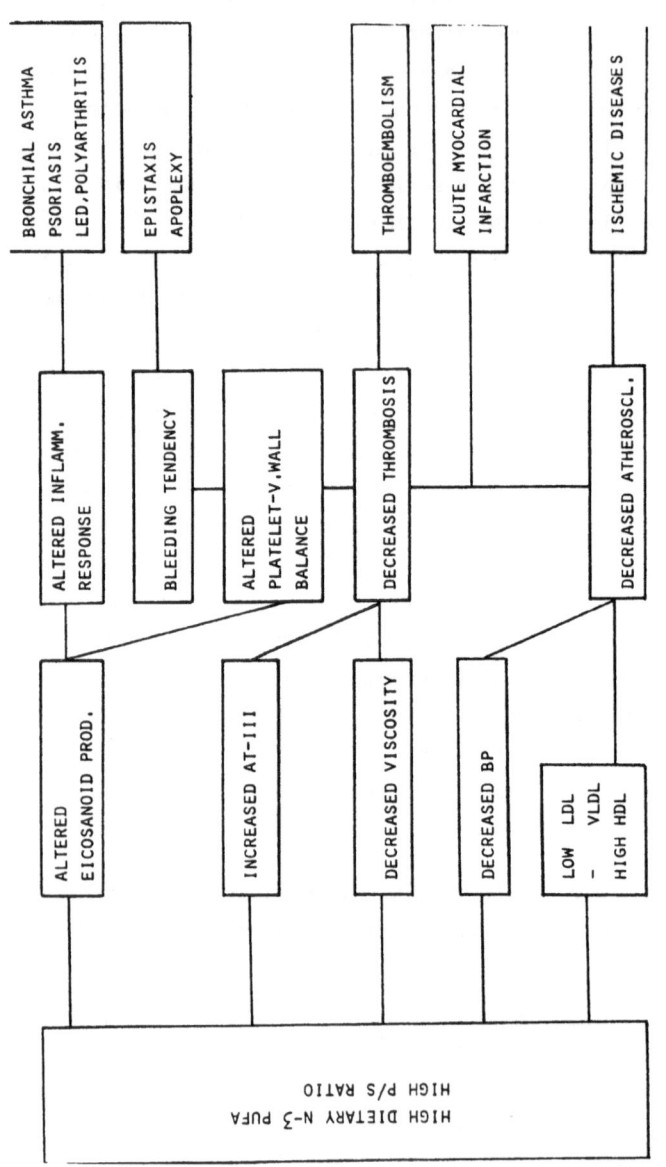

Fig. I.

Suggestions of interactive pathways for dietary n-3 PUFAs as mediators of morbidity expression.

118

Practical Nutritional Relevance

From what has been reviewed it is evident that fats from marine animals play an important role in modulation of human metabolism and have a potential in the prevention and treatment of different diseases. This raises a series of questions and considerations of practical nutritional relevance. The first is, whether n-3 PUFAs should be included in dietary recommandations, and if so, at what level, and should that be related to the intake of n-6 PUFAs? As mentioned in the introduction, the recommandation of 10% of the energy intake as polyunsaturated fat has been accepted as equivalent to n-6 PUFAs. There is certainly no absolute truth in the 10% figure, but more specific: should the 10% be devided into 5% each to the two PUFA families? In considerations like these, the relative strength of any beneficial effect of the two PUFA families must be evaluated. This is certainly not an easy task. n-3 PUFAs may excert more profound metabolic effects, when added to the diet than a concomitant reduction in n-6 PUFAs may have.

The next practical problem that arises is to translate any recommandation of n-3 PUFA enrichment of the diet into practical advices and guidelines for food processing. Long-chained n-3 PUFAs are not easily handled nor well liked by the public. On the other hand it seems an intolerable vaste that the bulk af marine oils is hydrogenated to nutritionally less valuable material for the manufacture of margarine and shortening, which is the major commercial use of fish oil presently. The food industry faces a hard task to solve these problems. Untill they are solved the pharmaceutical industry and health food manufactorers will certainly explore the possibilities of filling the gap by offering different types of concentrated food additives, comprising fish oil concentrates or derivatives.

Finally it must be checked whether n-3 PUFAs as part of natural food items or as concentrates or in prosessed food may have any adverse effects. Uncritically used, they certainly may have, as the n-6 PUFA family may! Our present status does not allow us to advocate more agressively for the use of n-3 PUFAs, than to encourage the inclusion of sea food in the dietary habits to ensure a well balanced diet.

References

1. AHA Committee Report: Risk Factors and Coronary Disease, _Circulation_ 62:449A (1980).
2. Select Committee on Nutrition and Human Needs, United States Senate, _Dietary Goals for the United States_, 2.ed., (1977).

3. B. M. Rifkind, R.S. Goor, and R. I. Levy, Current status of the role of dietary treatment in the presentation and management of coronary hart disease, Med.Clin.N.Amer., 63:911 (1979).

4. G. O. Burr, and M. M. Burr, On the nature and role of the fatty acids essential in nutrition, J.Biol.Chem., 86:587 (1930).

5. T. P. Hilditch and P. N. Williams, Chemical Constitution of Natural fats, pp 172. Chapman and Hall, London (1964).

6. R. T. Holman, Prog.Chem.Fats Other lipids, 9:279 (1968).

7. M. Neuringer, W. E. Connor, C. V. Petten, and L. Baistad, Dietary Omega-3 fatty acid deficiency and visual loss in infant rhesus monkeys, J.Clin.Invest., 73:272 (1984).

8. R. G. Ackmann, New Scources of Fats and Oils. E. H. Pryde, L. H. Princen, and Mukhergee Eds. pp. 340. American Oil Chemist.Scoiety Campain, Ilinois (1981).

9. J. Dyerberg, H. O. Bang, and O. Aagaard, α -linolenic acid and eicosapentaenoic acid, Lancet, 1: 199 (1980).

10. S. Renaud and A. Nordøy, "Small is beautiful": α -linolenic acid and eicosapentaenoic acid in man, Lancet, 1:1169 (1983).

11. A. Strøm and R. A. Jensen, Mortality from circulatory diseases in Norway 1940-1950, Lancet, 1:126 (1951).

12. A. M. Nelson, Diet therapy in coronary disease, Geriatrics, 27:103 (1972).

13. H. O. Bang, J. Dyerberg, and Aa. Nielsen, Plasma lipid and lipoprotein pattern i Greenlandic west-coast Eskimos, Lancet, 1:1143 (1971).

14. J. Dyerberg, and H. O. Bang, Hæmostatic function and platelet polyunsaturated fatty acids in Eskimos, Lancet, 2:433 (1979).

15. A. Hirai, T. Hamazaki, T. Terano, T. Nishikawa, Y. Tamura, A. Kumagai, and J. Sajiki, Eicosapentaenoic acid and platelet function in Japanese, Lancet, 2:1132 (1980).

16. W. S. Harris, W. E. Connor, and M. P. McMurry, The comparative reduction of the plasma lipids and lipoproteins by dietary polyunsaturated fats: Salmon oil versus vegetable oil, Metabolism, 32: 179 (1983)

17. B. E. Phillipson, D. W. Rothrock, W. E. Connor, W. S. Harris, and D. R. Illingworth, Reduction of plasma lipids, lipoproteins and apoproteins by dietary fish oils in patients with hypertriglyceridemia, N.Engl.J.Med., 312:1210 (1985).

18. P. J. Nestel, W. E. Connor, F. Reardon, S. Connor, S. Wong, and R. Boston, Suppression by diets rich in fish oil of very low density lipoprotein production in man, J.Clin.Invest., 74:82 (1984).

19. M. F. Readon, P. J. Nestle, I. H. Craig, and R. W. Harper, Lipoprotein predictors of the severity of coronary artery diseases in men and women. Circulation, 71:881 (1985).

20. S. Fischer and P. C. Weber, Thromboxane A3(TXA3) is formed in human platelets after dietary eicosapentaenoic acid (20:5w3). Biochim. Biophys. Res. Comm., 116:1091 (1983).

21. S. Fischer and P. C. Weber. Prostaglandin I3 is formed in vivo in man after dietary eicosapentaenoic acid. Nature, 307:165 (1984).

22. J. Z. Mortensen, E. B. Schmidt, A. H. Nielsen, and J. Dyerberg, The effect of n-6 and n-3 polyunsaturated fatty acids on hemostasis, blood lipids and blood pressure, Thromb. Hemostasis (Stuttgart), 50:543 (1983).

23. P. Singer, M. Wirth, S. Voigt, S. Zimoutkowski, W. Gödicke, and H. Heine, Clinical studies on lipids and blood pressure lowering effect of eicosapentaenoic acid - rich diet, Biomed. Biochim.Acta, 43:421 (1984).

24. R. Lorenz, V. Spengler, S. Fischer, J. Duhm, and P. Weber, Platelet function, thromboxane formation and blood pressure control during supplementation of western diet with cod liver oil, Circulation, 67: 504 (1983).

25. T. Terano, A. Hirai, T. Hamazaki, S. Kobayashi, T. Fujita, Y. Tamura, and A. Kumagai, Effect of oral administration of highly purified eicosapentaenoic acid on platelet function, blood viscosity and red cell deformability in healthy human subjects, Atherosclerosis, 46: 321 (1983).

26. D. Kromhout, E. B. Bosschieter and C. Coulander, The inverse relation between fish consemption and 20-year mortality from coronary heart disease, N.Engl.J.Med. 312:1205 (1985).

27. T. H. Lee, J. M. Mencia-Husta, C. Shih, E. J. Corey, R. A. Lewis and K. F. Austen, Effects of exogenous arachidonic, eicosapentaenoic, and docosahexaenoic acid on the generation of 5-lipoxygenase pathway products by ionophore-activated human neutrophils, J.Clin.Invest., 74: 1922 (1984).

28. J. D. Picket, D. R. Robinson and A. D. Steinberg, Effetcs of dietary enrichment with eicosapentaenoic acid upon autoimmune nephritis in femal NZB X NZW/F1 mice, Athritis,Rheum. 26:133 (1983).

29. J. J. Voorhees, Leukotrienes and other lipoxygenase products in the pathogenesis and therapy of psoriasis and other dermatoses, Arch.Dermatol. 119:541 (1983).

30. J. M. Kremer, J. Bigaueette, A. V. Michalek, M. A. Timchalk, L. Lininger, R. I. Rynes, C. Huyck, J. Zieminski and Bartholomew. Effects

of manipulation of dietary fatty acids on clinical manifestations of rheumatoid arthritis, <u>Lancet</u> 1:184 (1985).

31. N. Kromann, and A. Green, Epidemiological studies in the Upernavik district, Greenland, <u>Acta Med.Scand</u>. 208:401 (1980).

n-3 FATTY ACIDS AND THE EICOSANOID SYSTEM

Peter C. Weber

Department of Preventive Medicine and Clinical Epidemiology
Harvard Medical School, Massachusetts General Hospital
Boston   MA 02114

BACKGROUND

Cardiovascular diseases are rare among Eskimos living on their traditional maritime diet even though this diet is high in calories, rich in fat and relatively low in "classical" polyunsaturated fatty acids of the n-6 type, but rich in polyunsaturated n-3 fatty acids (2,6,13). Furthermore, although it had already been observed almost 30 years ago that fish oil reduces blood fat levels even more than n-6 fatty acid rich vegetable oil do (1), the potential of the n-3 fatty acids - contained neither in butter nor margarine in any relevant amount - in atherothrombotic and chronic inflammatory diseases was not recognized for a long time.

The ORIGIN of n-3 FATTY ACIDS

In the n-3 polyunsaturated fatty acids the cis double bonds begin at the 3rd carbon atom counting from the methyl end of the molecule; in the n-6 fatty acids, the better known fatty acids of the linoleic acid family, they start at the 6th carbon atom.

The desaturation step from the n-6 fatty acids to the n-3 fatty acids seems to be carried out exclusively at the level of linoleic acid, $C18:2n-6$ to form the α-linolenic acid, $C18:3n-3$ in the chloroplasts of e.g. green leaves and phytoplancton. Following chain elongation, further desaturation and concentration in the food chain, the polyunsaturated, long chain n-3 fatty acids eicosapentaenoic acid (EPA, $C20:5n-3$) and docosahexaenoic acid (DCHA, $C22:6n-3$) are found in high concentrations in cold water fish such as mackerel, salmon and trout (29).

The EICOSANOID SYSTEM

Physiological and pathophysiological reactions like vascular resistance, thrombosis, wound healing, inflammation and allergy are modulated by oxygenated metabolites of polyunsaturated fatty acids with 20 carbon atoms, such as arachidonic acid (AA; C20:4n-6). These metabolites are collectively termed eicosanoids. Interference with eicosanoid synthesis is one mechanism of action of various therapeutic agents, among them antithrombotic compounds, antihypertensives and antiinflammatory drugs (32). A change of eicosanoid production and eicosanoid-dependent cellular functions also may be achieved by altering eicosanoid precursor availability (6,30,31,32).

In contrast to the nutrition of our remote ancestors, which has been always low in fat but relatively rich in n-3 fatty acids (7), the predominance of the n-6 fatty acids in our current diet and, conse-quently, in the cell membranes, leads to a dominance of the eicosanoids derived from arachidonic acid. These include, for example, the highly pro-aggregatory and vasoconstrictive thromboxane A_2 (TXA$_2$) in the thrombocytes, as "antagonist" the antiaggregatory and vasodilating prostacyclin (PGI$_2$) in the endothelium, and the highly chemotactic and pro-inflammatory leukotriene B_4 (LTB$_4$) in granulocytes, monocytes and macrophages.

SUBSTITUTION of n-6 by n-3 FATTY ACIDS

In an early study we demonstrated that it is possible to induce in Caucasians less reactive platelets and reduced formation of pro-aggregatory and vasoconstrictive thromboxane by substituting mackerel, rich in n-3 polyunsaturated fatty acids, as the sole source of their dietary fat (24). In plasma and platelet membrane phospholipids n-3 fatty acids increased at the expense of arachidonic acid inducing a pattern of fatty acids which was characterised by a low content of AA and high contents of EPA and DCHA.

SUPPLEMENTATION of the WESTERN DIET with n-3 FATTY ACIDS

In subsequent studies (17,26,27) the Western diet, which supplies saturated, monounsaturated and - almost exclusively - n-6 polyunsaturated fatty acids, was supplemented with 10-40 ml/day of cod liver oil (CLO) providing per day about 1-5 grams of EPA and about 1-6 grams of DCHA. EPA and DCHA were time- and dose-dependently incorporated into plasma-, platelet- and erythrocyte membrane phospholipids at the expense of n-6 polyunsaturated fatty acids C18:2n-6 and C20:4n-6. Bleeding time increased, platelet count, platelet aggregation upon ADP and collagen and associated thromboxane formation decreased. Blood pressure and pressure responses to norepinephrine and angiotensin II fell, without major changes in plasma catecholamines and red cell cation fluxes, but slight decreases in renin, urinary aldosterone, kallikrein and prostaglandins E2 and F2α (Table1). Biochemical and functional changes were rapid in onset and reversed 4-8 weeks after cessation of the fish oil supplement.

FORMATION of EICOSANOIDS from n-3 FATTY ACIDS

In vitro, cell incubates form eicosanoids from exogenous eicosa-pentaenoic acid that differ chemically only in the presence of an additional n-3 double bond, but biologically are vastly different (18) and more favorable than the eicosanoids derived from arachidonic acid (Table 2).

Table 1

Bleeding time, platelet aggregability, thromboxane formation and vascular reactivity after dietary supply of n-3 fatty acids for four weeks (40 ml of cod liver oil (CLO) per day corresponding to 4 grams of EPA and 6 grams of DCHA per day).

	Western diet	CLO Supplement
Bleeding time (seconds)	104 ± 34	145 ± 52*
Platelet aggregation on collagen (0.75 ug/ml) (in % L.T.)	70 ± 5	53 ± 15 *
Associated TXB formation (ug/ml)	40 ± 13	15 ± 2.5*
Blood pressure (BP) (mmHg, systolic, upright)	122 ± 4	115 ± 6**
BP after Norepinephrine-infusion (5 ug/min)	137 ± 9	127 ± 8*
BP after Angiotensin II-infusion (1 ug/min)	148 ± 15	141 ± 10

* : $P < 0.01$; ** : $P < 0.05$ (table modified from ref. 30).

Table 2

Biological effects of major eicosanoids derived from arachidonic acid (C20:4n-6) and eicosapentaenoic acid (C20:5n-3).

	C20:4n-6	C20:5n-3
Platelet	$TX A_2$ pro-aggregatory vasoconstrictory	$TX A_3$ non pro-aggregatory non vasoconstrictory
Endothelial Cell	$PG I_2$ anti-aggregatory vasodilatory	$PG I_3$ anti-aggregatory vasodilatory
Peripheral Granulocytes, Macrophages, Monocytes	$LT B_4$ strongly chemotactic	$LT B_5$ weakly chemotactic

For instance, thromboxane A_3 was found to be practically non-proaggregatory and non-vasoconstricting, but prostacyclin from EPA (PGI_3) is biologically as active as PGI_2. The leukotriene B_5 (LTB_5) from eicosapentanoic acid is only weakly chemotactic and pro-inflammatory as compared to LTB_4 (for ref. see 28).

In order to relate beneficial functional effects (4,6,17,24) of a dietary modification of the eicosanoid precursor fatty acids from the n-6 to the n-3 class to an alteration in the spectrum of eicosanoids produced, it is necessary to analyze in vivo/ex vivo the spectrum of eicosanoids after manipulation of the dietary supply with n-6 or n-3 fatty acids. We measured by GC, GC-MS, HPLC and RIA, the kinetic changes in plasma- and cellular phospholipid fatty acid composition in relation to the formation of TXB_3 and TBX_2 in activated platelets and LTB_4 and LTB_5 in stimulated granulocytes (PMNL) ex vivo. We also determined the excretion of the major metabolites, PGI_2-M and PGI_2-M and $TXA_{2/3}$-M of endogenously formed PGI_3, PGI_2 and TXA_3/TXA_2, respectively. The experiments were performed after dietary substitution (mackerel for 1 to 7 days; corresponds to 30-35 g EPA+DCHA/day) or supplementation (fish oil for 1 to 5 months; corresponding to 2.5-11g EPA+DCHA/day) of the Western diet in male Europeans (8,9,26,28).

As shown in Table 3, after dietary fish or fish oil, rich in the n-3 PUFAs, EPA and DCHA, TXB_3 was formed from EPA in stimulated platelets at only 5-15% of TXB_2 which was reduced to 30-50% of control. In volunteers with high basal excretion rates of the major thromboxane metabolite, TXB_2-M, the excretion of $TXB_{2/3}$-M after the n-3 PUFA enriched diet was reduced to control (26), results not shown. Importantly, and in contrast to the pattern of thromboxane formation, PGI_3-M increased dose- and time-dependently up to 60% of an <u>unchanged</u> PGI_2-M excretion (9,26). PGI_2-M excretion even increased when the Western diet was substituted by mackerel (9). Kinetic and dose response studies indicated that the changes of plasma fatty acid spectra and in eicosanoid formation occurred rapidly within hours, lasted for the entire supplementation periods and were dose related to the amounts of n-3 PUFAs provided in the diet (9,26). This indicates a biologically important relationship.

After dietary supply with n-3 PUFAs, peripheral PMNL formed LTB_5 from incorporated cellular EPA after stimulation with the CA^{++} ionophore A23187 (28). The conversion rates of cellular EPA and AA to LTB_5 and LTB_4, respectively, were similar. The products formed were quantitatively related to their parent fatty acids in the cellular lipids. At high concentrations of exogenous EPA, the formation of the highly chemotactic LTB_4 from endogenous AA was completely suppressed. In long term supplementation experiments the formation of LTB_4 also fell and granulocyte and monocyte functions were dampened (16).

Altogether, the findings demonstrate cellular differences in the cyclooxygenation and lipoxygenation of EPA after dietary supply and time- and dose-dependent effects of dietary n-3 PUFAs on the formation of eicosanoids from endogenous AA in different cells (e.g. platelets, endothelial cells, PMNL). In addition, we found important differences in the metabolism of the two n-3 PUFA, EPA and DCHA (10) and in the in vivo versus in vitro metabolism of EPA in platelets (25). Very recently, we demonstrated that dietary DCHA can be retroconverted to EPA in man (27) and that DCHA can reduce platelet aggregability by a mechanism probably not related to eicosanoid formation.

Table 3

Eicosanoid formation in man (in vivo, ex vivo) after dietary supply of n-3 fatty acids.

	Control diet "Western"	n-3 fatty acid-enriched diet CLO[1]	n-3 fatty acid-enriched diet Mackerel
Thromboxane (as percent of control)			
TXB_2	100 %	40-70	30-50
TXB_3	n.d.	10	-
Prostacyclin, major urinary metabolite (ng/g creat.)			
PGI_2-M	146 ± 39	162 ± 52	236 ± 32 *
PGI_3-M	n.d.	83 ± 25	134 ± 48
Leukotrienes ($pmol/10^7$ PMNL[2])			
LTB_4	218 ± 89	253 ± 19	-
LTB_5	n.d.	70 ± 19	-

[1] CLO = cod liver oil

[2] PMNL = polymorphonuclear leukocytes

n.d. = not detectable

- = not determined

* = P < 0.05 vs control diet

(table modified from ref. 31).

The functional effects of diets enriched in both EPA and DCHA include reduced platelet aggregation and leukocyte function, a prolonged bleeding time, reduced blood pressure and reduced blood pressure response to pressor hormones, and blunted inflammatory reactions (22). The results of our studies suggest that these effects may be induced by:

- reduced TXA_2-formation in platelets with little synthesis of inactive TXA_3 ;

- synthesis of active PGI_3 in addition to unaltered production of PGI_2 ; and

- formation of biologically less active LTB_5 in PMNL associated with unaltered or reduced synthesis of LTB_4 .

These effects of n-3 fatty acids on eicosanoid formation and cell function might also provide the basis for the inverse relationship between fish consumption and myocardial infarction and death that have been reported recently in two large longitudinal population studies of risk factors and chronic diseases (14,23).

OTHER EFFECTS of n-3 FATTY ACIDS

The relationship of other biochemical and functional characteristics observed after dietary n-3 fatty acids - including a reduction of triglyceride- and cholesterol-concentrations (11,12,21,26), reduced VLDL formation (20,21), decreased intimal proliferation (15), and an increase in fibrinolytic activity (3) or even proper neuronal function (5,19) - to changes in n-3 and n-6 fatty acid composition of cell membranes and in plasma, or to the altered spectrum of eicosanoids remains to be determined.

Acknowledgements

Supported by Deutsche Forschungsgemeinschaft and by Wilhelm-Sander -Stiftung.

REFERENCES

1. Ahrens E.H., W.Insull, J.Hirsch, W.Stoffel, M.L.Peterson, J.W. Farquhar, T.Miller, and H.J.Thomasson, The effect on human serum-lipids of a dietary fat, highly unsaturated, but poor in essential fatty acids, Lancet 1:115 (1959)
2. Bang H.O., J.Dyerberg, and A.Nielsen, Plasma lipid and lipoprotein pattern in Greenlandic west coast Eskimos, Lancet 1:1143 (1971)
3. Barcelli U., P.Glas-Greenwalt, and V.E.Pollak, Enhancing effect of dietary supplementation with n-3 fatty acids on plasma fibrinolysis in normal subjects, Thromb.Res. 39:307 (1985)
4. Brox J.H., J.E.Killie, S.Gunnes, and A.Nordoy, The effect of cod liver oil and corn oil on platelets and vessel wall in man, Thromb. Haemostasis 46:604 (1981)
5. Crawford M.A., A.G.Hassam, and G.Williams, Essential fatty acids and fetal brain growth, Lancet 1:452 (1976)
6. Dyerberg J., H.O.Bang, E.Stofferson, S. Moncada, and J.R.Vane, Eicosapentaenoic acid and prevention of thrombosis and athero-sclerosis, Lancet 1:117 (1978)
7. Eaton S.B., and M.Konner, Paleolithic nutrition. A consideration of its nature and current implications, N.Engl.J.Med. 312:283 (1985)
8. Fischer S., and P.C.Weber, Thromboxane A_3 (TXA$_3$) is formed in human platelets after dietary eicosapentaenoic acid (C20:5n-3), Biochem.Biophys.Res.Comm. 116:1091 (1983)
9. Fischer S., and P.C.Weber, Prostaglandin I_3 is formed in vivo in man after dietary eicosapentaenoic acid, Nature (Lond.) 307: 165 (1984)
10. Fischer S., C.von Schacky, W.Siess, T.Strasser, and P.C.Weber, Uptake, release and metabolism of docosahexaenoic acid (DHA, C22:6n-3) in human platelets and neutrophils, Biochem.Biophys. Res.Comm. 120:907 (1984)
11. Goodnight S.H., W.S.Harris, W.E.Connor, and D.R.Illingworth, Poly-unsaturated fatty acids, hyperlipidemia, and thrombosis, Arterio-sclerosis 2:87 (1982)

12. Illingworth D.R., W.S.Harris, and W.E.Connor. Inhibition of low density lipoprotein synthesis by dietary omega-3 fatty acids in humans, Arteriosclerosis 4:270 (1984)

13. Kromann N. and A.Green, Epidemiological studies in the Upernavik district,Greenland, Acta Med.Scand. 208:401 (1980)

14. Kromhout D., E.B.Bosschieter, and C.D.L.Coulander, The inverse relations between fish consumption and 20 year mortality from coronary heart disease, N.Eng.J.Med. 312:1205 (1985)

15. Landymore R.W., C.E.Kinley, J.H.Cooper, M.MacAulay, B.Sheridan, and C.Cameron, Cod-liver oil in the prevention of intimal hyperplasia in autogenous vein grafts used for arterial bypass, J.Thorac. Cardiovasc.Surg. 89:351 (1985)

16. Lee T.H., R.L.Hoover, J.D.Williams, R.I.Sperling, J.Ravalese, B.W.Spur, D.R.Robinson, E.J.Corey, R.A.Lewis, and K.F.Austen, Effect of dietary enrichment with eicosapentaenoic and docosa-hexaenoic acids on in vitro neutrophil and monocyte leukotriene generation and neutrophil function, N.Engl.J.Med. 312:1217 (1985)

17. Lorenz R., U.Spengler, S.Fischer, J.Duhm, and P.C.Weber, Platelet function, thromboxane formation and blood pressure control during supplementation of the western diet with cod liver oil, Circulation 67:504 (1983)

18. Needleman P., A.Raz, M.S. Minkes, J.A.Ferrendelli, and M.Sprecher, Triene prostaglandins: prostacyclin and thromboxane biosynthesis and unique biological properties, Proc.Natl.Acad.Sci.,U.S.A., 76:944 (1979)

19. Neuringer M., W.E.Connor, C.VanPetten, and L.Barstadt, Dietary omega-3 fatty acid deficiency and visual loss in infant rhesus monkeys, J.Clin.Invest. 73:272 (1984)

20. Nestel P.J., W.E.Connor, M.R.Reardon, S.Connor, S.Wong, and R. Boston, Suppression by diets rich in fish oil of very low density lipoprotein production in man, J.Clin.Invest. 74:82 (1984)

21. Phillipson B.E., D.W.Rothrock, W.E.Connor, N.S.Harris, and R. Illingworth, Reduction of plasma lipids, lipoproteins, and apo-proteins by dietary fish oils in patients with hypertriglyceridemia, N.Engl.J.Med. 312:1210 (1985)

22. Prickett J.D., D.R.Robinson, and A.D.Steinberg, Dietary enrichment with the polyunsaturated fatty acid eicosapentaenoic acid prevents proteinuria and prolongs survival in NZBxNZW F_1 mice, J.Clin. Invest. 68:556 (1981)

23. Shekelle R.B., L.V.Missell, P.Oglesby, A.MacMillan Shryock, and J.Stamler, Fish consumption and mortality from coronary heart disease (letter), N.Engl.J.Med. 313:820 (1985)

24. Siess W., P.Roth, B.Scherer, I.Kurzmann, B.Bohlig, and P.C.Weber, Platelet membrane fatty acids, platelet aggregation, and thromboxane formation during a mackerel diet, Lancet 1:441 (1980)

25. von Schacky C., W.Siess, S.Fischer, and P.C.Weber, A comparative study of eicosapentaenoic acid metabolism by human platelets in vivo and in vitro, J.Lipid Res. 26:457 (1985)

26. von Schacky C., S.Fischer, and P.C.Weber, Long term effects of dietary marine n-3 fatty acids upon plasma and cellular lipids, platelet function and eicosanoid formation in humans, J.Clin. Invest. 76:1626 (1985)

27. von Schacky C., and P.C.Weber, Metabolism and effects on platelet function of the purified eicosapentaenoic and docosahexaenoic acids in humans, J.Clin.Invest. 76:2446 (1985)

28. Strasser T., S.Fischer, and P.C.Weber, Leukotriene B$_5$ is formed in human neutrophils after dietary eicosapentaenoic acid, Proc. Natl.Acad.Sci.,U.S.A. 82:1540 (1985)

29. Tinoco J., Dietary requirements and functions of alpha-linolenic acid in animals (Review), Prog.Lipid Res. 21:1 (1982)

30. Weber P.C., S.Fischer, C. von Schacky, R.Lorenz, and T.Strasser, Dietary eicosapentaenoic acid and eicosanoid formation, in "Health effects of polyunsaturated fatty acids in seafoods", ed. A.P. Simopoulos,R.R.Kifer, Raven Press (in press)(1986a)

31. Weber P.C., S.Fischer, C. von Schacky, R.Lorenz, and T.Strasser, The conversion of dietary eicosapentaenoic acid to prostanoids and leukotrienes in man, Prog.Lipid Res. (in press) (1986b)

32. Weber P.C., Vascular resistance and platelet function on cardio-vascular drugs and diets affecting prostanoid formation, in "Atherosclerosis", eds. F.G.Schettler, A.M.Gotto, G.Middelhoff, A.J.R.Habenicht, K.R.Jurutka, pp.715-719, Springer-Verlag,Berlin, (1983)

EFFECT OF EICOSAPENTAENOIC ACID ON LEUKOTRIENE B FORMATION BY HUMAN NEUTROPHILS AND THE RELEVANCE TO INFLAMMATION

John A. Salmon and Takashi Terano*

Department of Mediator Pharmacology, Wellcome Research Laboratories, Langley Court, Beckenham, Kent, BR3 3BS U.K.

INTRODUCTION

Epidemiological surveys suggest a causal relationship between high intake of eicosapentaenoic acid (EPA) in the diet and a reduced incidence of thrombo-emobolic episodes (1-3). Subsequent studies with animals and human volunteers fed an EPA-rich diet confirmed that EPA reduces platelet aggregation, increases cutaneous bleeding times, lowers plasma lipid levels and reduces whole blood viscosity (4-7). Some of these changes have been attributed to the synthesis of metabolites of EPA which are less biologically active than similar compounds formed from arachidonic acid (AA) (8,9). However, most investigators acknowledge that EPA is a poor substrate for the cyclo-oxygenase enzyme (9,10) and therefore one of the main mechanisms by which EPA exerts its anti-thrombotic activity is probably by direct competition of the conversion of AA by the cyclo-oxygenase. Thus, high intake of EPA leads to a decrease in the amount of pro-thrombotic thromboxane A_2 (TXA_2) formed during platelet aggregation.

However, EPA is a good substrate for the lipoxygenase enzymes (11,12) which convert some polyunsaturated fatty acids to hydroperoxy intermediates. One such product of AA metabolism, 5-HPETE, is of particular interest because it can be further metabolised to leukotrienes (13,14). The leukotrienes C_4 and D_4 (LTC_4, LTD_4) are the major constituents of the "slow reacting substance of anaphylaxis" (SRS-A) and probably contribute to the development of symptoms associated with asthma and similar hypersensitivity reactions (15,16). Another leukotriene, LTB_4, is a potent chemokinetic and chemotactic agent for polymorphonuclear leukocytes (PMN) (17-19) and it has been suggested that LTB_4 plays a role in the recruitment of PMN to sites of inflammation. Both LTB_5 and LTC_5 can be biosynthesized by murine mastocytoma cells from EPA-rich oils (20): indeed, Murphy et al., (20) indicated that the efficiency of this conversion to LTB_5 is similar to the synthesis of LTB_4 from AA.

---

* Present address, 2nd Department of Internal Medicine, Chiba University, Chiba City, Japan.

In our experiments we initially investigated whether EPA is converted to leukotrienes by human PMN and whether it modified the metabolism of AA. Additionally we assessed the biological consequence of biochemical changes in vitro and in vivo.

METABOLISM OF EPA BY HUMAN PMN

Calcium ionophore (A23187)-stimulated human neutrophils are able to transform EPA into LTB_5. As illustrated in Fig. 1 the metabolism of $[^{14}C]$-EPA is qualitatively similar to that of $[^{14}C]$-AA; the major products are 5-hydroxy-eicosapentaenoic acid (5-HEPE), LTB_5 and a product with the same mobility as TXB_3 in the thin layer chromatograpy (TLC) solvent system employed. Further investigations of the "TXB_3-like" compound indicated that it was a mixture of 20-COOH- and 20-OH-LTB_5 which are metabolites of LTB_5 (see below).

After 5 min incubation of human neutrophils labelled with $[^{14}C]$-fatty acid with A23187 there was greater formation of 5-HEPE than of 5-hydroxy-eicosatetraenoic acid (5-HETE; the analogous product derived from AA). This can be appreciated by comparing the density of respective zones on the autoradiogram but it is more apparent after scraping the areas from the TLC plate and subjecting them to liquid scintillation counting. The hydroxy metabolites of both AA and EPA are derived from the corresponding hydroperoxy-acids which are the initial products of 5-lipoxygenase action. Thus, these data confirm that EPA is as good a substrate for the 5-lipoxygenase in human PMN as AA. Previously, Jakschik et al, (11) and Ochi et al, (21) showed that EPA was a slightly better substrate for 5-lipoxygenase (partially purfied enzyme) than AA.

Lower amounts of both 5-HETE and 5-HEPE were recovered at longer periods of incubation which probably reflects incorporation into phospholipids (see 22,23).

The product with the Rf similar to that of LTB_4 which was formed from EPA was confirmed as LTB_5 by reverse-phase high performance liquid chromatography (RP-HPLC) and gas chromatography-mass spectrometry (GC-MS) (See reference 24 for details of these procedures). Although the conversion of EPA to 5-HEPE was greater than that of AA to 5-HETE, the biosynthesis of LTB_5 in human PMN was less than that of LTB_4 (see Fig. 1).

Further quantitative differences in the biosynthesis of the leukotrienes from the two fatty acids were exposed when the products were subjected to RP-HPLC. These studies revealed that wheras the enzymic formation of LTB_4 was always greater than that of the diastereomers of 6-trans-LTB_4 (the latter are formed by non-enzymic hydrolysis of the epoxide intermediate, LTA_4) the formation of the 6-trans diastereomers of LTB_5 was greater than that of LTB_5 itself (see Fig. 2).

These data imply that although the respective 5-hydroperoxy compounds formed by 5-lipoxygenase activity are good substrates for the dehydrase, leukotriene A-synthase, the resultant epoxides are handled differentially by the LTA-hydrolase: LTA_4 is far more efficiently converted to LTB_4 than is LTA_5 to LTB_5. The diastereomers of 6-trans-LTB_4 are much less biologically active than LTB_4 itself (25) and therefore this implies that the formation of biologically active compounds from EPA is not as high as from AA. The metabolic pathways of AA and EPA in human PMN are summarised in Fig. 4; similar data have been reported by Prescott (26).

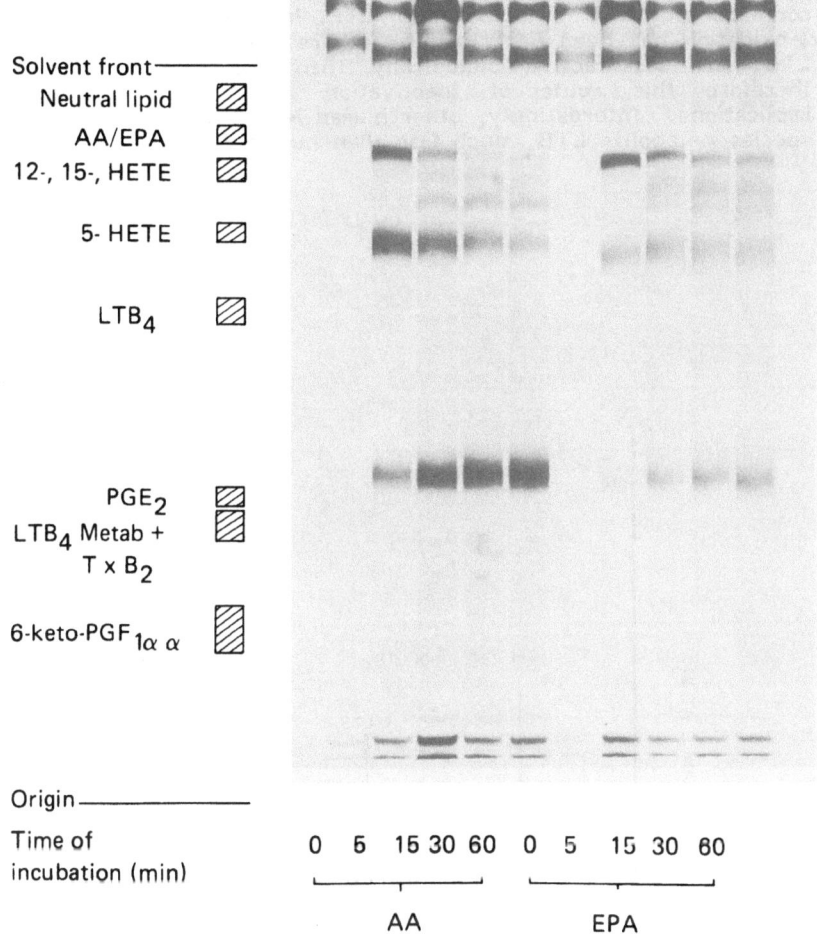

Solvent front ———
Neutral lipid ▨
AA/EPA ▨
12-, 15-, HETE ▨

5- HETE ▨

LTB$_4$ ▨

PGE$_2$ ▨
LTB$_4$ Metab + ▨
T x B$_2$

6-keto-PGF$_{1\alpha}$ α ▨

Origin ———

Time of
incubation (min)   0   5   15 30 60   0   5   15 30 60

                          └──────┬──────┘ └──────┬──────┘

                                 AA   EPA

Fig. 1.   Autoradiogram of radioactive compounds separated by TLC,
which were obtained after A23187-stimulation of human PMN
pre-labelled with either [^{14}C]-AA or [^{14}C]-EPA. Fatty acid
(100nCi) was incorporated into cells (approx. 1ml of 5 x
10^6/ml) which were incubated with A23187 (1 µg/ml) for the
times indicated. After acidification (pH 4), compounds were
extracted into ethyl acetate and then subjected to TLC
(Solvent system: The organic phase of ethyl acetate - iso -
octane - water - acetic acid; 110 : 50 ; 100 : 20).

Incubation of human PMN with A23187 for longer than 5 min produced lower amounts of both LTB$_4$ and LTB$_5$ (see Fig. 1) but there was a corresponding increase in the formation of the product with an Rf similar to TXB$_2$. Subsequent RP-HPLC studies indicated that this product derived from both AA and EPA was actually a mixture of two polar metabolites, namely the 20-OH and 20-COOH derivatives of the corresponding LTB. Metabolism of LTB$_4$ in human neutrophils by ω-oxidation has been described by others (27-29). Both metabolites of LTB$_4$ are less active biologically than the parent compound and therefore this route of inactivation may have pathophysiological implications. Interestingly, other human leukocytes and PMN from other species metabolize LTB$_4$ much less than human PMN (28).

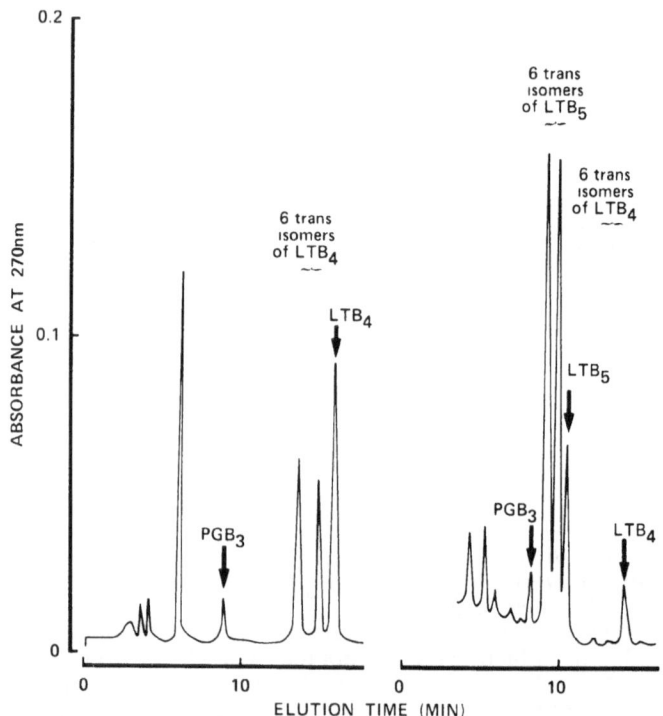

Fig. 2. RP-HPLC of products obtained after incubating either AA (Left Hand Panel) or EPA (Right Hand Panel) with human PMN. Cells (3 x 10^7/ml) were stimulated with A23187 (1 µg/ml) in the presence of 50 µg/ml fatty acid. The products were extracted using C18-silica mini-columns (Bond-Elut) and separated using RP-HPLC (5µ C18-Spherisorb eluted at 1 ml/min with methanol – water – acetic acid, 70 : 30 : 0.01 v/v, adjusted to an apparent pH of 5.9). Prostaglandin B$_3$ was added as an internal standard.

The metabolism of the leukotrienes can be more thoroughly monitored using RP-HPLC. Inspection of Figure 3 reveals that the ratios of the most polar peak (peak I;20-COOH-LTB) to that of peak II (20-OH-LTB) was higher for the LTB_4 metabolites than for the LTB_5 products. This was a consistent finding which indicates that conversion of 20-OH-LTB_5 to 20-COOH-LTB_5 was slower than the analogous oxidation of 20-OH-LTB_4. Indeed the overall metabolism of LTB_5 was slower than that of LTB_4 when measured by RP-HPLC (see Fig. 4) or by radioimmunoassay (30); the half-lives for the disappearance of LTB_4 and LTB_5 (20ng of each) during incubation with human neutrophils (5×10^6) were approximately 20 and 40 minutes respectively.

The approximate efficiencies of the major transformation of AA and EPA in human neutrophils are summarized in Fig. 4.

The addition of unlabelled EPA to human PMN pre-labelled with [^{14}C]-AA reduced the formation of all metabolites of AA at high doses (31) but, of the 5-lipoxygenase products, the synthesis of LTB_4 was particularly vulnerable (30% inhibition at 10 µg/ml) suggesting that transformation by the LTA-hydrolase was a rate limiting step which has also been proposed by Jakschik and Kuo (32). A similar interpretation of experimental data with EPA was reported by Prescott (26) and Lee et al, (33).

BIOLOGICAL PROPERTIES OF LTB_4 AND LTB_5

In order to compare the biological properties of LTB_4 and LTB_5, we initially had to prepare LTB_5; this was achieved by biosynthesis from EPA in A23187-stimulated rabbit peritoneal PMN (24). The product was extracted, purified by RP-HPLC, identified by GC-MS and quantified by UV spectrophotometry. The LTB_4 used in the following experiments was obtained from Professor E.J. Corey, Harvard Universty.

Leukotiene B_5 was 30-100 times less active than LTB_4 in causing degranulation and chemokinesis of human PMN in vitro; also, it induced less aggregation of rat PMN (24). In vivo, LTB_5 was at least 10 times less active than LTB_4 in potentiating bradykinin-induced extravasation in rabbit skin (24); this property is believed to be a reflection of the leukotatic activity of LTB (see 34).

Other investigators have reported similar lower activity of LTB_5 compared to LTB_4 (25,35-37).

EFFECT OF EPA ON FORMATION OF LTB_4 AND LTB_5 IN VIVO

The effect of oral administration of EPA to rats on the biosynthesis of LTB_4 and LTB_5 by leukocytes stimulated with A23187 has been evaluated (38). The concentration of LTB_4 was determined by radioimmunoassay (RIA) and RP-HPLC using PGB_2 as internal standard. Supplementation of a normal rat diet with EPA (240 mg/kg/day as the ethyl ester which had a purity of 75%) for 4 weeks caused a significant increase in the formation of LTB_5 and a decrease in the synthesis of LTB_4 by stimulated leukocytes. The ratio of EPA:AA in leukocyte phospholipids correlated ($r = 0.795$, $p < 0.001$) with the $LTB_5 : LTB_4$ ratio produced after stimulation of the leukocytes (38).

Similar qualitative changes have been observed (39-41) after feeding animals and human subjects with diets rich in EPA (using MaxEPA, menhaden oil or unprocessed mackeral).

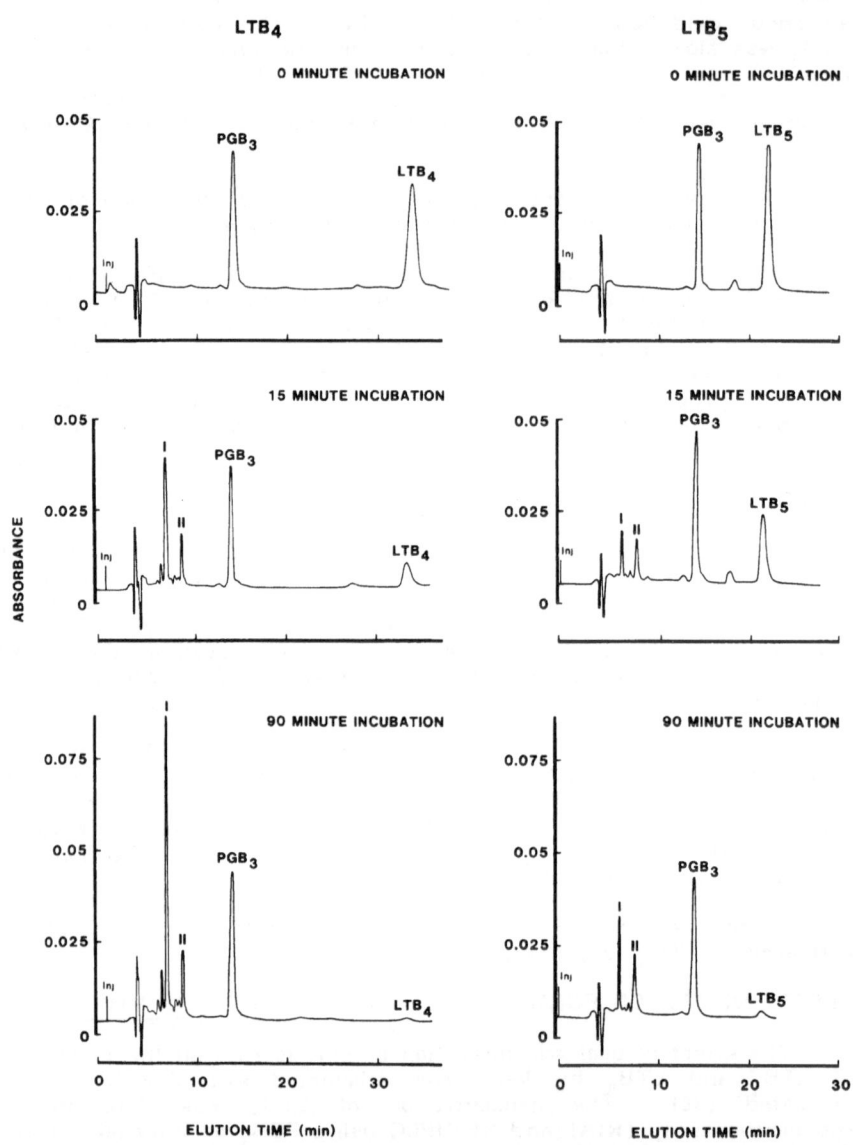

Fig. 3.   RP-HPLC separation of metabolites of LTB$_4$ and LTB$_5$ formed in human PMN LTB$_4$ and LTB$_5$ were incubated with cells for the times indicated and then the products were extracted and subjected to HPLC as described in the legend to Fig. 2.

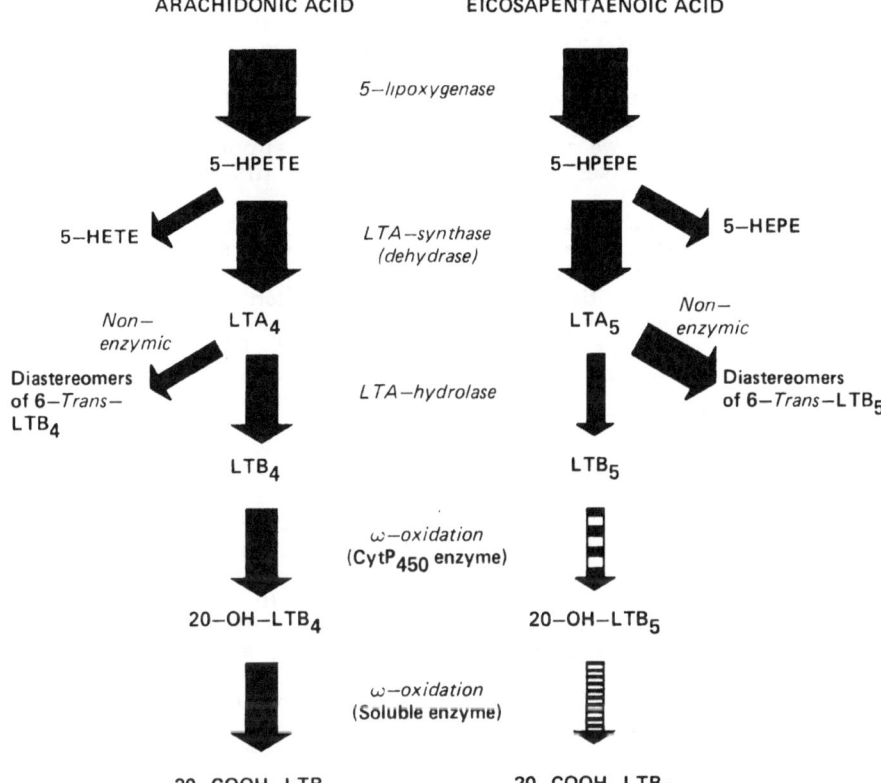

ARACHIDONIC ACID   EICOSAPENTAENOIC ACID

5–lipoxygenase

5–HPETE   5–HPEPE

5–HETE   *LTA–synthase*
(dehydrase)   5–HEPE

*Non–
enzymic* LTA_4 LTA_5 *Non–
enzymic*

Diastereomers
of 6–*Trans*–
LTB_4 *LTA–hydrolase*   Diastereomers
of 6–*Trans*–LTB_5

LTB_4 LTB_5

ω–*oxidation*
(CytP$_{450}$ enzyme)

20–OH–LTB_4 20–OH–LTB_5

ω–*oxidation*
(Soluble enzyme)

20–COOH–LTB_4 20–COOH–LTB_5

Fig. 4.   Metabolism of arachidonic acid and eicosapentaenoic acid in human neutrophils. The width of the arrows provides an approximate guide to the relative yields from the reactions which are controlled by the enzymes indicated in italics or, as is the case in the formation of the diastereomers of 6-trans-LTB, occur non-enzymically. The oxidation of LTB_5 and 20-OH-LTB_5 occur more slowly than the analogous transformation of LTB_4 and 20-OH-LTB_4 and this is represented by the broken arrows.

Tissue injury induced by a variety of factors triggers an inflammatory type reaction with infiltration of PMN. The invading cells are capable of synthesizing and releasing a variety of pro-inflammatory mediators which can exacerbate the damage produced by the injurious stimulus. These mediators include AA metabolites, oxygen radicals (e.g. superoxide), platelet activating factor (PAF) and lysosomal enzymes. Thus, agents which prevent the activation and infiltration of PMN should prevent the extent of the damage; this is relevant to chronic inflammatory disease (e.g. rheumatoid arthritis) and also to myocardial ischemia, pulmonary embolism and endotoxin-induced injury (42-45). Therefore, if LTB_4 has a chemotactic role, the data discussed in the previous sections suggest that an EPA-rich diet could reduce tissue damage by suppressing the formation of LTB_4 and thereby decreasing the leukocyte infiltrate at sites of inflammation. In order to evaluate this potential we assessed the effects of supplementing the diet with EPA on the responses in two models of acute inflammation.

Supplementation of a standard rat diet with 240 mg/kg/day EPA-ethyl ester for 4 weeks reduced, but not significantly, the concentration of immunoreactive LTB_4 (i-LTB_4) and leukocytes in inflammatory exudate derived from implantation of carrageenin-impregnated sponges (46; Table 1). Leukotriene B_5 cross-reacts (approx. 17%) with the antibody used for the RIA of LTB_4 and additional examination by combined RP-HPLC and RIA demonstrated that concentrations of LTB_5 were increased after the dietary supplement. Thus, the value of i-LTB_4 recorded in Table 1 is the sum of the LTB_4 and LTB_5 present; it is possible that the synthesis of LTB_4 may have been depressed significantly.

The evidence that prostaglandins, particularly PGE_2, mediate some cardinal signs of inflammation (erythema, oedema, hyperthermia, hyperalgesia) is convincing (47) and therefore the effect that EPA has on prostaglandin formation was also considered. In the model of acute inflammation described above, supplementation of the diet with EPA decreased significantly the concentrations of PGE_2 and TXB_2 (46; see Table 1). There were no detectable trienoic cyclo-oxygenase products thereby confirming that EPA is a poor substrate for this enzyme.

Table 1. Effect of supplementing a normal rat diet with EPA on eicosanoid concentrations measured by RIA and leukocyte numbers in inflammatory exudate obtained by s/c implant of sponges impregnated with 2% carrageenin for 4 hr

Diet	Concentration of immunoreactive (i-) eicosanoids (ng/ml)			Leukocyte count (x 10^6/ml)
	i-PGE_2	i-TXB_2	i-LTB_4	
Control	13.9 ± 1.9	9.8 ± 2.4	4.1 ± 1.2	9.5 ± 2.1
EPA-enriched	5.9 ± 1.9***	2.3 ± 0.28*	1.9 ± 0.3	7.1 ± 1.1

Values are the mean ± S.E.M. of data from five animals; *P < 0.05, *** P < 0.01; Data from reference 45

In the second model of acute inflammation, oedema formation 2-4h after injection of carrageenin into rat paws was significantly reduced in animals fed an EPA-rich diet (46; Fig. 5). Oedema formation during the first hour has been attributed to the release of vasoactive amines, such as histamine and serotonin, whereas the swelling between 2-4h is considered to be due to the presence of prostaglandins, particularly PGE_2 (48). This hypothesis is supported by the evidence that inhibitors of prostaglandin synthesis, such as indomethacin, are not effective at 1h but do reduce swelling 2-4h. The data obtained in animals with the dietary supplement suggest that EPA affects the prostaglandin phase of oedema formation and this is supported by the observed reduction of PGE_2 in exudate obtained after implanting carrageenin-impregnated sponges as discussed above. The suppression of LTB_4-synthesis after EPA could also contribute to the lower oedema formation.

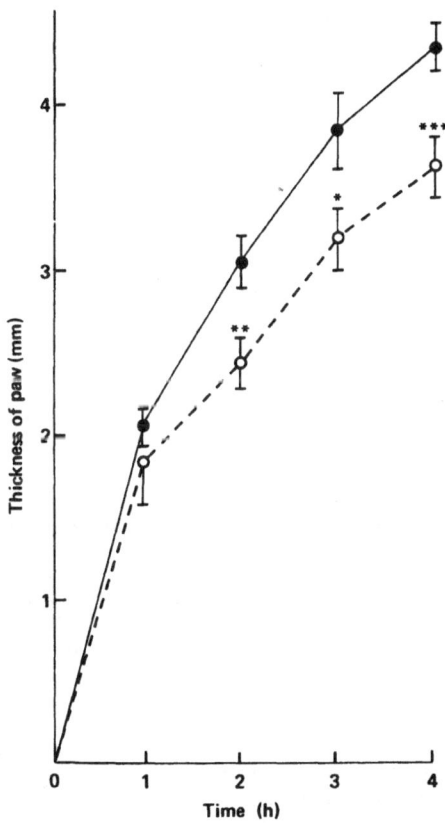

Fig. 5.   Comparison between control (—●—) and EPA-fed rats (--o--) in paw oedema response induced by sub-plantar injection of carrageenin.   Each point is the mean ± S.E.M. of the mean (n=15), * P < 0.005, ** P < 002, *** P < 0.01.   Data from reference 45.

New Zealand black x New Zealand white (NZB x NZW/F1) mice develop a glomerulonephritis that has served as a model for systemic lupus erythematosus. Prickett and colleagues (49,50) reported that a diet enriched with EPA (given as menhaden oil) delayed the development of proteinuria and prolonged survival of mice.

In another model of inflammation, Prickett et al, (51) noted that an EPA-rich diet increased the incidence, although not the severity, of collagen-induced arthritis in rats. The authors suggested that the detrimental effect of EPA in this model could be attributable to the decrease of PGE_2 synthesis which has been reported to exert suppressor influences on a variety of immune responses. However, in a similar model, Leslie et al, (52) reported that a fish-oil diet increased the time of onset, and reduced the incidence and severity of the arthritis.

Lee et al, (53) have recently reported that a fish-oil enriched diet augmented the decrement in dynamic compliance during anaphylaxis in guinea-pigs when the histamine-mediated component of the response was eliminated. These authors considered that the most likely explanation for their data was that EPA in the fish-oil competitively inhibited metabolism of AA by the cyclo-oxygenase leading to lower amounts of bronchodilator prostaglandins (e.g. PGE_2). Thus, asthmatic subjects – at least those sensitive to non-steroidal anti-inflammatory agents (NSAID) – could experience an exacerbation of symptoms if they ingested EPA. However, bronchoconstriction in asthmatics can also be due to mediators released from recruited acute inflammatory cells and therefore EPA could decrease the severity of airway responses by reducing the cellular influx.

CONCLUSIONS

An EPA-enriched diet may reduce the severity of acute inflammatory reactions by decreasing the synthesis of dienoic prostanoids and tetraenoic leukotrienes. Also, EPA promotes the formation of LTB_5 which is considerably less active biologically than LTB_4 and may antagonize the pro-inflammatory action of LTB_4. Our results suggest that a nutritional approach to anti-inflammatory therapy is an interesting possibility which deserves further study.

Epidemiological surveys do indicate that populations that consume a high amount of EPA in their diet (e.g. Greenland eskimos) have a lower incidence of several inflammatory disorders when compared to a matched population with a much lower intake of EPA (see 54). However, such surveys do not confirm whether this is a causal or casual relationship.

Supplementation of the diet with EPA may not always be beneficial because it suppresses the synthesis prostaglandins, which have important modulatory activities in some situations; for example, in chronic inflammation and asthma the reduced formation of PGE_2 may be detrimental.

We support the suggestion of Lands and Bimbo (55) that an altered dietary intake of EPA can be expected to act in a long-term manner suited to preventative or prophylactic approaches rather than the short-term activity associated with pharmacological agents. Thus experiments in animal models of human disease, which are necessarily compressed into a short time-scale, are probably unsuitable for assessing the effects of EPA. The benefits or shortcomings of adding EPA to the diet will only be clarified by conducting well-controlled clinical trials. In a recently published limited and open trial, a daily supplement of EPA (1.8g) to rheumatoid arthritis patients improved morning stiffness and the number of tender joints and these benefits disappeared after stopping the diet (56). Clearly, further longer-term trials are required to substantiate the potential anti-inflammatory effect of EPA in patients.

REFERENCES

1. J. Dyerberg, H.O. Bang and N. Horne. Fatty acid composition of the plasma lipids in Greenland Eskimos. Am. J. Clin. Nutr. 28:958 (1975).

2. J. Dyerberg, and H.P. Bang. Hemostatic function and platelet polyunsaturated fatty acid in Eskimos. Lancet ii:433 (1979).

3. A. Hirai, T. Hamazaki, T. Terano, T. Nishikawa, Y. Tamura, A. Kumagai, and J. Sajiki. Eicosapentaenoic acid and platelet function in Japanese. Lancet ii:1132 (1980).

4. W. Siess, P. Roth, B. Schere, T. Kurzman, B. Bohling and P.C. Weber. Platelet membrane fatty acids, platelet aggregation and thromboxane formation during a mackerel diet. Lancet i:441 (1980).

5. J.H. Brox, J.E. Kille, S. Gunnes and A. Nordoy. The effects of cod liver oil and corn oil on platelet and vessel wall in man. Thromb. Haem. 46:604 (1981)

6. T. Terano, A. Hirai, T. Hamazaki, S. Kobayashi, T. Fujita, Y. Tamura and A. Kumagai. Effect of oral administration of highly purified eicosapentaenoic acid on platelet function, blood viscosity and red cell deformability in healthy human subjects. Atherosclerosis 46:321 (1983).

7. Editorial. Eskimo diets and diseases. Lancet i:1139 (1983).

8. R.J. Gryglewski, J.A. Salmon, F.B. Ubatuba, B.C. Weatherley, S. Moncada and J.R. Vane. Effects of all cis 5,8,11,14,17 EPA and PGH_3 on platelet aggregation. Prostaglandins 18:453 (1979).

9. P. Needleman, A. Raz, M. Minkes, J.A. Ferrendelli and H. Sprecher. Triene prostaglandins; prostacyclin and thromboxane biosynthesis and unique biological properties. Proc. Natl. Acad. Sci. 76:944 (1979).

10. D.A. van Dorp. Aspects of the biosynthesis of prostaglandins. Progr. Biochem. Pharmacol. 3:71 (1967).

11. B.A. Jakschik, A.R. Sams, H. Sprecher and P. Needleman. Fatty acid structural requirement for leukotriene biosynthesis. Prostaglandins 20:401 (1980).

12. C. Yokoyama, K. Mizuno, H. Mitachi, T. Yoshimoto, S. Yamamoto and C.R. Pace-Asciak. Partial purification and characterization of arachidonate 12-lipoxygenase from rat lung. Biochem. Biophys. Acta. 750:237 (1983).

13. R.C. Murphy, S. Hammarstrom and B. Samuelsson. Leukotriene C: A slow reacting substance from murine mastocytoma cells. Proc. Natl. Acad. Sci. 76:275 (1979).

14. P. Borgeat and B. Samuelsson. Transformation of arachidonic acid by rabbit polymorphonuclear leukocytes. Formation of a novel dihydroxyeicosatetraenoic acid. J. Biol. Chem. 254:2643 (1979).

15. B. Samuelsson. Leukotrienes: A novel group of compounds including SRS-A. Prog. Lipid Res. 20:23 (1981).

16. E.J. Goetzl. Oxygenation of products of arachidonic acid as mediators of hypersensitivity and inflammation. Med. Clin. North Am. 55:809 (1981).

17. A.W. Ford-Hutchinson, M.A. Bray, M.V. Doig, M.E. Shipley and M.J.H. Smith. Leukotriene B, a potent chemokinetic and aggregating substance released from polymorphonuclear leukocytes. Nature (Lond). 286:264 (1980).

18. M.J.H. Smith, A.W. Ford-Hutchinson and M.A. Bray. Leukotriene B: a potential mediator of inflammation. J. Pharmacol. 32:517 (1980).

19. R.M.J. Palmer, R.J. Stepney, G.A. Higgs and K.E. Eakins. Chemokinetic activity of arachidonic acid lipoxygenase products on leukocytes of different species. Prostaglandins 20:411 (1980).

20. R.C. Murphy, W.C. Pickett, B.R. Culp and W.E.M. Lands. Tetraene and pentaene leukotrienes; selective production from murine mastocytoma cells after dietary manipulation. Prostaglandins 22:613 (1981).

21. K. Ochi, T. Yoshimoto, S. Yamamoto, K. Taniguchi and T. Miyamoto. Arachidonate 5-lipoxygenase of guinea-pig peritoneal polymorphonuclear leukocytes. J. Biol. Chem. 258:5754 (1983).

22. W.F. Stenson and C.W. Parker. 12-L-Hydroxy-5,8,10,14-eicosa-tetraenoic acid, a chemotactic fatty acid is incorporated into neutrophil phospholipids and triglyceride. Prostaglandins, 18:285 (1979).

23. R.W. Bonser, M.I. Siegel, S.M. Chung, R.T. McConnell and P. Cuatrecasas. Esterification of an endogenously synthesized lipoxygenase product into granulocyte cellular lipids. Biochemistry, 20:5297 (1981).

24. T. Terano, J.A. Salmon and S. Moncada. Biosynthesis and biological activity of leukotriene B_5. Prostaglandins 27:217 (1984).

25. A.W. Ford-Hutchinson, M.A. Bray, F.M. Cunningham, E.M. Davison and M.J.H. Smith. Isomers of leukotriene B_4 possess different biological potencies. Prostaglandins, 21:143 (1981).

26. S.M. Prescott. The effect of eicosapentaenoic acid on leukotriene B production by human neutrophils. J. Biol. Chem. 259:7615 (1984).

27. G. Hansson, J.A. Lindgren, S-E. Dahlen, P. Hedqvist and B. Samuelsson. Identification and biological activity of novel ω-oxidised metabolites of leukotriene B_4 from human leukocytes. FEBS Lett, 130:107 (1981).

28. W.S. Powell. Properties of leukotriene B_4 20-hydroxylase from polymorphonuclear leukocytes. J. Biol. Chem. 259;3082 (1984).

29. R.D.R. Camp, P.M. Woollard, A.I. Mallet, N.J. Fincham, A.W. Ford-Hutchinson and M.A. Bray. Neutrophil aggregating and chemokinetic properties of 5,12,20-trihydroxy-6,8,10-14-eicosatetraenoic acid isolated from human leukocytes. Prostaglandins, 23;631 (1982).

30. T. Terano, H. Saito, J.A. Salmon and S. Moncada. Metabolism of leukotrienes B_4 and B_5 by human polymorphonuclear leucocytes. Prog. Lipid Res. (In Press).

31. T. Terano, A. Hirai, Y. Tamura, S. Yoshida, J.A. Salmon and S. Moncada. Effect of eicosapentaenoic acid on eicosanoid formation by stimulated human polymorphonuclear leukocytes. Prog. Lipid Res. (In Press).

32. B.A. Jakschik and C.G. Kuo. Characterization of leukotriene A_4 and B_4 biosynthesis. Prostaglandins 25:767 (1983).

33. T.K. Lee, J.M. Mencia-Huerta, C. Shih, E.J. Corey, R.A. Lewis and K.F. Austen. Effects of exogenous arachidonic, eicosapentaenoic and docosahexaenoic acids on the generation of 5-lipoxygenase pathway products by ionophore-activated neutrophils. J. Clin. Invest. 74:1922 (1984).

34. C.V. Wedmore and J.T. Williams. Control of vascular permeability by polymorphonuclear leukocytes in inflammation. Nature, Lond. 289:646 (1981).

35. D.W. Goldman, W.C. Pickett and E.J. Goetzl. Human neutrophil chemotactic and degranulating activities of leukotriene B_5 (LTB_5) derived from eicosapentaenoic acid. Biochem. Biophys Res. Commun. 117:282 (1983).

36. T.H. Lee, J.M. Mencia-Huerta, C. Shih, E.J. Corey, R.A. Lewis and K.F. Austen. Characterization and biologic properties of 5,12-dihydroxy derivatives of eicosapentaenoic acid including leukotriene B_5 and the double lipoxygenase product. J. Biol. Chem. 259:2383 (1984).

37. A.G. Leitch, T.H. Lee, E.W. Ringel, J.D. Prickett, D.R. Robinson, S.G. Pyne, E.J. Corey, J.M. Drazen, K.F. Austen and R.A. Lewis. Immunologically induced generation of tetraene and pentaene leukotrienes in the peritoneal cavities of menhaden-fed rats. J. Immunol. 132:2559 (1984).

38. T. Terano, J.A. Salmon and S. Moncada. Effect of orally administered eicosapentenoic acid (EPA) on the formation of leukotriene B_4 and leukotriene B_5 by rat leukocytes. Biochem. Pharmacol. 33;3071 (1984).

39. T. Strasser, S. Fischer and P.C. Weber. Leukotriene B_5 is formed in human neutrophils after dietary supplementation with icosapentaenoic acid. Proc. Natl. Acad. Sci. USA, 82:1540 (1985).

40. T.H. Lee, R.L. Hoover, J.D. Williams, R.I. Sperling, J. Ravelese, B.W. Spur, D.R. Robinson, E.J. Corey, R.A. Lewis and K.F. Austen. Effect of dietary enrichment with eicosapentaenoic and docosahexaenoic acids on in vitro neutrophil and monocyte leukotriene generation and neutrophil function. N. Engl. Med, 312:1217 (1985).

41. S.M. Prescott, G.A. Zimmerman and A.R. Morrison. The effects of a diet rich in fish oil on human neutrophils: identification of leukotriene B_5 as a metabolite. Prostaglandins, 30:209 (1985).

42. A.R. Faggiotto, R. Ross and L. Harker. Studies on hypercholesterolemia in the non-human primate. Arteriosclerosis 4:323 (1984).

43. P.G. Hazen and B. Michel. Management of Necrotizing vasculitis with colchicine. Arch. Dermatol. 115:1303 (1979).

44. B.R. Luchesi, S.R. Jolly, M.B. Baslie and G.D. Abrams. Protection of ischemic myocardium by BW755C. Fed. Proc. 41:1737 (1982).

45. K.M. Mullane, and S. Moncada. The salvage of ischaemic myocardium by BW755C in anaesthetised dogs. Prostaglandins 24:255 (1982).

46. T. Terano, J.A. Salmon, G.A. Higgs and S. Moncada. Eicosapentenoic acid as a modulator of inflammation: effect on prostaglandin and leukotriene synthesis. Biochem. Pharmacol. 35:779 (1986).

47. S. Moncada, J.R. Vane. Mode of action of aspirin-like drugs. Adv. Int. Med 24:1 (1979).

48. M. Di Rosa, J.P. Giroud and D.A. Willoughby. Studies of the mediators of the acute inflammatory response induced in rats by carrageenan and turpentine. J. Pathol. 104:15 (1971).

49. J.D. Prickett, D.R. Robinson and A.D. Steinberg. Dietary enrichment with the polyunsaturated fatty acid eicosa-pentaenoic acid prevents proteinuria and prolongs survival in NZB x NZW/F1 mice. J. Clin. Invest. 68:556 (1981).

50. J.D. Prickett, D.R. Robinson and A.D. Steinberg. Effects of dietary enrichment with eicosapentaenoic acid upon autoimmune nephritis in female NZB x NZW/F1 mice. Arth. Rheum. 26:133 (1983).

51. J.D. Prickett, D.E. Trentham and D.R. Robinson. Dietary fish oil augments the induction of arthritis in rats immunized with type II collagen. J. Immunol. 132:725 (1984).

52. C.A. Leslie, W.A. Gonnerman, M.D. Ullman, K.C. Hayes, C. Franzblau and E.S. Cathcart. Dietary fish oil modulates macrophage fatty acids and decreases arthritis susceptability in mice. J. Exp. Med, 162;1336, (1985).

53. T.H. Lee, K.F. Austen, A.G. Leitch, E. Israel, D.R. Robinson, R.A. Lewis, E.J. Corey and J.M. Drazen. The effects of a fish-oil-enriched diet on pulmonary mechanics during anaphylaxis. Am. Rev. Respir. Dis. 132:1204 (1985).

54. N. Kromann and A. Green. Epidemiological studies in the Upernavik district, Greenland. Acta. Med. Scand. 208:401 (1980).

55. W.E.M. Lands and A.P. Bimbo. in "Possible beneficial effects of polyunsaturated fatty acids in maritime foods" International Association of Fish Meal Manufacturers, Potters Bar (1983).

56. J.M. Kremer, A.V. Michalek, L. Liningek, C. Huyck, J. Bigaudette, M.A. Timchalk, R.I. Rynes, J. Zieminski and L.E. Bartholomew. Effects of manipulation of dietary fatty acids on clinical manifestation of rheumatoid arthritis. Lancet i:184 (1985)

A BIOLOGICAL ANTIOXIDANT FUNCTION FOR VITAMIN E:

ELECTRON SHUTTLING FOR A MEMBRANE-BOUND "FREE RADICAL REDUCTASE"

Paul B. McCay, Edward K. Lai, Gemma Brueggemann, and
Saul R. Powell

Biomembrane Research Laboratory, Oklahoma Medical
Research Foundation, Oklahoma City, OK

INTRODUCTION

In the last several years, several investigators have reported
that the tocoperoxy radical, the initial product of the reaction of
vitamin E (alpha-tocopherol), is recycled back to alpha-tocopherol in
the course of its function as an antioxidant in biological membranes
(1-3). However, direct evidence for such a process has not been
provided, and the mechanism by which such a reaction could be mediated
was not known. Some of the evidence which tends to support this idea
was based on observations that lipid peroxidation in liver microsomes
from tocopherol-deficient rats is poorly inhibited by glutathione
(GSH), but is markedly inhibited by GSH in microsomes from tocopherol-
sufficient rats (4). Burk and coworkers (5-7) have shown that a heat-
labile factor is present in microsomes which, in the presence of GSH,
prevents lipid peroxidation in these membranous particles under
conditions which promote vigorous lipid peroxidation in the absence of
GSH. The activity of this heat-labile, membrane-bound factor was shown
to be heat-labile and destroyed by trypsin, and, therefore, presumably
a constitutive protein in the microsomal membrane.

An activity of this type would constitute a potent protective
mechanism for components of membranes against oxidative stress. The
various findings led us to investigate the mechanism of the GSH-
dependent inhibition of lipid peroxidation in microsomal membranes,
with the particular goal of determining whether or not this system
actually involved the recycling of tocoperoxy radicals. We wished to
determine if the GSH-dependent, heat-labile factor was involved in the
reduction of tocoperoxy radicals formed by the reaction of tocopherol
with free radicals which are known to be generated in membranes by
certain enzymic functions or by environmental factors.

The results of these experiments provide substantial support for
the conclusion that alpha-tocopherol functions as an electron shuttle
in the membrane by quenching reactive radicals produced either by
enzyme-catalyzed processes or environmental factors. The resulting
relatively stable tocoperoxy radicals will be shown to be rapidly
quenched by a membrane-bound protein utilizing GSH as a source of
electrons. This report presents data which indicate that the
protective effect of the GSH-dependent factor is completely dependent

on a catalytic quantity of tocopherol being present in the membrane. An electron shuttling capacity of this type forges an efficient link between the excellent radical quenching properties of tocopherol in a hydrophobic domain (4) with the substantial capacity of cytosolic enzymes to maintain GSH in the reduced form (8).

In essence, the system comprises a potent mechanism for stabilizing membrane structure and function when the membrane is subjected to conditions favoring free radical formation. The regeneration of the very small amount of tocopherol required for the function of this system by a membrane-bound "free radical reductase" may explain not only the potent antioxidant properties of this substance in in cell membranes, but also the low requirement for alpha-tocopherol in mature animals as well as the difficulty in causing tocopherol deficiency symptoms in fully grown animals. In addition, the findings may shed light on the reason for the serious consequences of tocopherol deficiency in the rapidly growing young animal species observed, including humans (9, 10). It may also explain some of the membrane perturbations observed in animals depleted of GSH (11).

EXPERIMENTAL PROCEDURES AND METHODS

Chemical components used in this work were either reagent grade or of the best quality available commercially. dl-Alpha-tocopherol was obtained from Sigma Chemical Co., St. Louis, MO. Glutathione (GSH) was obtained from Boehringer Mannheim Biochemicals, Indianapolis, IN. Male Sprague-Dawley rats were purchased from Sasco, Inc, Omaha, NE. The tocopherol-deficient diet was purchased from ICN Life Sciences, Cleveland, OH.

Incubation Systems

Rat liver microsomes were prepared as by differential centrifugation from livers homogenized in 0.15 M potassium phosphate buffer, pH 7.4, (1:3 w/v) as described previously (12). Each 0.1 ml of this microsomal suspension contained approximately 2.0_{mg} of protein.

Three types of peroxidation systems were used to assay the effectiveness of the microsomal factor. System No. 1 (NADPH oxidase-catalyzed lipid peroxidation) contained 0.1 ml of the microsomal suspension, 0.1 ml of either 0.3 mM NADPH or an NADPH-generating system (containing 0.3 mM NADPH, 5.0 Kornberg units of glucose-6-phosphate dehydrogenase, and 5.0 mM glucose-6-phosphate), 0.1 ml of an ADP-Fe^{3+} complex (composed of 4.0 mM ADP and 0.12 mM $FeCl_3$), and 0.7 ml of 0.15 M potassium phosphate buffer, pH 7.4. System No. 2 (non-enzymic lipid peroxidation catalyzed by ascorbic acid) contained 0.1 ml of microsomal suspension, 0.1 ml of 6.60 mM ascorbic acid, 0.1 ml of the ADP-Fe^{3+} complex, and 0.7 ml of phosphate buffer. System No. 3 (lipid peroxidation linked to carbon tetrachloride metabolism) contained 0.1 ml of microsomal suspension, 0.1 ml of the NADPH-generating system, 20 μl of CCl_4, and 0.78 ml of phosphate buffer. All systems were incubated at 37° for 30 minutes unless specified otherwise.

Index of Extent of Lipid Peroxidation

The extent of free radical-mediated peroxidation of membrane lipids can be estimated by the thiobarbituric acid reaction in which some of the products of lipid peroxidation react with this acid to form a pigment absorbing at 532 nm, and which can be converted to equivalents of malondialdehyde (13).

146

Determination of microsomal alpha-tocopherol.

Tocopherol was extracted from microsomes according to the procedure of Burton et al'(14). HPLC analyses of tocopherol in the extracts were carried out as described by Hatam and Kayden (15). Processing known amounts of alpha-tocopherol through the extraction and analytical procedures showed that no significant loss of tocopherol occurs in the processing.

Tocopherol-Deficient Rats

Rats were fed a tocopherol-deficient diet ad libitum for at least 8 weeks before use in the preparation of liver microsomes as described above. The microsomes from these animals still contained low, but detectable amounts of tocopherol.

Preparation of microsomes with extremely low tocopherol content

In order to establish whether or not there was an absolute requirement for tocopherol by the GSH-dependent microsomal factor, microsomes free of measureable tocopherol content were prepared by cold (-70°)acetone extraction of these particles. The extracted particle still contained most of the phospholipids as well as enzymic activity. In some cases, tocopherol was included in the acetone extraction medium so that the treated microsomes still contained normal amounts of tocopherol. This control was performed to demonstrate that acetone extraction per se did not inactivate the heat-labile, glutathione-dependent "free radical reductase" activity. The experimental data shown below indicate that the factor tolerated this treatment well.

Determination of Glutathione Utilization

The oxidation of GSH in the varioussystems was determined as described by Sedlak and Lindsay using Ellman's reagent (16).

Production and isolation of the alpha-Tocopheroxy Radical

The relatively stable tocopheroxy radical was produced by reacting dl-alpha-tocopherol with diphenyl dipycryl hydrazyl according to the procedure described by Boguth and Niemann (17). The tocopheroxy radical fraction was separated from the reaction products by the method of Glavind and Holmer (18). The radicals slowly decay by a second order disproportionation reaction in a hydrophobic environment, but we observed very little formation of dimers in the time scale involved in these experiments.

Incorporation of the Tocopheroxy Radical Preparation into Microsomes

Tocopheroxy radicals were incorporated into liver microsomes by mixing the preparation with microsomal pellets and incubating the mixture. Sample of these microsomes which, after resusupension and centrifugation in buffer, exhibited the EPR signal for the tocopheroxy radicals were employed in certain experiments descibed below.

RESULTS

Free radical processes can initiate lipid peroxidation (which is itself a free radical process) resulting in the the loss of membrane integrity and function. We employed three different type of lipid peroxidation-promoting systems in these investigations. The systems which initiate lipid peroxidation were investigated: the enzymic

TABLE I

INHIBITION OF LIPID PEROXIDATION BY A LIVER MICROSOMAL
HEAT-LABILE FACTOR REQUIRING GLUTATHIONE

The data shown are averages for four experiments. The
experimental procedures for assaying malondialdehyde formation
in each system are described in the methods section of the
text. The concentrations of GSH and BME in these experiments was
10 mM. The concentration of DTT was 5.0 mM. Incubations were
carried out for 30 min at 37°. The heat lability studies
on the microsomal factor were carried out using microsomes which
had been heated to 80° for 10 minutes prior to assembling
the incubation systems.

SYSTEM	THIOBARBITURIC ACID VALUE (O.D 532 nm/mg micr. protein)
1. ENZYMIC LIPID PEROXIDATION ASSOCIATED WITH CCl_4 METABOLISM):	
MICROSOMES ----------------------------	0.04 ± 0.01
MICROSOMES + NADPH + CCl_4 -------------	0.68 ± 0.28
MICROSOMES + NADPH + CCl_4 + GSH -------	0.14 ± 0.03
MICROSOMES + NADPH + CCl_4 + BME -------	0.68 ± 0.09
2. NON-ENZYMIC LIPID PEROXIDATION CATALYZED BY ASCORBATE-IRON:	
MICROSOMES ----------------------	0.05 ± 0.01
MICROSOMES + ADP-Fe^{3+} + AA -------------	0.63 ± 0.27
MICROSOMES + ADP-Fe^{3+} + AA + GSH -------	0.11 ± 0.05
MICROSOMES + ADP-Fe^{3+} + AA + BME -------	0.95 ± 0.02
HEATED MICR. + ADP-Fe^{3+} + AA ---------	0.83 ± 0.15
HEATED MICR. + ADP-Fe^{3+} + AA + GSH -----	0.83 ± 0.14
HEATED MICR. + ADP-Fe^{3+} + AA + CYS-SH --	0.19 ± 0.01

oxidation of NADPH in the presence of small amounts of ferric iron
(peroxidation system 1); during non-enzymic ascorbate-catalyzed lipid
peroxidation (peroxidation system 2); and during the enzymic metabolism
of $\cdot CCl_3$ radicals during CCl_4 metabolism (12) (peroxidation system 3).
These different systems were studied in order to establish whether or
not the effects of the GSH-dependent factor was limited to a particular
type of peroxidation (metal-catalyzed, for example). The results shown
below demonstrate that the factor appears to be effective in various
types of free radical-mediated processes.

GSH-dependency of the inhibition of lipid peroxidation

Table I shows that the GSH-dependent inhibition of lipid
peroxidation in microsomal membranes is effective in protecting the

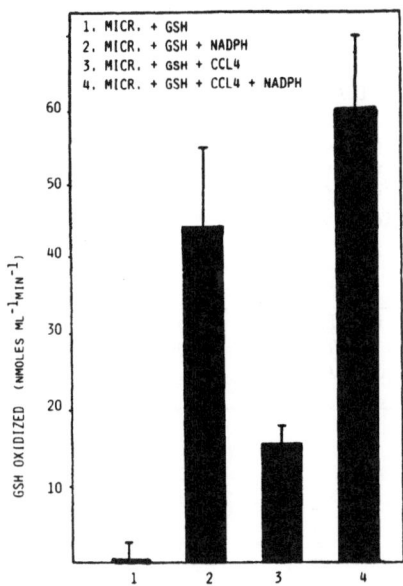

Fig. 1. Inhibition by a heat-labile, GSH-dependent factor of lipid peroxidation initiated by NADPH oxidation and by CCl4 metabolism in liver microsomes.

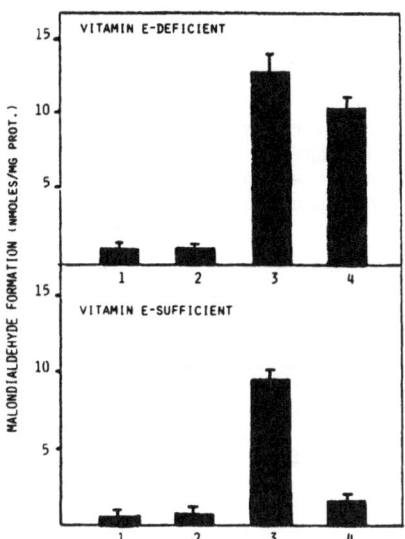

Fig. 2. Impairement of the GSH-dependent inhibition of lipid peroxidation in liver microsomes from tocopherol-deficient rats. 1. Microsomes + GSH; 2. Microsomes + GSH + NADPH; 3. Microsomes + NADPH + CCl4; 4. Microsomes + GSH + CCl4 + NADPH.

membranes against both non-enzymic metal-catalyzed lipid peroxidation as well as lipid peroxidation catalyzed by formation of a radical during the enzymic metabolism of a halocarbon. The latter process is not dependent on traces of metal ions in the system to observe extensive peroxidation. In addition, the effect is shown to be heat-labile. types of peroxidizing systems, and that it is heat-labile. There is a considerable specificity for glutathione in this system. peroxidation by the microsomal heat-labile factor. Neither beta-mercaptoethanol nor dithiothreitol at equivalent concentrations of thiol groups could replace GSH. The inhibitory effect of cysteine is shown in System 2 to be due to a heat-stable, non-enzymic effect on lipid peroxidation while the effect of GSH required the heat-labile microsomal factor (Table I). GSH was effective at concentrations of 1.0 mM while other thiols, including cysteine had no effect.

Figure 1 shows that GSH is consumed during the suppression of lipid peroxidation by the heat-labile factor. The oxidation of GSH occurred in all three types of lipid peroxidation systems. When all GSH was oxidized, the protection against lipid peroxidation is lost (data not shown). GSH oxidation does not occur if the heat-labile factor in the microsomes is inactivated. GSH consumption only occurred when conditions existed that would have produced lipid peroxidation if GSH were not in the system.

Preservation of tocopherol by GSH during oxidative stress

Table II shows that when rat liver microsomes were incubated in

TABLE II

PRESERVATION OF THE TOCOPHEROL CONTENT OF MICROSOMES BY A GSH-
DEPENDENT FACTOR DURING PEROXIDATIVE STRESS

System compositions and incubation conditions are described in
the methods section of the text. The values represent the
averages $\pm$ the standard deviations (n = 4 for the NADPH
oxidase- and CCl_4-dependent peroxidation systems; n = 3 for the
ascorbate-ADP-Fe^{3+} system).

SYSTEM	INCUBATION TIME (min)	TOCOPHEROL CONTENT (nmoles/mg prot.)	MALONDIALDEHYDE EQUIVALENTS (O.D. 532 nm)
1. NADPH OXIDASE-DEPENDENT LIPID PEROXIDATON:			
MICROSOMES	0	455 $\pm$ 56	0.40 $\pm$ 0.08
MICROSOMES	30	432 $\pm$ 58	0.41 $\pm$ 0.07
MICROSOMES + ADP-Fe^{3+} + NADPH	30	142 $\pm$ 35	7.98 $\pm$ 1.65
MICROSOMES + ADP-Fe^{3+} + NADPH + GSH (10mM)	30	350 $\pm$ 37	2.12 $\pm$ 0.79
MICROSOMES + ADP-Fe^{3+} + NADPH + BME (10 mM)	30	170 $\pm$ 51	8.20 $\pm$ 1.47
2. ASCORBATE-ADP-Fe^{3+}-DEPENDENT LIPID PEROXIDATION:			
MICROSOMES	0	503 $\pm$ 48	0.35 $\pm$ 0.35
MICROSOMES	30	442 $\pm$ 39	0.31 $\pm$ 0.09
MICROSOMES + ADP-Fe^{3+} + ASCORBATE	30	343 $\pm$ 19	4.73 $\pm$ 0.32
MICROSOMES + ADP-Fe^{3+} + ASCORBATE + GSH (10 mM)	30	438 $\pm$ 3	0.78 $\pm$ 0.63
3. CCl_4-DEPENDENT LIPID PEROXIDATION:			
MICROSOMES	0	471 $\pm$ 61	0.41 $\pm$ 0.13
MICROSOMES	30	454 $\pm$ 35	0.50 $\pm$ 0.09
MICROSOMES + CCl_4 + NADPH	30	121 $\pm$ 50	6.60 $\pm$ 2.76
MICROSOMES + CCL_4 + NADPH + GSH (10 mM)	30	334 $\pm$ 63	6.63 $\pm$ 1.72

any of the three systems which exert oxidative stress on membranes,
extensive lipid peroxidation occurs in the membrane along with loss of
a major portion of the its tocopherol content. If GSH is included in
the systems, however, the peroxidation is markedly inhibited as mentioned
above, and most of the tocopherol in the membrane is preserved. GSH
levels as low as 1.0 mM are effective. Other thiol compounds (such as
ß-mercaptoethanol) cannot function in the place of GSH (Table II).

Effect of depletion of microsomal tocopherol on the GSH-dependent inhibition of lipid peroxidation

The earlier work of Reddy et al (4) indicated that the GSH-dependent-inhibition of lipid peroxidation was not effective in microsomes from tocopherol-deficient rats. This suggested that tocopherol might play an essential role in the inhibition process, perhaps by functioning as an electron shuttle in which it quenches free radicals that may be generated in the hydrophobic zone of the membrane, forming the relatively stable tocopheroxy radical. The latter might then be reduced back to tocopherol by the heat-labile membrane-bound factor, using cytosolic GSH as the reducing agent. To explore this possibility further, liver microsomes were prepared from tocopherol-deficient and tocopherol-sufficient rats. Figure 2 shows that microsomes from the deficient animals have little capacity to inhibit lipid peroxidation in the presence of GSH. The microsomes from the tocopherol supplemented animals show a marked ability to inhibit peroxidation in the presence of GSH.

Even more suggestive evidence of a requirement for tocopherol in the GSH-dependent inhibition of peroxidation was acquired in experiments in which microsomes were totally depleted of vitamin E by extraction in cold acetone. The extraction removes neutral lipids, including

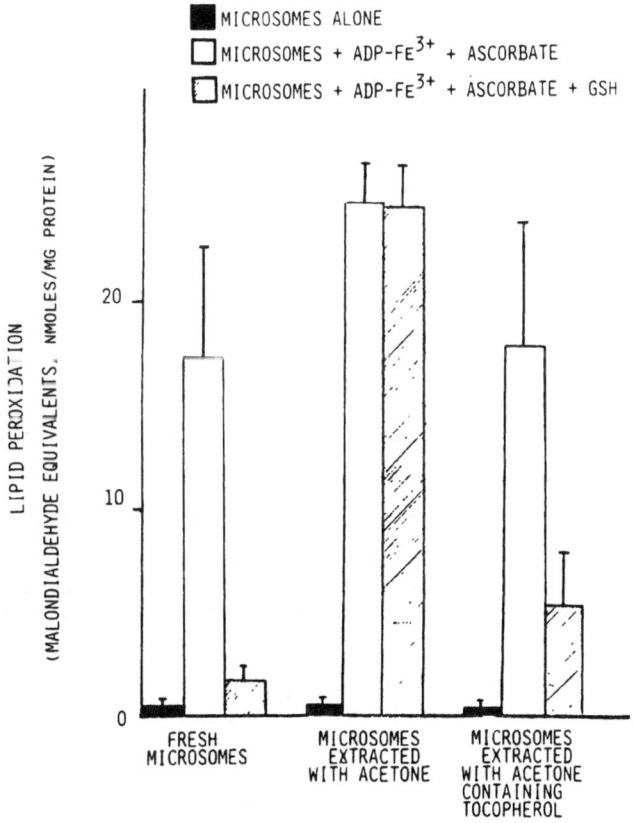

Fig. 3. Loss of the capacity of microsomes depleted of tocopherol to effect GSH-dependent inhibition of lipid peroxidation. The microsomes were extracted with cold acetone or with cold acetone containing tocopherol as described under Experimental Procedures.

TABLE III

QUENCHING OF THE TOCOPHEROXY RADICALS INCORPORATED INTO
MICROSOMAL MEMBRANES BY A MICROSOMAL FACTOR
REQUIRING GLUTATHIONE

The tocopheroxy radical was prepared and incorporated into
liver microsomes as described in the Eperimental Procedures
section. chromatography. Varian E-9 EPR spectrometer sectings
were: microwave power was 25 mW; receiver gain, 6.3 X 10^4;
time constant, microwave frequency, 9.109 GHz; 3 sec; scan
time, 16 min.; temperature, 25°. Signal intensity was
determined by measurement of center line height.

SYSTEM	SIGNAL INTENSITY
MICROSOMES CONTAINING TOCOPHEROXY RADICALS ---------------------	6.7 cm
MICROSOMES CONTAINING TOCOPHEROXY RADICALS I INCUBATED WITH GSH FOR 5 MIN ---------------------	0.9 cm
TOCOPHEROXY RADICALS INCUBATED WITH GSH FOR 5 MIN. (NO MICROSOMES) ----------------	7.5 cm

(VALUES REPRESENT RESULTS OF 3 EXPERIMENTS)

tocopherol, but leaves the membrane structure and most of the enzymic
activities largely intact. As a control for the possibility that the
solvent extraction may have inactivated the GSH-depended heat-labile
factor responsible for inhibition of peroxidation, microsomes were
extracted with acetone which contained sufficient tocopherol so that at
the end of the extraction procedure, the microsomes still contained
normal levels of tocopherol. Fig. 3 shows that acetone extraction
essentially eliminated the GSH-dependent supression of peroxidation.
The microsomes extracted with acetone plus tocopherol still displayed
substantial inhibition of lipid peroxidation in the presence of GSH.
The data support the possibility that tocopherol plays an obligatory
role in the functioning of this system.

Quenching of tocopheroxy radicals by the GSH-dependent factor

To obtain additional information that would implicate an electron
shuttling function for tocopherol in preventing lipid peroxidation,
tocopheroxy radicals were prepared and incorporated into liver

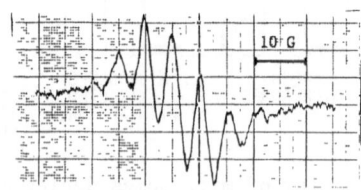

Fig. 4. Electron paramagnetic resonance scan of tocopheroxy radicals
prepared by the reaction of alpha-tocopherol with diphenyl picryl
hydrazyl.

TABLE IV

QUENCHING BY THE GSH-DEPENDENT MICROSOMAL FACTOR OF
TRICHLOROMETHYL RADICALS GENERATED DURING THE METABOLISM OF
CCl_4 BY MICROSOMES

Trichloromethyl radical linked to carbon tetrachloride
metabolism were generated in a system containing 0.1 ml of
microsomal suspension (approx. 2.0 mg protein), 0.1 ml of
the NADPH-generating system, 20 ul of CCl_4, and 0.78 ml of
phosphate buffer. The systems also contained 0.1 mM phenyl-
t- butyl nitrone to trap the trichloromethyl radicals as
they are formed by the drug metabolizing system. All
systems were incubated at 37° for 30 minutes unless
specified otherwise. At the end of the incubation period,
the samples were placed in a Pasteur pipette, sealing the
tip end, and centrifuging the microsomal pellet into the
capillary portion. The capillary part of the pipette was
placed in the EPR cavity and assayed resonance signals.

SYSTEM	SIGNAL INTENSITY
MICR. + NADPH + CCl_4 ----------------------	6.8
MICR. + NADPH + CCl_4 + GSH ----------------	0.9
MICR. + NADPH + CCl_4, THEN ADDING GSH AT THE END OF THE INCUBATION PERIOD AND INCUBATING ANOTHER 30 MIN. (CONTROL FOR POSSIBLE REDUCTION OF THE SPIN ADDUCT BY GSH) ----------------------	6.5

microsomes as described under Experimental Procedures. After washing,
these microsomes displayed an EPR signal characteristic of the
tocopheroxy radical (Fig. 4). When these microsomes were incubated in
0.15 M phosphate buffer, pH 7.4, in the absence of GSH, the center line
height of the signal was determined to be 6.7 cm at the spectrometer
setting indicated (Table III). When these same microsomes were
incubated in buffer in the presence of 10 mM GSH, the signal was
immediately quench (line height of 0.9 cm). That the quenching was not
due to a direct reaction between the tocopheroxy radicals and GSH was
shown by incubating an emulsion of tocopheroxy radicals with GSH under
the same conditions. No quenching of the tocopheroxy radical was
observed (line height 7.5 cm). Since the quenching effect was
eliminated by prior heating of the microsomes, the results indicate that
there is a GSH-dependent "free radical reductase" activity in
microsomes.

Additional evidence indicating that tocopherol functions as an electron
shuttle for quenching free radicals in microsomal membranes

If this system does function to quench free radicals which may
form in the microsomal membrane as a result of drug metabolism or
environmental factors, it should be possible to test this possibility
by observing the effect of the system on the generation of
trichloromethyl radicals during CCl_4 metabolism. Peroxidation System
No. 3 was for this study. Phenyl-t-butyl nitrone was added to trap the
·CCl_3 radicals which are formed. Table IV shows that a strong EPR

153

signal was obtained for the latter. However, when GSH was also
included in the system, the signal strength was greatly reduced. To
demonstrate that this loss of the signal for the $\cdot CCl_3$ radical adduct
was not due to direct reduction of the spin adduct by GSH, the system
was incubated without GSH first, after the signal intensity reached its
maximum, 10 mM GSH was added to the system. No quenching occurred,
and, hence, the loss of signal was not due to reduction of the spin
adduct by GSH.

DISCUSSION

When considering the data as a whole, the results are consistent with
the hypothesis that tocopherol functions in the microsomal membrane to
quench free radicals resulting in formation of the relatively stable
tocopheroxy radical which is then immediately reduced back to tocopherol by
a GSH-dependent ", membrane-bound free radical reductase".   Such radicals
may be produced during the metabolism of xenobiotic compounds or which may
be formed as a result of environmental factors (radiation, iron overload,
ischemia-reperfusion, excess polyunsaturated fat intake, etc.), and may
account for the untoward effects of tocopherol deficiency.

The structure, stability, lipid solubility, chemical properties,
and ready availability in food sources make tocopherol an ideal
molecule for a free radical quenching functioning. The stability of
the radical intermediate would be an essential feature of a molecule
functioning in such a cyclic system.

A recycling function of this type would be an extremely efficient
mechanism for protecting cell membranes against oxidative stress because
the system's capacity to quench free radicals would be limited only by the
reaction rate of the "free radical reductase" and the capacity of the
cytosolic enzymes to maintain GSH in the reduced state (which is
considerable).

Although ascorbic acid has been shown to reduce tocopheroxy radicals
to tocopherol non-enzymatically, it probably does not play a significant
biological role in the "free radical reductase" system due to the fact that
the concentration of GSH in most cell is several fold greater than that of
ascorbate (19). In addition, the symptoms of ascorbic acid deficiency do
not appear to mimic those of tocopherol deficiency as one might expect
would be the case if ascorbic acid was essential to recycle tocopheroxy
radicals back to tocopherol. The symptoms of GSH depletion, however, do
resemble those of an oxidatively stressed organism, but it is not known if
GSH depletion is accompanied by a conversion of tocopherol to the quinone
or dimer. It would be helpful in confirming the biological role for
tocopherol as an electron shuttle for a "free radical reductase" if the
effect of GSH depletion on membrane tocopherol content were determined.

The distribution of this "free radical reductase" activity in animal
tissues is not known at present, although we have evidence this it appears
to be present in other organelles (kidney mitochondria, for example). If
the vitamin's primary function is to link the reducing power of systems
which maintain GSH in the reduced state to the quenching of free radicals
in membranes, it would explain why adult animals (including humans) seldom
show symptoms of vitamin E deficiency. That is, when young animals are
born (including humans, and especially premature babies), vitamin E levels
tend to be low, yet they are rapidly creating new cells which must be
supplied with vitamin E which can only come from the diet. If the diet is
deficient, the young usually show rapid development of symptoms and may die
unless supplemented. Mature animals, however, are not producing net

numbers of new cells and would be recycling the tocopherol present in their cell membranes. In addition, body stores of tocopherol in adipose tissue would be available to replace the little tocopherol that is metabolized (21).

If other antioxidant compounds can function in a similar manner as tocopherol, an explanation of the ability of some of these compound to prevent tocopherol deficiency symptioms would be obvious. Investigations on the capacity of various antioxidants to substitute for tocopherol in the "free radical reductase" system are under way in the authors' laboratory.

REFERENCES

1. H.-W. Leung, M.J. Vang, and R.D. Mavis, The Cooperative Interaction Between Vitamin E and Vitamin C in Suppression of Peroxidation of Membrane Phospholipids, Biochim. Biophys. Acta 664:266-272 (1981).
2. M. Scarpa, K.A. Rigo, M. Maiorino, F. Ursini, and C. Gregolin, Formation of α-Tocopherol Radical and Recycling of α-Tocopherol by Ascorbate During Peroxidation of Phosphatidylcholine Liposomes, Biochim. Biophys. Acta 801:215-219 (1984).
3. E. Bascetta, F.D. Gunstone, and J.C. Walton, Electron Spin Resonance Study of the Role of Vitamin E and Vitamin C in the Inhibition of Fatty Acid Oxidation in a Model Membrane, Chem. Phys. Lipids 33:207-210 (1983).
4. C.C. Reddy, R.W. Scholz, C.E. Thomas, and E.J. Massaro, Vitamin E-dependent Reduced Glutathione Inhibition of Rat Liver Microsomal Lipid Peroxidation, Life Sci. 31:571-576 (1982).
5. R.F. Burk, M.J., Trumble, and R.A. Lawrence, Rat Hepatic Cytosolic Glutathione-dependent Enzyme Protection Against Lipid Peroxidation in NADPH-Microsomal Lipid Peroxidation System, Biochem. Biophys. Acta 618:35-41 (1980).
6. R.F. Burk, K. Patel, and J.M. Lane, Reduced Glutathione Protection Against Rat Liver Microsomal Injury by Carbon Tetrachloride. Dependence on 02, Biochem. J. 215:441-445 (1983).
7. K.E. Hill and R.F. Burk, Influence of Vitamin E and Selenium on Glutathione-Dependent Protection Against Microsomal Lipid Peroxidation, Biochem. Pharmacol. 33:1065-1068 (1984).
8. N.S. Kosower and E.M. Kosower, The Glutathione Status of Cells, International Review of Cytology 54:109-160 (1978).
9. M.L. Scott, Advances in Our Understanding of Vitamin E, Fed. Proc. 39:2736-2739 (1980).
10. J.G. Bieri, L. Corash, and V.S. Hubbard, Medical Uses of Vitamin E, N. Engl. J. Med. 308:1063-1071 (1983).
11. I.A. Arias and W.B. Jakoby, "Glutathione: Metabolism and Function", Raven Press, N.Y. (1976).
12. P.B. McCay, E.K. Lai, J.L. Poyer, C.M. DuBose, and E.G. Janzen, Oxygen- and Carbon-centered Free Radical Formation During Carbon Tetrachloride Metabolism. Observation of Lipid Radicals In Vivo and In Vitro, J. Biol. Chem. 259:2135-2143 (1984).
13. H.E. May and P.B. McCay, Reduced Triphosphopyridine Nucleotide Oxidase-Catalyzed Alterations of Membrane Phospholipids. I. Nature of the Lipid Alterations, J. Biol. Chem. 243:2288-2295 (1968).
14. G.W. Burton, A. Webb, and K.U. Ingold, A Mild, Rapid, and Efficient Method of Lipid Extraction for Use in Determining Vitamin E/Lipid Ratios, Lipids 20:29-39 (1985).
15. L. Hatam and H.J. Kayden, A High-Performance Liquid Chromatographic Method for the Determination of Tocopherol in Plasma and Cellular Elements of the Blood, J. Lipid Res. 20:639-645 (1979).

16. J. Sedlak and R.H. Lindsay, Estimation of Total Protein Bound and Non-protein Bound Sulfhydryl Groups in Tissues with Ellman's Reagent, Anal. Biochem. 25:192-205 (1968).

17. W. Boguth and H. Niemann, Electron Spin Resonance of Chromanoxy Free Radicals From Tocopherol and Tocol, Biochim. Biophys. Acta 248:121-130 (1971).

18. J. Glavind and G. Holmer, Thin-Layer Chromatographic Determination of Antioxidants by the Stable Free Radical α,α-Diphenyl-ß-Picryl Hydrazyl, J. Amer. Oil Chem. Soc. 44:539-542 (1967).

19. D. Hornig, Distribution of Ascorbic Acid Metabolites and Analogs in Man and Animals, Ann. N.Y. Acad. Sci. 258:103-118 (1975).

20. S. Kasparek, Chemistry of Tocopherols and Tocotrienols, in: "Vitamin E: a Comprehensive Treatise," L.J. Machlin, ed., Marcel Dekker, Inc., New York, N.Y. (1980).

21. H.E. Gallo-Torres, Vitamin E: Transport and Metabolism in: "Vitamin E: A Comprehensive Treatise," L.J. Machlin, ed., Marcel Dekker, Inc., New York, N.Y. (1980).

BIOLOGICAL EFFECTS OF TRANS FATTY ACIDS

J.L. Zevenbergen

Unilever Research Laboratorium
P.O. Box 114
3120 AC Vlaardingen, The Netherlands

1. INTRODUCTION

Hydrogenation is widely applied in the edible fats industry. The aim of this process is to shift the melting range of edible oils to higher temperatures to obtain solid fats (hardstock) for production of margarines and shortenings. It also improves the flavour stability of oils, especially of those containing substantial amounts of linolenic acid, such as soybean oil.

During the hydrogenation process, saturated fatty acids and a variety of geometric (i.e., trans) and positional isomers of unsaturated fatty acids are formed in varying amounts, depending on the initial fatty acid composition of the oils and the process conditions. The most commonly used oils for hydrogenation are of vegetable origin, like soybean oil (1). It is estimated that in the U.S. 2.5 million tonnes of vegetable oil (44% of the total edible fat production) are hydrogenated per annum (2).

Vegetable oils contain unsaturated fatty acids predominantly with 18 carbon atoms (C18): oleic and linoleic acid with some linolenic acid. Hydrogenation of these oils can thus lead to the formation of monoenoic and dienoic trans fatty acids with the double bond(s) between carbon atoms 6 to 14 from the carboxyl end. The main isomeric fatty acids of partially hydrogenated vegetable oils are monoenoic trans fatty acids with the double bond at positions 9 to 12; furthermore much smaller amounts of trans dienoic acids, cis,trans- and trans,cis- and some trans,trans-linoleic acid isomers are present (3,4).

Partially hydrogenated marine oils contain C20 and C22 trans fatty acids. Very long chain fatty acids, irrespective of their geometrical configuration, show specific biological effects, which are caused by their chain length (5). These fatty acids will not be discussed in this review.

In this report the biological effects of trans fatty acids, as present in partially hydrogenated vegetable oils, will be reviewed. Data from the literature and our own work will be used to evaluate the safety of trans fatty acids for human consumption.

2. CONSUMPTION OF TRANS FATTY ACIDS

The content of trans fatty acids in commercially available edible products has been reported by several investigators (6-10). The trans fatty acids content in U.S. margarines ranges from almost 0 to 40%; as

a guideline the average content of trans fatty acids in stick (hard)
margarines amounts to 25% and in tub (soft) margarines to 17% (9,10).
Shortenings contain ca. 16% on average. The average levels of trans
fatty acids in European margarines differ from country to country.
(6,7,11-14). Shortenings in Europe generally contain a higher level of
trans fatty acids than in the U.S.

Trans fatty acids are not only present in partially hydrogenated
vegetable oils, but also occur in ruminant fats and dairy products.
These fatty acids are mainly formed in a bio-hydrogenation process by
microorganisms in the rumen. Ruminant milkfat may contain up to 8% (15)
and meat up to 6% (10) of trans fatty acids. All the trans fatty acid
isomers present in partially hydrogenated vegetable oils are also found
in dairy products (16).

Because of the wide use of margarines, shortenings and ruminant- or
milk-fat, trans fatty acids may be present in many edible products
(9,13). Values for the total per capita daily consumption of trans
fatty acids in the U.S. vary between 6.8 and 12.1 g, depending on the
method of estimation (see 17 for recent review). However, on careful
analysis of the available data, around 8 g seems to be the best
estimate; this is about 6% of the total fat consumption (around 2.4%
of energy). In Germany the estimated intake of trans fatty acids is
4.5-6.4 g per capita per day (18); in Sweden 5 g (19) and in the U.K.
12 g of trans fatty acids per capita per day (11).

3. NUTRITIONAL EVALUATION OF TRANS FATTY ACIDS

Since 1945 extensive long-term animal feeding studies of partially
hydrogenated fats have been performed. Deuel et al. (20,21) and later
Alfin-Slater et al. (22-24) fed 46 generations of rats a diet
containing margarine as the sole source of dietary fat. This fat
contained 35% trans fatty acids. No adverse effects were found in
reproductive performance, longevity and in the histopathological
examination of many organs. Two additional 25-generation feeding
studies with rats fed similar diets as in her previous experiments,
were reported by Alfin-Slater et al. (25). Again no significant adverse
effects could be attributed to the dietary margarines.
Vles and Gottenbos (26,27) reported long-term studies involving large
groups of male and female, newly weaned, SPF Swiss mice and SPF Wistar
rats fed diets containing soybean oil, three hydrogenated soybean oils
containing different levels of trans fatty acids, coconut oil or
butterfat. In each diet, 54% of the energy was provided by the
experimental fat and 6 % by soybean oil. The soybean oil ensured that
the content of essential fatty acids was adequate. In both mice and
rats, no significant differences were observed in mortality, lifespan,
growth, organ weights or histopathology except for higher liver weights
in mice fed the more extensively hydrogenated soybean oil containing
60% trans fatty acids. Histopathological examination of these livers
revealed no abnormalities. No systematic differences in total tumor
frequency were found between the dietary groups. Neither did obvious
differences occur in the type and site of the neoplasms. In mice,
leukemia and lung papiloma and, in rats, adenoma of the pituitary
formed the most frequently occurring type of tumors. Other pathological
(non-neoplastic) changes were randomly distributed among the groups.
In another experiment Vles et al. (28) fed female Viennese x Alaska
rabbits diets containing the same fats (except butterfat) used in the
rat and mice studies just described. The diets contained 22.5 energy
percent experimental fat plus 2.5 energy percent soybean oil as a
source of EFA. Diets were fed to the rabbits during their entire
lifespan - up to 10 years. Differences in lifespan among the dietary
groups were not significant. There were no systematic differences among

the dietary groups in total tumor incidence. Atherosclerotic lesions were more severe in rabbits fed the diets with the lowest amounts of linoleic acid, i.e. coconut oil or hydrogenated soybean oil.

Nolen et al. (29) performed a 2-year chronic toxicity study of used frying fats (including a soybean oil and a partially hydrogenated soybean oil). Each fat was fed at the 28% of energy level in a semipurified diet to 50 male and 50 female, weanling Sprague-Dawley rats. Histopathological examination did not show significant differences attributable to diet.

The short term human studies with partially hydrogenated vegetable fats, to be discussed in the next section, did not reveal any adverse effects either. Therefore it can be concluded that neither partially hydrogenated fats nor the individual components present (isomeric fatty acids and the unsaponifiable part of the fats) show specific physiological or pathological effects. It has to be emphasized that in these long term experiments the dietary fats contained at least 5% cis,cis-linoleic acid to cover the requirement for EFA.

4. TRANS FATTY ACIDS, ATHEROSCLEROSIS AND CANCER

The effects of hydrogenated fats on serum cholesterol has been studied in both animals and man. Saturated fatty acids, especially lauric, myristic and palmitic acid increase serum cholesterol levels in man and animals, whereas linoleic acid has a serum cholesterol lowering effect. Cis-monounsaturated fatty acids, particularly oleic acid, appear to occupy an intermediate position in this respect.

Therefore when evaluating the effects of trans fatty acids on serum cholesterol levels one should study these effects relative to those of other comparable fatty acids, such as saturated or cis-monounsaturated fatty acids of the same chain length. Moreover, the experimental and control diet should be nutritionally adequate and contain equal amounts of linoleic acid and cholesterol.

Only a few human studies with well balanced diets containing hydrogenated fats have been performed. Mattson et al.(30) fed liquid formula diets containing a high level of either trans fatty acids or cis-monounsaturated fatty acids. After 21 days no significant difference in serum cholesterol levels was detected. Vergroesen and Gottenbos (31) reported two studies in which they investigated the effect of trans fatty acids (elaidic acid) on serum cholesterol levels. They concluded that trans fatty acids (elaidic acid) induced a total serum cholesterol level lying between that induced by saturated and that by cis-monounsaturated fatty acids.

A rabbit experiment performed in our laboratory (32) was aimed at comparing the potential atherosclerotic effects of trans fatty acids with those of saturated fatty acids at the same linoleic acid level (10%). This experiment showed no significant difference in atherogenicity between trans fatty acids and saturated fatty acids. In a second experiment Hornstra and Vles (32) showed that linoleic acid also reduced the atherogenic effects of trans fatty acids as it does the atherogenicity of saturated fatty acids.

In two other animal studies, one with swine (33) and the other with Vervet monkeys (34), no significant differences in incidence and severity of atherosclerosis were found between animals fed partially hydrogenated fats and those fed control fats (oleic acid rich fats).

In conclusion, when trans fatty acids are compared with other fatty acids in well designed studies, it appears that they are certainly no more atherogenic than saturated fatty acids. The atherogenic effects of trans fatty acids are reduced by linoleic acid as are those of saturated fats.

The long-term experiments described in an earlier section never gave
any indications of carcinogenic effects of trans fatty acids. Moreover,
four additional studies on the tumor promoting effects of hydrogenated
fats have been performed. These recent studies, two with mice (35,36)
and two with rats (37,38), fed well balanced diets, showed no more
tumors in the animals fed trans fatty acids than in those fed either
cis-monounsaturated or saturated fatty acids.

5. BIOCHEMICAL ASPECTS OF TRANS FATTY ACIDS

In the first part of this review it has been discussed that the
results of the long-term studies show that trans fatty acids have no
specific effects. For the evaluation of biochemical or physiological
aspects of trans fatty acids, they can therefore be regarded as members
of the fatty acid family. However, to regard trans fatty acids as one
single group of fatty acids, as is often done in discussions on
partially hydrogenated oils, is an oversimplification. A distinction
has to be made, from a biochemical point of view, between different
types of trans fatty acids. During hydrogenation of vegetable oils
mainly C18 trans fatty acids are formed.
Two types can be differentiated: trans fatty acids with one double bond
(isomers of oleic acid) and trans fatty acids with two double bonds
(isomers of linoleic acid). The latter group consists of fatty acids
with one trans double bond (cis,trans or trans,cis) or two trans double
bonds (trans,trans, eg. 9t,12t linoleic acid = linolelaidic acid).

Emken (39,40) recently reviewed the biochemistry of trans fatty
acids. In vivo investigations have indicated that hydrogenated fats and
specific fatty acid isomers can influence the activity of the desaturases, elongases, acyltransferases, oxygenases and prostaglandin
synthetases. This is often interpreted as an indication of undesirable
effects of trans fatty acids on essential fatty acid metabolism
(41,42). Indeed, in essential fatty acid deficiency trans fatty acids
aggravate the characteristic symptoms of this disease (43). For proper
lipid metabolism, however, the organism needs to have ample choice of
fatty acids; the more the dietary composition restricts the variation
of fatty acids available to the organism, the more limited the possibilities to carry out proper metabolic processes. For example, in the
absence of essential fatty acids dietary stearic and oleic acid can be
readily converted into a polyunsaturated fatty acid: Δ9,12,15-eico-
satrienoic acid. Elaidic acid can only be converted marginally to a
polyunsaturated fatty acid (44). Because large amounts of dietary fatty
acids (any fatty acid) inhibit the fatty acid synthesis, trans fatty
acids fed to EFA-deficient animals in considerable amounts, decrease
the animals' ability to produce polyunsaturated fatty acids and as a
result will aggravate the EFA-deficiency symptoms. However, if sufficient amounts of linoleic acid are fed, trans monoenoic fatty acids do
not adversely influence these lipid metabolizing enzymes (45).
The effects of 9-trans, 12-trans 18:2 appears to be more potent than
those of other trans fatty acids, but this fatty acid is of no nutritional importance because it is present in the human diet in trace
amounts only (46).

Trans fatty acids can be incorporated in biomembranes of all the
investigated human and animal tissues. Extensive data is reported on
the incorporation of both monoenoic trans fatty acids (47-51) and
dienoic trans fatty acids (41,52) in lipid classes of many organs. The
amount of trans fatty acids present in biomembranes largely depends on
the amount in the diet, but even when very high amounts are fed, no
excessive deposition occurs.
Emken states in his review (40) that large amounts of partially
hydrogenated vegetable oils generally decrease the arachidonic acid

content in tissue phospholipids somewhat. So trans fatty acids are
capable of altering the fatty acid composition of biomembranes. But,
even in much smaller amounts, some other fatty acids like α-linolenic
acid and eicosapentaenoic acid have a much more profound effect on
membrane arachidonic acid content than trans fatty acids (53,54).

More important than the fact that the fatty acid composition of
biomembrane can be changed, is the question whether the function of
those membranes is also altered. The mere fact that the fatty acid
compostion of a biomembrane is altered does not neccesarily imply that
its function is disturbed. The fatty acid composition of the mitochondrial membrane for example can be changed by feeding trans fatty
acids (48,55). Despite these changes, the oxidative phosphorylation of
mitochondria, which is strongly dependent on a high structural integrity of the inner mitochondrial membrane, is not influenced (55,56).

6. LINOLEIC ACID REQUIREMENT IN TRANS FATTY ACIDS-FED RATS

As is apparent from the previous part of this paper and from recent
reviews dealing with the biological effects of trans fatty acids
(4,11,39,46) no undesirable effects occurred provided a sufficient
amount of essential fatty acids (linoleic acid) was present in the
diet. To define the minimum requirement for linoleic acid necessary to
prevent specific or adverse effects (like aggravation of essential
fatty acid deficiency) of trans fatty acids, we performed two rat
feeding studies.

In the first study 8 groups of weanling, male SPF Wistar rats were
fed semi-synthetic diets (with 40 % of energy from fat) for 3 months.
Six of those groups received a diet containing large amounts of trans
fatty acids (around 50 % of total fat) as present in a partially hydrogenated soybean oil. By using variable amounts of sunflower seed oil,
olive oil or coconut oil, the linoleic acid content of the diets varied
from 0.4 to 7.1% of total energy. One group was fed a diet containing
40 % of energy as olive oil (giving 2 energy% linoleic acid); and
finally one group received a diet with a mixture of coconut oil and
olive oil (giving 5 energy% linoleic acid content). The last two groups
served as references.

Growth, food- and water- consumption were not systematically different and neither did (histo-) pathological examination revealed any
abnormalities in any of the groups. After the feeding period the fatty
acid composition of phospholipids of heart and liver mitochondria,
blood platelets and segments of aorta were investigated. It was
observed that trans fatty acids were incorporated in all investigated
phospholipids and that the content of trans monoenoic acids was not
influenced by the amount of linoleic acid in the diet. The level of
trans isomers of linoleic acid was low, and tended to decrease with
increasing amounts of linoleic acid in the diet. The linoleic acid
content in these phospholipids hardly changed with the increase of its
content in the diet.

The arachidonic acid level in the phospholipids generally increased
with increasing amounts of dietary linoleic acid, and reached a constant level at a dietary linoleic acid level between 2 and 5 energy%,
depending on the tissue and phospholipid class. Surprisingly, the
arachidonic acid level in some phospholipids was lower in the reference-group fed olive oil than in the reference-group fed the mixture of
coconut and olive oil, despite the higher linoleic acid level in the
first diet (5 and 2 energy% respectively).

Trans fatty acids increased the linoleic acid level of all investigated phospholipids. The arachidonic acid level in most phospholipids
was lower in the group fed trans fatty acids and 2 energy% linoleic
acid than in the reference-group fed the same amount of linoleic acid.
With 5 energy% linoleic acid in the diet the effect of trans fatty

acids on phospholipid arachidonic acid levels had disappeared or had become significantly smaller. This confirms other investigations in which it was found that trans fatty acids increase the linoleic acid level and decrease the arachidonic acid level in phospholipids (45,57). This effect decreases when the linoleic acid content of the diet is increased.

As stated before, much more important than the observation that trans fatty acids can influence the essential fatty acid composition of membrane phospholipids is whether trans fatty acids also influence the membrane functions. For this purpose we measured the respiration capacity and ATP-synthesis of liver and heart mitochondria. These functions are closely linked to the mitochondrial membrane and are therefore sensitive to changes in the composition of that membrane. Despite significant changes in the fatty acid composition of the mitochondrial membranes, the respiration and ATP-synthesis were not significantly influenced by the dietary trans fatty acids at any level of dietary linoleic acid investigated. This is in agreement with other investigations, in which it was found that dietary trans fatty acids in the presence of sufficient linoleic acid did not decrease the mitochondrial function (55,56).

The effect of trans fatty acids on eicosanoid synthesis was investigated by measuring the production of both lipoxygenase and cycloxygenase products by stimulated blood platelets and aorta-segments, as described by Hornsta et al. (54). The production of both hydroxy eicosatetraenoic acid and hydroxy heptadecatrienoic acid (indicative of lipoxygenase and cyclooxygenase activity, respectively) by blood platelets and prostacyclin (measured in a bio-assay) by segments of aorta was linear with the arachidonic acid level in the phospholipid fraction. Because trans fatty acids lowered the arachidonic acid level, they also reduced the production of these eicosanoids. However, this decrease was not greater than could be predicted from the decrease in the arachidonic acid level, indicating that trans fatty acids do not have a direct effect on the prostaglandin or hydroxy fatty acid production. Neither did Blomstrand et al. (45) find any interference of high amounts of trans fatty acids with platelet cycloxygenase and lipoxygenase activity provided sufficient amounts of linoleic acid were available.

In a following experiment we wanted to compare the effects of trans fatty acids with those of saturated or cis-monounsaturated fatty acids at the level of 2 energy% linoleic acid. In that experiment we investigated the same parameters as described above, in four groups of 40 male, weanling Wistar rats. The animals received semi-synthetic diets containing 40 % of energy as fat. Three of the four dietary fat blends were composed mainly of a partially hydrogenated soybean oil (containing 50 % of total fat as trans fatty acids), olive oil (special variety with only 5.5 % linoleic acid and nearly 80 % oleic acid) or cocoa-butter (containing a high amount of saturated fatty acids), respectively. The fourth fat blend was a mixture of the other three, in order to obtain a diet with half the amount of trans fatty acids (25 % of total fat) compared with the first blend. All blends were prepared in such a way as to give diets containing approximately 2 energy% linoleic acid.

We found that trans fatty acids decreased the arachidonic acid level in phospholipids when compared with saturated fatty acids; compared with cis-monounsaturated fatty acids the arachidonic acid level was equal or only slightly lower. Linoleic acid levels were generally increased by trans fatty acids. No systematic effects of trans fatty acids were found on the mitochondrial function or on the eicosanoid production by blood platelets and segments of aorta. From the results of these experiments we concluded that 2 energy% of linoleic acid is sufficient to prevent undesirable effects of dietary trans fatty acids.

7. CONCLUSIONS

From the recent reviews dealing with trans fatty acids and our own work in this field, we conclude that in view of the level of linoleic acid in the average diet (normally more than 2 energy%) trans fatty acids (average amount in the human diet less than 5 energy%) do not present a nutritional problem. The recent review of the ad-hoc panel of the FASEB (Federation of American Societies for Experimental Biology) (17) supports this conclusion by stating: "The available scientific information suggests little reason for concern with the safety of dietary trans fatty acids both at their present and expected levels of consumption and at the present and expected levels of consumption of linoleic acid".

8. REFERENCES

1. Applewhite, T.H. (1981) Nutritional effects of hydrogenated soya oil. J. Am. Oil Chem. Soc. 58, 260-269.

2. USDA (1980) US Department of Agriculture, Science and Education Administration: Food and Nutrient Intake of Individuals in 1 day in the United States, Spring 1977. Nationwide food consumption survey 1977-1978, Preliminary Report no. 2, September, Washington, DC, US Department of Agriculture.

3. Dutton, H.J. (1979) Hydrogenation of fats and its significance, in: Geometric and Positional Fatty Acid Isomers (eds. E.A. Emken and H.J. Dutton), pp. 1-16, American Oil Chemists' Society, Champaign, Illinois.

4. Gottenbos, J.J. (1983) Biological effects of trans fatty acids, in: Dietary fats and health. (eds. Perkins, E.G. and Visek, W.J.) pp. 375-390, American Oil Chemists' Society, Champaign, Illinois.

5. Sauer, F.D. and Kramer, J.K.G. (1983) In: High and low erucic acid rapeseed oils (eds. Kramer, J.K., Sauer, F.D. and Pigden, W.J.), pp. 253-292, Academic Press, New York.

6. Heckers, H. and Melcher F.W. (1978) Trans-isomeric fatty acids present in West German margarines, shortenings, frying and cooking fats. Am. J. Clin. Nutr. 32, 1041-1049.

7. Thomas, L.H., Jones, P.R., Winter, J.A. and Smith, H. (1981) Hydrogenated oils and fats: the presence of chemically modified fatty acids in human adipose tissue. Am. J. Clin. Nutr. 34, 877-886.

8. Davignon, J., Holub, B. Little, J.A., McDonald, B.E. and Spence, M. (1980) Report of the Ad Hoc Committee on the Composition of Special Margarines. Dept. of Health and Welfare, Minister of Supply and Services, Ottawa, Canada.

9. Enig, M.G., Pallansch, L.A., Sampugna, J. and Keeney, M. (1983) Fatty acid composition of the fat in selected food items with emphasis on trans fatty acids. J. Am. Oil Chem. Soc. 60, 1788-1795.

10. Slover, H.T., Thompson, R.H. Jr., Davis, C.S. and Merola, G.V. (1985) Lipids in margarines and margarine-like foods. J. Am. Oil Chem. Soc. 62, 775-786.

11. Gurr, M.I. (1983) Trans fatty acids. International Dairy Federartion Document 166, 5-18.

12. Katan, M.B., van Bovenkamp, P. and Brussaard, J.H. (1983) Fatty Acid Composition, Trans Fatty Acid and Cholesterol Content of Margarines and other Edible Fats, In: Voedingsmiddelenanalyses van de Vakgroep Humane Voeding, Landbouwhogeschool, Wageningen, The Netherlands.

13. Kochar, S.P. and Matsui, T. (1984) Essential fatty acids and trans content of some oils, margarine and other food fats. Food Chemistry 13, 85-101.

14. Druckrey, F., Høy, C.-E. and Hølmer, G. (1985) Fatty acid composition of Danish margarines. Fette Seifen Anstrichm. 87, 350-355.

15. Parodi, P.W. (1976) Distribution of isomeric octadecenoic fatty acids in milk fat. J. Dairy Sci. 59, 1870-1873.

16. Sommerfeld, M. (1983) Trans unsaturated fatty acids in natural products and processed foods. Prog. Lipid Res. 22, 221-233.

17. FASEB Report of the ad-hoc Committee of the Federation of American Societies for Experimental Biology (1985) Health Aspects of Dietary Trans Fatty Acids (ed. F.R. Senti) Contract Number FDA 223-83-2020.

18. Heckers, H., Melcher, F.W. and Dittmar, K. (1979) Zum Taeglichen Verzehr transisomere Fettsaeuren. Eine Kalkulation unter Zurgrundelegung der Zusammensetzung handelsueblicher Fette und verschiedener menschlicher Depotfette. Fette Seife Anstrichm. 6, 217-226.

19. Åkesson, B., Johansson, B.M., Eng, M., Svensson, M. and Ockerman, P.A. (1981) Content of trans-octadecenoic acid in vegetarian and normal diets in Sweden, analyzed by the Duplicate Portion technique. Am. J. Clin. Nutr. 34, 2517-1520.

20. Deuel, H.J., Hallman, L.F. and Movitt, E. (1945) Studies on the comparative nutritive value of fats. VI. Growth and reproduction over 10 generations on Sherman diet B where butterfat was replaced by a maragrine fat. J. Nutr. 29, 309-320.

21. Deuel, H.J., Greenberg, S.M., Savage. E.E. and Bavetta, L.A, (1950) Studies on the comparative nutritive value of fats. J. Nutr. 42, 239-255.

22. Alfin-Slater, R.B., Wells, A.F., Aftergood, L. and Deuel, H.J. (1957) Nutritive value and safety of hydrogenated vegetable fats as evaluated by long-term feeding experiments with rats. J. Nutr. 63, 241-261.

23. Alfin-Slater, R.B., Wells, A.F. and Aftergood, L. (1970) Longevity and multigeneration studies in rats fed hydrogenated fats of increasing polyunsaturated fat content. J. Am. Oil Chem. Soc. 47, 274.

24. Alfin-Slater, R.B., Aftergood, L. and Whitten, T. (1976) Nutritional value of trans fatty acids. J. Am. Oil Chem. Soc. 53, 658A.

25. Alfin-Slater, R.B., Wells, A.F. and Aftergood, L. (1973) Dietary fat composition and tocopherol requirement IV. Safety of polyunsaturated fats. J. Am. Oil Chem. Soc. 50, 479-484.

26. Vles, R.O. and Gottenbos, J.J. (1972a) Long-term effects of feeding butterfat, coconut oil, and hydrogenated or non hydrogenated soyabean oils. I. Eighteen-month experiment in mice. Voeding 33, 428-433.

27. Vles, R.O. and Gottenbos, J.J. (1972b) Long-term effects of feeding butterfat, coconut oil, and hydrogenated or non hydrogenated soyabean oils. II. Life-span experiment in rats. Voeding 33, 455-465.

28. Vles, R.O., Gottenbos, J.J. and van Pijpen, P.L. (1977a) Aspects nutritionels des huiles de soja hydrogenees et de leur acides gras insatures isomeriques. Bibl. Nutr. Dieta 25, 186-196.

29. Nolen, G.A., Alexander, J.C. and Atman, N.R. (1967) Long-term rat feeding study with used frying fats. J. Nutr. 93, 337-348.

30. Mattson, F.H., Hollenbach, E.J. and Klingman, A.M. (1975) Effect of hydrogenated fat on the plasma cholesterol and triglyceride levels in man. Am. J. Clin. Nutr. 28, 726-731.

31. Vergroesen, A.J. and Gottenbos, J.J. (1975) The role of fats in human nutrition: an introduction. in: The role of fats in human nutrition. (Ed. Vergroesen, A.J.), pp. 1-41, Academic press, New York.

32. Hornstra, G. and Vles, R. (1978) Effects of dietary fat on atherosclerosis and thrombosis. in: International Conference on Atherosclerosis (eds. Carlson, L.A. et al.), Raven Press, New York.

33. Royce, S.M., Holmes, R.P., Takagi, T. and Kummerow, F.A. (1984) The influence of dietary isomeric and saturated fatty acids on atherosclerosis and eicosanoid synthesis in swine. Am. J. Clin. Nutr. 39, 215-222.

34. Kritchevsky, D., Davidson, L.M, Weight, M., Kriek, N.P.J. and du Plessis, J.L. (1984) Effect of trans-unsaturated fats on experimental atherosclerosis in vervet monkeys. Atherosclerosis 51, 123-133.

35. Brown, R.R. (1981) Effects of dietary fat on incidence of spontaneous and induced cancer in mice. Cancer Res. 41, 3741-3742.

36. Erickson, K.L., Schlanger, D.S., Adams, D.A., Fregeau, D.R. and Stern, J.S. (1984) Influence of dietary fatty acid concentration and geometric configuration on murine mammary tumorigenesis and experimental metastasis. J. Nutr. 114, 1834-1842.

37. Selenskas, S.L., Ip, M.M. and Ip, C. (1984) Similarity between trans fat and saturated fat in the modification of rat mammary carcinogenesis. Cancer Res. 44, 1321-1326.

38. Watanabe, M., Koga, T. and Sugano, M. (1985) Influence of dietary cis- and trans- fat on 1,2-dimethylhydrazine-induced colon tumors and fecal steroid excretion in Fischer 344 rats. Am. J. Clin. Nutr. 42, 475-484.

39. Emken, E.A. (1983) Biochemistry of unsaturated fatty acid isomers. J. Am. Oil. Chem. Soc. 60, 995-1004.

40. Emken, E.A. (1984) Nutrition and biochemistry of trans and positional fatty acid isomers in hydrogenated oils. Annu. Rev. Nutr. 4, 339-376.

41. Kinsella, J.E., Bruckner, G. Mai. J. and Shimp, J. (1981) Metabolism of trans fatty acids with emphasis on the effects of trans,trans-octadecadienoate on lipid composition, essential fatty acid, and prostaglandins: an overview. Am. J. Clin. Nutr. 34, 2307-2318.

42. Holman, R.T., Mahfouz, M.M., Lawson, L.D. and Hill, E.G. (1983) Metabolic effects of isomeric octadecenoic acids. in: Dietary fats and health. (eds. Perkins, E.G. and Visek, W.J.), pp. 320- 340, American Oil Chemists' Society, Champaign, Illinois.

43. Holman, R.T. and Aaes-Jorgensen, E. (1956) Effects of trans fatty acid isomers upon essential fatty acid deficiency in rats. Proc. Soc. Exp. Biol. Med. 93, 175-179.

44. Mahfouz, M.M., Valicenti, A.M. and Holman, R.T. (1980) Desaturation of isomeric trans-octadecenoic acids by rat liver microsomes. Biochim. Biophys. Acta 618, 1-12.

45. Blomstrand, R., Diczfalusy, U., Sisfontes, L. and Svensson, L. (1985) Influence of dietary partially hydrogenated vegetable and marine oils on membrane composition and function of liver microsomes and platelets in the rat. Lipids 20, 283-295.

46. Beare-Rogers, J.L. (1983) Trans- and positional isomers of common fatty acids, in: Advances in nutritional research. (ed. Draper, H.) Vol. 5., pp. 171-200, Plenum Publishing Corporation, New York.

47. Alfin-Slater, R.B. and Aftergood, L. (1979) Nutritional role of hydrogenated fats (in rats), In: Geometric and Positional Fatty Acid Isomers (eds. E.A. Emken and H.J. Dutton), pp. 53-74, American Oil Chemists' Society, Champaign, Illinois.

48. Høy, C.-E. and Hølmer, G. (1979) Incorporation of cis- and trans-octadecenoic acids into the membranes of rat liver mitochondria. Lipids 4, 717-733.

49. Wood, R. (1979) Incorporation of dietary cis and trans octadecenoate isomers in the lipid classes of various rat tissues. Lipids 14, 975-982.

50. Emken, E.A. (1981) Influence of trans-9-, trans-12-, and cis-12-octadecenoic acid isomers on fatty acid composition of human plasma lipids. Prog. Lipid Res. 20, 135-141.

51. Ohlrogge, J.B., Gulley, R.M. and Emken, E.A. (1982) Occurence of octadecenoic fatty acid isomers from hydrogenated fats in human tissue lipid classes. Lipids 17, 552-557.

52. Anderson, R.L., Fullmer, C.S. and Hollenbach, E.J. (1975) Effects of the trans isomers of linoleic acid on the metabolism of linoleic acid in rats. J. Nutr. 105, 393-400.

53. Høy, C.-E, Hølmer, G., Kauer, N., Byrjalsen, I. and Kirstein, D. (1983) Acyl group distributions in tissue lipids of rats fed evening promrose oil (γ-linolenic plus linoleic acid) or soybean oil (α-linolenic plus linoleic acid). <u>Lipids</u> 18, 760-771.

54. Hornstra, G., Christ-Hazelhoff, E., Haddeman, E., ten Hoor F. and Nugteren, D.H. (1981) Fish oil feeding lowers thromboxane- and prostacyclin production by rat platelets and aorta and does not result in the formation of prostaglandin I3. <u>Prostaglandins</u> 21, 727-737.

55. Blomstrand, R. and Svensson, L. (1983) The effects of partially hydrogenated marine oils on the mitochondrial function and membrane phospholipid fatty acids in rat heart. <u>Lipids</u> 18, 151-170.

56. Royce, S.M. and Holmes, R.P. (1984) The saturation and isomerization of dietary fatty acids and the respiratory properties of rat heart mitochondria. <u>Biochim. Biophys. Acta</u> 792, 371-375.

57. Lawson, L.D., Hill, E.G, and Holman, R.T. (1985) Dietary fats containing concentrates of cis or trans octadecenoates and the patterns of polyunsaturated fatty acids of liver phosphatidyl-choline and phosphatidylethanolamine. <u>Lipids</u> 20, 262-267.

AGRICULTURE - INDUSTRY RELATIONSHIP

Giorgio Bonaga and Umberto Pallotta

Cattedra di Industrie Agrarie, Universita' di Bologna

via S. Giacomo 7- 40126 Bologna, Italy

INTRODUCTION

In 1984 the balance of foreign trade in the Italian agro-alimentary sector showed a deficit of 10,126 billion lire, the highest recorded in Italy,if we exclude the figure for 1985, announced recently, which amounts to 11,500 billion lire. Our first question is : how much weight does the item "vegetable oils and oilseeds" have in such an alarming balance ? The Table 1 will make it easier to answer. It sums up, in monetary value, the imports and exports of the vegetable oils and oilseeds in the year 1984. From the data presented we conclude that the Italian balance of trade in this sector records a deficit of 1,300 billion lire, equal to 12.5% of the total agro-alimentary deficit. The deficit has therefore increased in spite of the fact that imports have decreased in quantity (- 13%) and exports have increased (+ 57%) with respect to the previous year. These variations indicate an unfavourable trend in the prices of oil products for the Italian market. However, the figure is in itself sufficient to stress the importance of the oil sector in the Italian economy. Other questions which one might ask, are : why is Italy a country which imports considerable quantities of vegetable oils and oilseeds ? Secondly, what is the present agronomical, industrial and commercial state of the oil sector, if this does not even garantee self-sufficiency for domestic requirements ? Finally, what are the trends of the Italian oil sector for the next few years, considering that every decision on domestic affairs must also come to terms with the big world markets ? We shall attempt to answer these questions referring to statistics and forecasts updated to 1984, which are the only complete figures available (2,3,5,6,7,8).

CONSUMPTION OF FOODS AND CONSUMPTION OF EDIBLE FATS

The Table 2 shows the annual consumption "pro capita" of the main foods in the EEC-10. The figures for 1984 confirm that Italy is the major consumer of bread, pasta, fresh fruit and vegetable oils and is second to France in the consumption of wine, and to Greece in that of vegetables. This is confirmation of the stability of the "Mediterranean" type of diet.

Table 1. Import-Export Balance in the Vegetable Oil Sector: 1984

(a) I M P O R T S

Products	Value (million of lire)	% of imports	% of agro-alimentary deficit
1. OLIVE OIL (virgin, refined, high acidity)	339,942		
2. HUSK OIL (crude, and refined)	38,860		
Total (1+2)	378,802	2.0	3.7
3. SEEDOILS (crude, and not crude)	365,344		
4. OILSEEDS (various)	836,423		
Total (3+4)	1,201,767	6.1	11.9
Total (1+2+3+4)	1,580,569	8.1	15.6

(b) E X P O R T

Products	Value (million of lire)	% of imports
1. OLIVE OIL (virgin, refined, high acidity)	167,058	
2. HUSK OIL (crude, and refined)	31,928	
Total (1+2)	198,986	2.1
3. SEEDOILS (crude, and not crude)	116,269	
4. OILSEEDS (various)	2,742	
Total (3+4)	119,011	1.3
Total (1+2+3+4)	317,999	3.4

----------------------oooooooooooooooooooo--------------------

(a) - (b) 1,262,570 12,5

SOURCE: elaboration from ISTAT data(1984).

Any small fluctuations seem to be connected with a favourable or unfavourable trend in the prices of some products and with a temporary alarmist attitude to diet, rather than with permanent changes in tendency. Perhaps only wine seems likely to undergo a progressive decrease in consumption of a traditional kind, also because of the ever-increasing competition of other drinks. The Figure 1 shows the data concerning the consumption of edible fats in Italy, divided into the various types. The figures for 1984 show the annual consumption of edible fats to be 26.1 kilos pro capita, 70% of which are olive oil and seedoil.

Table 2. Annual Consumption of Main Foods in EEC-10 (kg pro capita)

(kg per head)

FOODS	West Germany	France	Italy	Holland	Belgium and Luxemburg	England	Ireland	Denmark	Greece
Wheat	50	73	124	55	69	66	79	44	77
Potatoes	86	77	38	81	98	101	103	68	70
Sugar	36	36	30	39	33	43	44	44	29
Fresh fruit (including "jams" and "fruit juices")	89	53	70	84	64	33	30	38	57
Citrus fruits	30	19	35	54	20	14	12	12	49
Vegetables (including "preserves")	71	119	162	87	77	75	83	56	232
Wine	25	94	87	12	20	7	3	13	44
Dairy products (excluding "cream")	84	90	80	136	87	137	190	157	27
Cheese	12	18	13	13	10	6	3	10	18
Beef	24	33	25	21	28	23	24	14	27
Pork	58	38	24	41	41	26	32	51	18
Mutton, lamb, and goat	1	4	2	1	2	7	8	1	14
Poultry	10	17	18	9	13	14	14	9	13
Eggs	17	15	11	11	14	14	12	13	12
OILS, AND FATS FOR COOKING	14.0	17.5	23.4	14.1	12.7	10.8	13.0	11.8	23.1
BUTTER	7.0	9.8	2.2	3.5	8.2	7.0	11.0	8.6	1.0
MARGARINE	8.0	3.6	0.5	15.2	11.8	7.0	4.0	16.6	-
Total edible fats (kg per head)	29.0	30.9	26.1	32.8	32.7	24.8	28.0	37.0	24.1

SOURCE: elaboration from EUROSTAT data (1984).

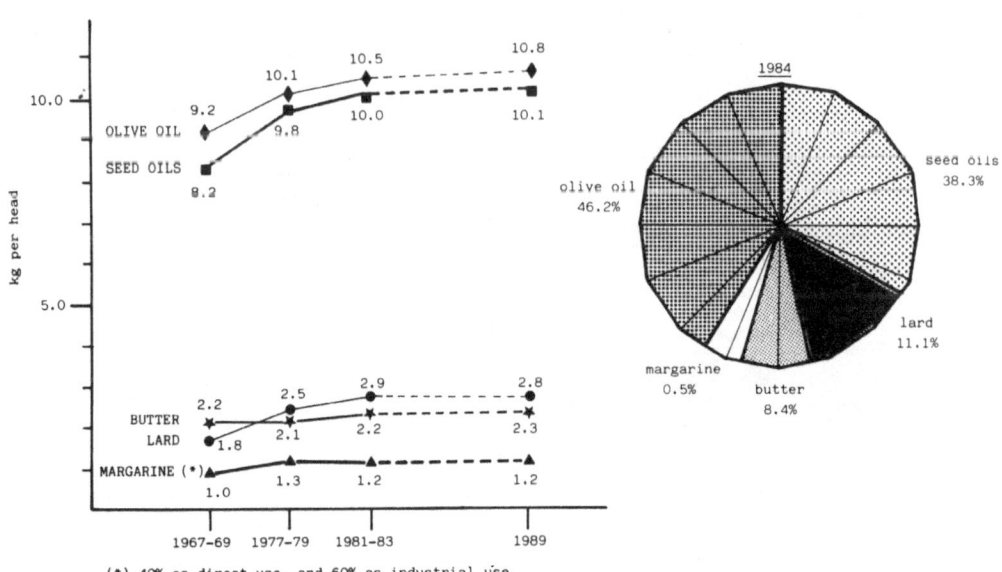

(*) 40% as direct use, and 60% as industrial use.

SOURCE: elaboration from ISTAT data (1984); forecasts from IRVAM data (1984).

Fig. 1. Annual Consumption of Edible Fats in Italy (and Forecasts)

The quantity has remained almost constant in the last five years and seems unlikely to increase much. The quality of the demand is expected to remain constant, except for minor fluctuations. In short, a slight increase in the consumption of olive oil and seedoil is forecast for the next few years, which is not compensated for by the increase in domestic production, and this will require the importation of about 800,000 quintals of olive oil (only 1% less than the average quantity imported in the three-year period 1981-83). No significant variations are forecast for margarine consumption, although this solid fat could occupy the space left free by the insufficient domestic supply of butter. A slight increase is also forecast in the consumption of butter, which is not compensated for by the increase in domestic production, so that the importation of about 500,000 quintals is forecast (4% less than the average quantity imported in the three-years period 1981-83). Finally, a slight decrease is forecast in the consumption of lard, which will leave a larger quantity for exportation.

THE OLIVE OIL SECTOR (10,11)

The Italian peninsula extends over 12 parallels and a length of 1,200 kilometres; about half of the cultivated area consists of hilly ground. The cultivation of olive-trees ranges over such a great variety of climates and soils that the number of types of oil which are obtained in Italy cannot be compared with those from other countries. As we shall explain later, this diversity of pedological and climatic situations has at the same time good and bad effects on Italian olive-growing. The Figure 2 shows the geographical distribution of the areas where olive-trees are grown and the quantity of olives harvested in the 1983 - 84 crop.

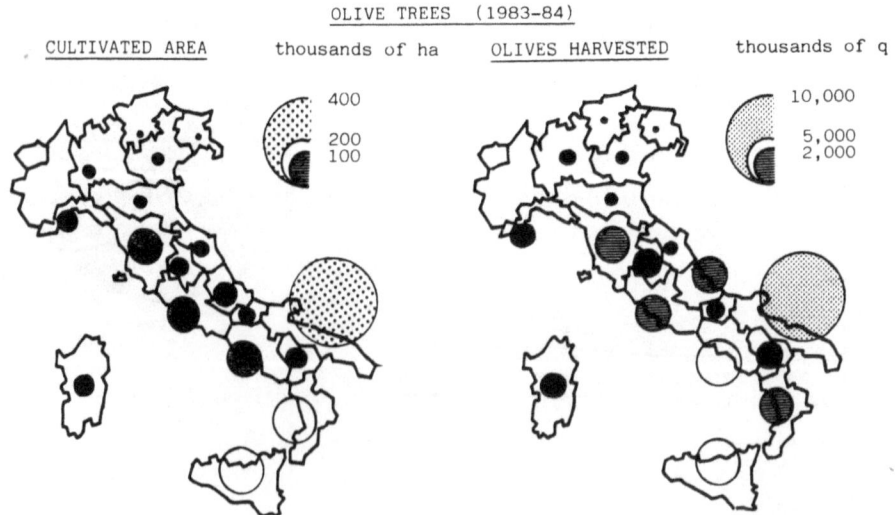

OLIVE TREES (1983-84)

CULTIVATED AREA   thousands of ha   OLIVES HARVESTED   thousands of q

400
200
100

10,000
5,000
2,000

Total cultivated area = 1,250,000 ha   Total olives harvested = 42,060,000 q

Figure 2 . Olive Trees (1983-84)

Source : elaboration from ISTAT data (1984)

The total area cultivated is about 1,250,000 hectares (97% of which are concentrated in central and southern Italy), but it is not difficult to predict that a new survey will find a considerable reduction in the area where olive-trees are grown, on account of the great damage caused by the frosts of January 1985 and the tendency to replace olive-trees with other crops bringing higher profits in those areas where it is possible. Although no definite statistics have been published yet, the olives harvested in the 1984 - 85, crop amount to 20,0 million quintals, which is the lowest quantity in the period 1980 - 85. The Figure 3 shows the trend of the last five crops; the olives harvested, the table olives, the olives for oil extraction and the oil extracted are shown separately. The scanty production of pressure oil in the 1984-85 crop caused about 450,000 quintals of A.I.M.A. (National Company for Intervention on Agricultural Markets) stocks to be put on the home market. The amount of imports and exports was also influenced by such a critical crop, as shown in Table 3.

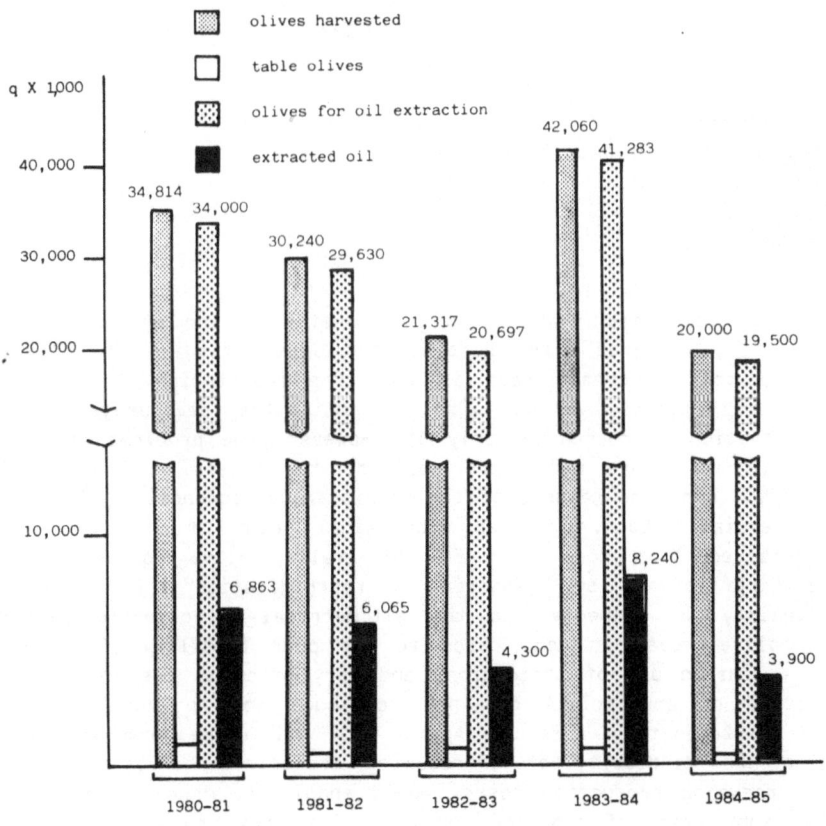

SOURCE: Annuario di Statistica Agraria. ISTAT: 1980, 1981, 1982, 1983, 1984, 1985.

Figure 3. Olives harvested, Table olives, Olives used for Oil extraction and Extracted Oil from 1980-81 to 1984-85

Table 3

IMPORTS AND EXPORTS OF VEGETABLE OILS IN ITALY

(Jenuary–October, 1985)

Products	Imports (Quintals)	Exports (Quintals)
1) Virgin olive oil	812,513	102,970
2) High acidity olive oil	907,643	1,323
3) Crude husk oil	201,200	52,206
4) Refined husk oil	4,717	56,200
5) Refined olive oil	125	371,058
Total	1,926,198	583,757
	(+83%)	(+1%)

SOURCE: ASSITOL data (1986).

It must be specified that the figures refer to the period January–October 1985, compared with those of the same period of the previous year. The most important fact is that the importation of olive oil and husk oil increased by 83% with respect to the previous year, while the exportation increased by only 1%. However, the problems of olive-growing in Italy are much more serious; the frosts of January 1985 accentuated them. The reasons of the crisis are of agronomical, technical and commercial nature. From an agronomical point of view it should be remembered that more than 50% of Italian olive-growing is held to be "inefficient" or "requiring a reorganization". This lack of functionality is connected to the altimetrical or climatic position of the olive-trees, in some cases to the poor fertility of the soil or to a limited use of irrigation, and very often to the insufficient application of appropriate growing techniques and to the lack of a correct phytosanitary defence. Apart from all these problems, there is also the difficulty of introducing mechanized harvesting which, besides reducing production costs, would enable the drupes to be picked at the right stage of ripening and also when they are whole, and thus able to provide oil at higher quality. As far as processing and marketing are concerned, it must be pointed out that the preservation, the transport and the system of carrying the olives to the mills are also in need of much improvement. In Italy the system of carrying the olives has favoured the multiplicity of the crushing structure.

As the Figure 4 shows, the number of oil-mills working in Italy in the 1983-84 crop was 9,346, 96% of which were concentrated in central and southern Italy. Only 589 of these (6% of the total number) have a cooperative organization and almost all of them are concentrated in Puglia Region. This large number of mills means that the system of carrying the olives is not at all functional and that each mill works at a reduced average daily capacity. This obviously results in a lower quality pressure oil and higher production costs. The geographical distribution of the industry which processes the virgin olive oil, the crude husk oil and, as we shall see later, the seedoil, is on the other hand completely different.

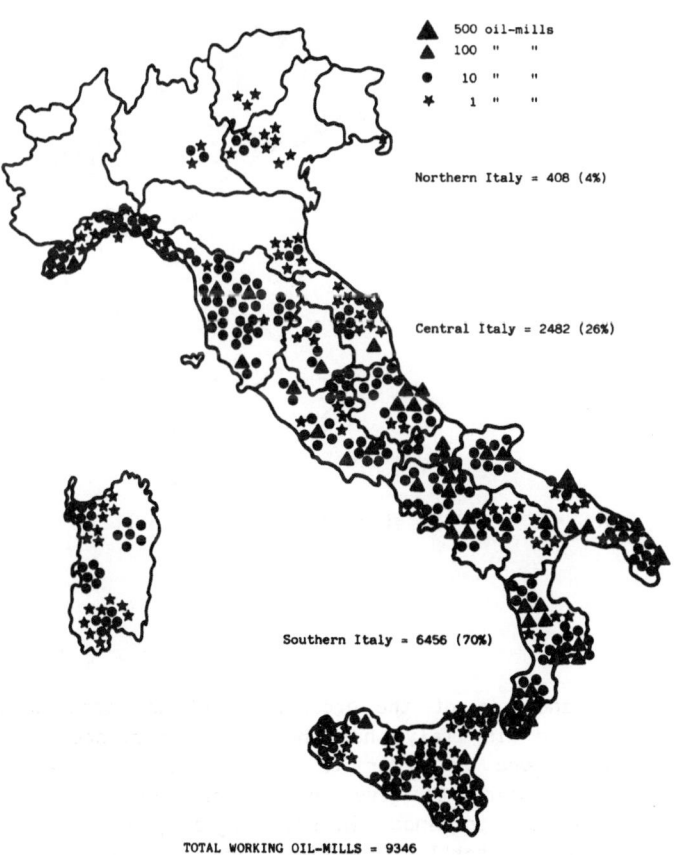

Figure 4 . Working Oil-Mills in Italy (1985)

The Figure 5 shows the distribution of the processing industry : 60% of the industries are concentrated in central and northern Italy, as are the bottling and marketing of the oil.

Finally, the Figure 6 shows the different lines of processing the olives for the production of the various types of olive oil on the Italian market ("extra virgin", "sopraffino virgin", "olive" and "husk and olive").

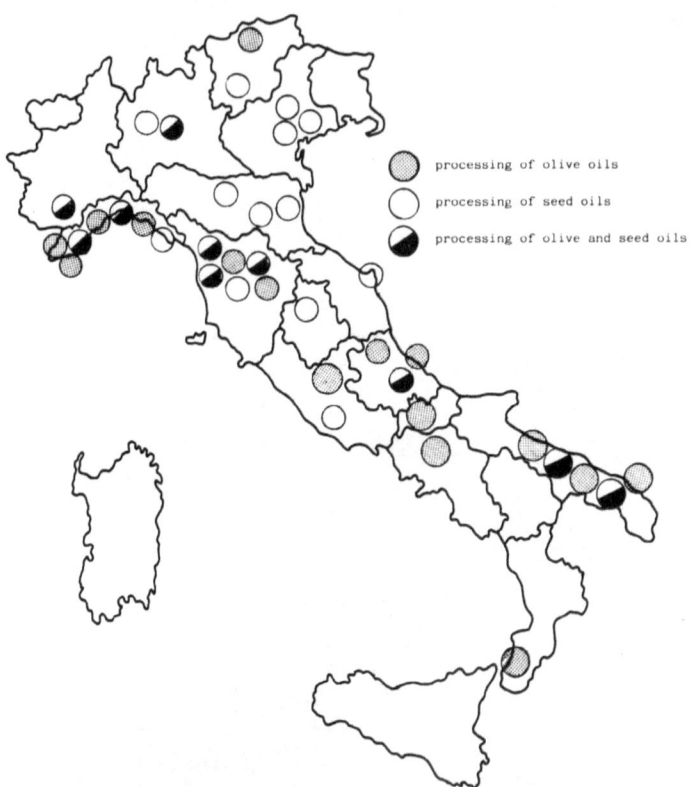

processing of olive oils

processing of seed oils

processing of olive and seed oils

Figure 5 . Processing of Edible Oil in Italy (1985)

This brief analysis of the organizational and productive structure of olive oil sector in Italy, has shown the importance of the problems which must be solved in order to give Italian olive-growing a new boost. It is necessary to give the sector new vitality on account of the considerable importance, in some regions, of the oil production with respect to the total production, and also the lack - in many areas of peninsula - of valid alternatives to olive-growing.

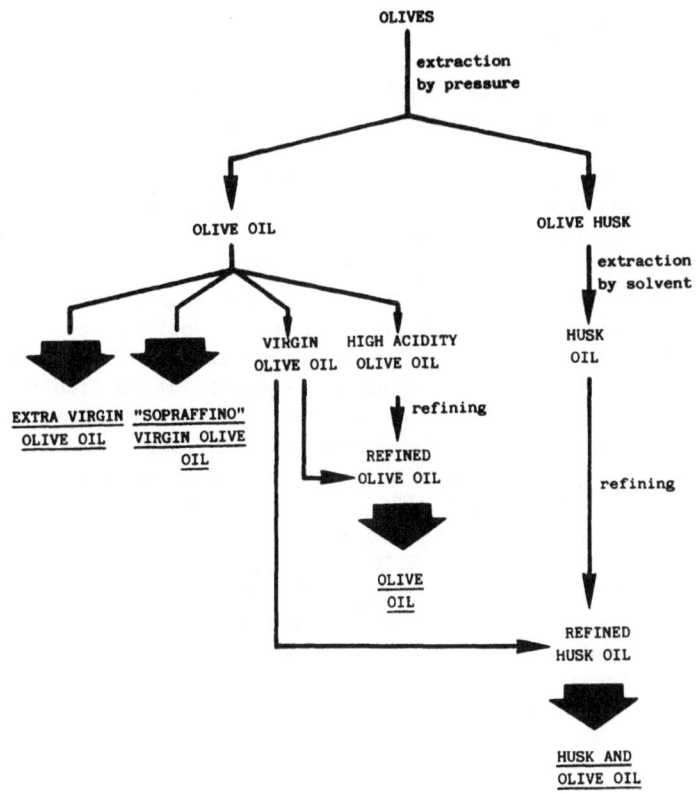

Figure 6

Within an EEC agricultural policy which seem set on penalizing olive oil, it is becoming clear that the only objective which the Italian olive oil sector must pursue is that of increasing the value of olive oil, and in particular that of defending a high quality product. The modernization of Italian olive-growing must have as its aim the production of "extra virgin olive oil"; it must be remembered that in 1985 the production of "extra virgin olive oil" was only the 20% of the total pressure oil produced. The goal can be reached by improving production (by reorganizing the olive-groves, increasing yields, reducing production costs) and processing (by expanding the cooperative system, making technological improvements to the plants, increasing industrial prod-uctivity and strengthening the distribution network). This is the only way in which Italian olive-growing can be competitive with those from other countries that have already begun the process of modernization, in a situation in which competition has become even fiencer, owing to the entry of Spain and Portugal in the EEC. The surplus olive oil of the Community will determine limits to production and to prices, so that only higher quality products will be competitive on international markets.

177

THE SEEDOIL SECTOR

In 1984 Italy imported seedoil and oilseeds for a total of 1,202 billion lire and exported the same products for 119 billion lire. The deficit therefore amounts to 1,000 billion lire and shows the gap which divides domestic production from self-sufficiency. The figure is even more alarming if one thinks of the recent expansion of oleaginous cultivations in Italy. The "success" of the oil crops is undoubtedly linked to changes in EEC agricultural policy, which has included in its new objectives the development of the production of proteins. This is a sector in which the deficit of the balance of trade of the EEC is particularly heavy. Another reason is that of finding alternative crops to those of which a surplus is produced, in particular, cereals. In order to give a general idea of the expansion of oleaginous culti- vations in Italy, we show the areas in which soybeans and sunflowers were grown in 1982 and 1984. The regions in which the two crops expanded most and the estimates for 1985 are also given in the Figure 7 and Figure 8. The areas in which soybeans were grown increased from 2,900 hectares in 1982 to about 35,000 in 1984 (a 1,070% increase), with an estimate for 1985 of 98,000 hectares (a 180% increase).
The areas in which sunflowers were grown increased from 54,000 hectares in 1982 to 85,000 in 1984 (a 58% increase), with an estimate for 1985 of 124,000 hectares (a 48% increase).

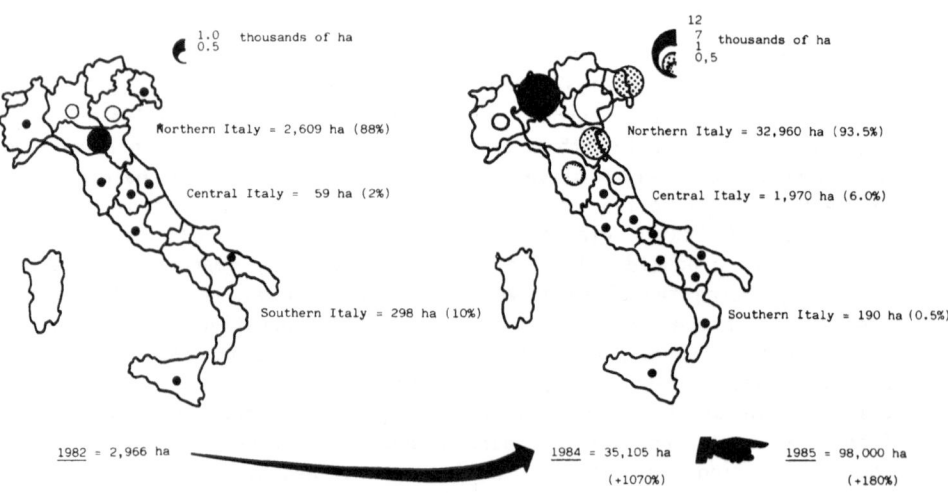

Figure 7 .   Soybean : cultivated area

Source : Annuario di Statistica Agraria, 1982 and 1984. ISTAT, 1985.

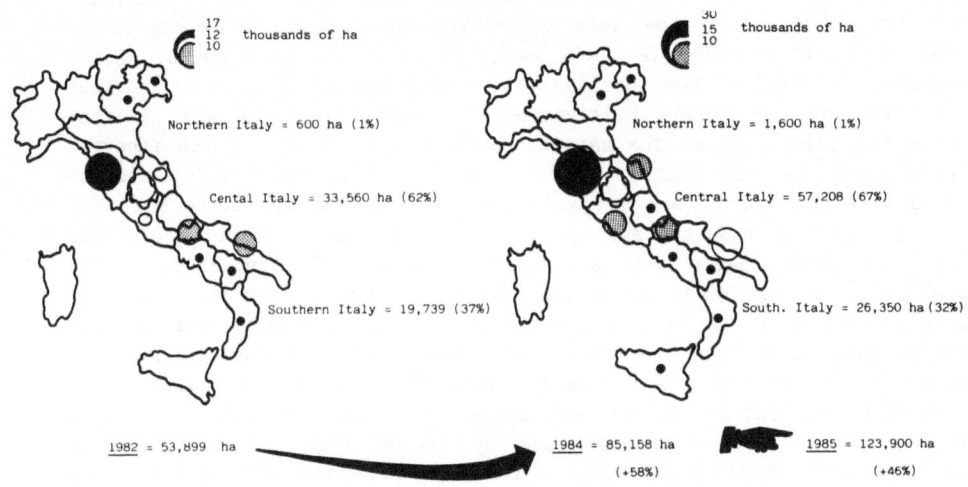

Figure 8 .   Sunflower : cultivated area

Source : Annuario di Statistica Agraria, 1982 and 1984. ISTAT, 1985.

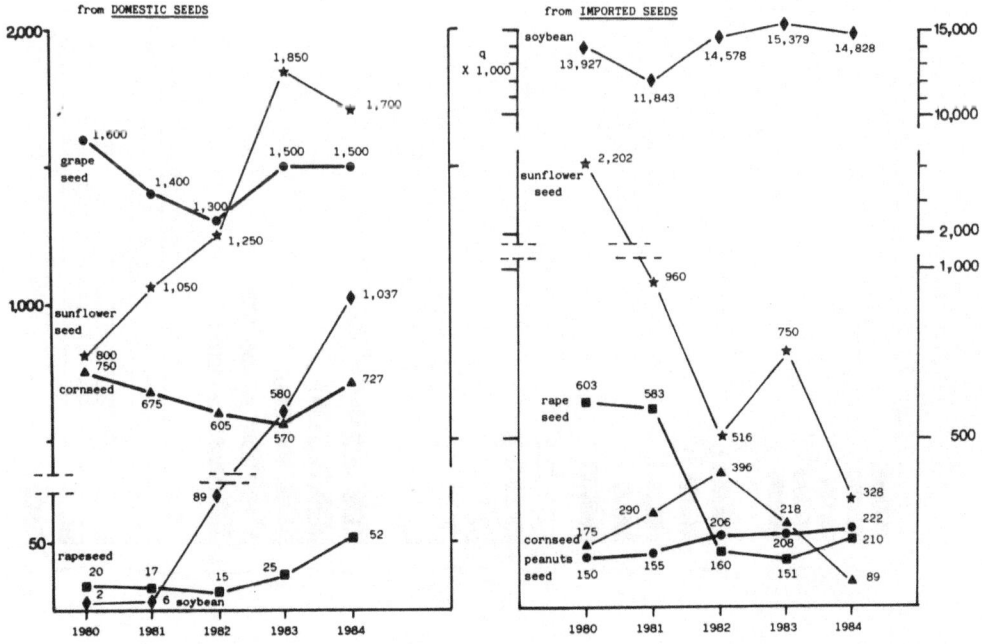

Figure 9. Main Oilseeds used for Oil Extraction from 1980 to 1984

Moreover, with the introduction of the "zero erucic acid" variety, the areas in which rape were grown increased from 800 hectares in 1982 to 8,000 in 1984 (a 900% increase), with an estimate of 15,000 hectares for 1985 (a 90% increase). As a confirmation of such an enormous expansion of oleaginous cultivations we present the Figure 9 that shows the quantities of the main home-produced oilseeds and the imported seeds used for oil extraction from 1980 to 1984. With the exclusion of grapeseeds, which are a typically Italian product connected with the distillation industry and not with the oil industry, the conclusions about the other seeds used are extremely clear. The increases in the production of sunflower seeds and rapeseeds have brought about considerable reductions in the quantities imported, although the average yields have sometimes been poor on account of unfavourable seasonal trends. In the case of soybeans, however, the home-produced seeds used for oil extraction do not exceed 6.5% of the total quantity. The same trends can be confirmed by taking into consideration the main seedoil obtained in Italy from home-produced and imported raw materials in the period from 1980 to 1984, as shown in Figure 10. The imporvement in the domestic supply of seedoil is considerable only for sunflower oil, it is not very significant for rapeseed oil and it is only an indicator of the trend for soybean oil, because the quantity of soybean oil obtained from home-produced raw material is negligible with respect to that obtained from imported seeds. The crude oil, obtained either by pressure or by means of solvents, must, however, be processed.

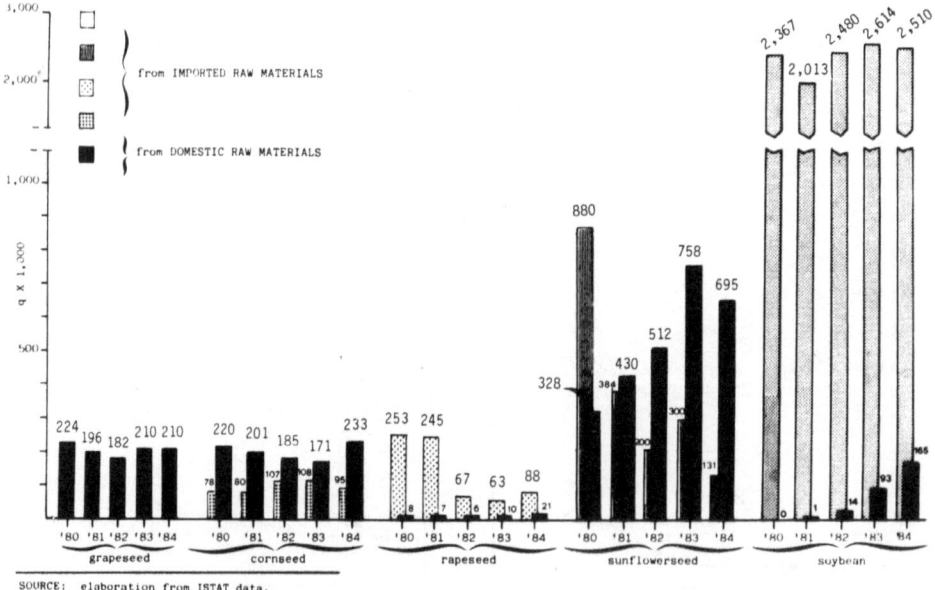

SOURCE: elaboration from ISTAT data.

Figure 10. Seedoils produced from 1980 to 1984 (grapeseed, cornseed, rapeseed, sunflowerseed, soybean)

In Italy not all oil-works carry out the processing themselfes; many of them purchase oil that has already been processed and take care only to the packing. There are about 200 firms dealing with oil, but more than 50% of the total production is concentrated in the 8 most important concerns in this sector. In Italy there has been a considerable expansion of sunflowers and soybeans only in the last few years, but the domestic requirements of oilseeds and their by-products still depend almost wholly on importation, as the Figure 10 shows. The item which has the greatest weight is soybeans (about 80% of the oilseed imported), not only for oil, but also for the proteinic fraction of the seed. In fact, the Italian import of proteinic soya flour amount to about the same monetary value as that of the seeds. The Italian Ministry of Agriculture has set up, for a five-year period, an "Oleaginous Project" which has as its aim the quantitative and qualitative improvement of oleaginous products by means of genetic and agro-technical inter-vention. The crops included in the project are : soybeans, sunflowers, rape, ricinus and safflowers. Aa far as the two main crops are concerned, it can be said that the results seem encouraging after five years' experimentation. In the case of sunflowers, experimentation on hybrids has identified new combinations which are able to produce more oil and more proteinic substances, so offering the possibility of a complete use of sunflower seeds. The process of decorticating the kernels is however fundamental in a technology that aims at improving the yield and the quality of the oil and also increasing the value of the by-products. In Italy, unlike the industries in U.S.A., France and West Germany, there are no extraction plants which decorticate sunflower seeds. This improvement in technology is one of the objective of a modern oil industry. As far as soybeans are concerned, the aims of experimentation were numerous : where and how to grow soybeans in Italy, how to increase the yield and the quality of the oil and proteins and also which technological operations are able to eliminate, render harmless or attenuate the many anti-nutritional factors present in soybean flour. Even though not all solutions have been supplied yet, it can be said that the project has brought considerable improvement to the growth of soybean in Italy.

THE NUTRITIONAL FEATURES OF THE VEGETABLE OILS (1,4,10)

We would like to conclude this report by briefly answering two questions of a certain interest. What are the nutritional features of olive oil and seedoil ? And how can technological processes influence the nutritional features of natural products on which they are carried out ? We must start by saying that it is generally believed that diet plays a significant role in health and that -within it- the lipidic diet has particular importance. The reasons for this area based on the fact that lipids: a) are one of the irreplaceable sources of energy; b) are the source of essential fatty acids; c) are the carrier of certain fundamental vitamins (liposoluble ones); d) are the main source of aromatic substances in foods, contributing to a great extent to the definition of their organoleptic properties; e) are regulators of the hunger centers, giving the sensation of satiety which is felt after a meal.

Olive oil is made up, for 98-99% of glycerides. The fatty acids that esterify the glycerine are oleic acid (60-80%), linoleic acid (5-15%), palmitic acid (9-15%), stearic acid (1.5-3.0%) and traces of other fatty acids. The remaining 0.5-1.5% consists of minor components which are called "unsaponifiable"; in this fraction are included hydrocarbons, waxes, aliphatic and triterpenic alcohols, sterols, pigments, tocopherols, phenols and vitamines. One can generally say that on account of its chemical composition olive oil is a food which is readily and rapidly difested and assimilated, also owing to the eupeptic action of the aromatic microelements which increase the secretory stimuli along the digestive tract and to the presence of other microelements (the chlorophyllic pigments) which ease absorption. From a physiological point of view oilve oil fulfils the requirements of the period of growth and the development of the bone structure; it is also a bearer of mono- and poly-unsaturated fatty acids useful for the development of the central nervous system. The positive effects of olive oil on plasma lipids and lipoproteins are suggested by the results of epi-demiological investigations carried out in various countries : the results have shown the low incidence of cardiovascular diseases in those populations which consume only olive oil.

How does the processing of the oil affect its original chemical com-position ? The effects of processing on the glyceride fraction are negligible, except for the possibility of a more or less intense formation of positional and geometrical isomers during the decolorizing. The consequences on the non-glyceride fraction are more significant : deodorizing produces the loss of almost all of the phenolic compounds and a considerable quantity of tocopherols, linear hydrocarbons, aliphatic and triterpenic alcohols; the decolorizing produces variations in the composition of the sterols and aromatic polycyclic hydrocarbons. The phospholipids, carotenoids, chlorophyll, pheophitins and ubiquinons are considerably reduced during the refining of the oil, as are metals and any possible pollutants. Finally, it can be said that only in virgin olive oil, and in particular "extra virgin" and "sopraffino virgin", the natural composition remains intact, while in the olive oils made up of mixture of "virgin" oil and "refined" oil, the final composition depends a great deal on the percentages of each oil used in the mixture and on intensity of the refining.

As far as seedoils are concerned, it is clear that the composition depends on the botanical origin of the seeds. In general, it can be said that these oils are mainly made up of polyunsaturated fatty acids; this composition makes seedoils easily oxidable, especially if they do not contain sufficient quantities of antioxidants (natural and added). The effect is not limited to the clear loss of "essential" fatty acids but also leads to the accumulation of neoformation compounds (free radicals, polymers, peroxides) responsible for the negative effects of some constituents of the oil.

It seems clear that apart from the original nutritional value of an oil - whether it is olive oil or seedoil (a feature which must be improved with suitable agronomic techniques) - the improvement of extraction and processing technologies, as well as the efficient preservation of the products, are the objectives of a modern oil industry.

It must therefore be able to subordinate the technical, economic and commercial criteria to considerations of bionutritional kind.

CONCLUSIONS

From what has been said, it can be concluded that the closer relationship between agriculture and industry has already produced a considerable improvement in the Italian oil sector. Agriculture has the task of producing raw materials (both olives or oilseeds) at the right price and of good quality. Industry must process them by means of technologies suitable for maintaining the quality of the natural product and possibly eliminate substances of doubtful nutritional value from them. The specific role of agriculture and industry, and also their reciprocal interdependence, are the most important aspects of a modern view of the oil sector, and of its development and success on domestic and international markets.

REFERENCES

1. Arrigo L., Tiscornia E.,(1984) Correlazioni tra trattamenti tecnologici e proprieta' nutrizionali dell'olio di oliva, Riv.Soc.It. Sci.Alim. 1, 83.
2. ASSITOL,(1984) various bulletins
3. ASSITOL (1985), various bulletins
4. Daghetta A., Grieco D.(1984) Considerazioni e relazioni in ordine alla esistenza o meno di differenze nel valore nutritiovo e biologico delle diverse categorie di olio di oliva, Riv.Soc.It.Sci. Alim. 1:91
5. IRVAM (1984), various issues
6. IRVAM (1985), various issues
7. ISTAT (1984), various issues
8. ISTAT (1985), various issues
9. Lattanzio V.,(1983) Relazione annuale dell'8[a] Assemblea Ordinaria UNAPROL, Roma, 29-30 novembre
10. Montedoro G., Garofalo L.,(1984) Caratteristiche qualitative degli olii vergini di oliva.Influenza di alcune variabili: varieta', ambiente, conservazione, estrazione, condizionamento del prodotto finito, Riv.It.Sost.Grasse 3, 157
11. Petruccioli G., Tombesi A.,(1984) Aspetti agronomici per la razionalizzazione dell'olivicoltura, Atti delle Giornate dell'Olio umbro, Foligno, 3-4 dicembre.

THE IMPACT OF CONVENTIONAL TECHNOLOGIES ON THE CHEMICAL COMPOSITION OF

FATS

Enzo Fedeli

Stazione Sperimentale per le Industrie degli Oli e dei Grassi

Milano, Italy

Before they are utilized as foodstruffs, fats and oils have to be subjected to technological processing in order that they be isolated from the natural raw materials and purified.

The product quality that may be achieved varies according to the previous history of the departure material and its morphology, the technology utilized and the chemical composition of the fat.

The above observation exclude from this lecture any technology applicable to fats other than those directly concerned in the extraction and purification of them.

Currently we are using both physical and chemical processes to achieve the preparation of edible fats and oils and the general flowchart that may be drawn for any of the fatty materials is shown in figure 1; there are very few exceptions, a very noticeable one being virgin olive oil, which is extracted by a physical process but which is not refined.

We shall discuss quality in terms of the appropriate variables but before doing so we have to define quality.

The quality of a fat or of an oil is the nutritional value thereof defined by equation 1; the higher value of Q, the higher the quality will be (Fig.2).

I am not a nutritionist even if I am aware of the most recent literature on the subject as a whole and I should not be able to define each of the variables, and therefore Q, but before discussing the part in which I am competent, I whish to introduce a few considerations.

In terms of calorific value in the nutritional framework, each fat or oil is more or less equivalent; the fatty acids content is however of very considerable significance as demonstrated by thousands of scientific works on the subject.

```
NATURAL SOURCE ---> RAW MATERIAL ---> SEEDS STORAGE

SEEDS  STORAGE ---> EXTRACTION   --->   OIL STORAGE

OIL   REFINING --->    CANNING    --->    UTILIZATION
```
Fig.1


```
VARIABLE                                             F(....)

Fatty acids composition                                FA
Minor components                                       MC
Contaminants                                           CN
Side-Products                                          SP
Keeping 1                                              K1
Keeping 2                                              K2

Q=F(FA)+F(MC)+F(CN)+F(SP)+F(K1)+F(K2)
```
Fig.2 Variables that define quality


```
+-OCOR1  enz.+H2O     +-OH
+-OCOR2  ---------->  +-OCOR2  +R1COOH
+-OCOR3               +-OCOR3

+-OH     enz.+H2O     +-OH
+-OCOR2  ---------->  +-OH      +R2COOH
+-OCOR3               +-OCOR3

+-OH     enz.+H2O     +-OH
+-OH     ---------->  +-OH      +R3COOH
+-OCOR3               +-OH
```
Fig.3 Hydrolysis of fats by lipases


```
+OCO-()-=-()-  enz.+O2   +-OCO-()-=-()-
+OCO-()-=-()-  -------->  +-OCO-()-=-()-
+OCO-()-=-()-             +-OCO-()+-==-()-
                              OOH
```
Fig.4 Oxydation of fats by lipoxydases

Nutritionists would be able to attribute a value to $\bar{\phi}$ (FA) better than I. Minor components, which are always present in fats and oils, have received less attention from the nutritionists except in a few cases, it would be very difficult to give value of $\bar{\phi}$ (MC) proportional to that attributable to fatty acids.

Let us consider, for example, virgin olive oil, which does not undergo refining. At least 100-150 mg of a mixture of simple and complex phenols, squalene, ß-carotene and triterpenic acids are introduced with the daily diet. Those quantities could have a pharmacological rather than a nutritional significance; the same might be true in the presence of harmful minor components.[1-4]

When we return to the principal theme and examine the influence of the source on the FA composition, it is common knowledge that the application of new technologies to agriculture can influence it at times and change completely the composition, as in the case of rapeseed.

We shall leave aside these standpoints, and consider the effects of storage. It is noteworthy that storage can affect the structure of triglycerides, which are hydrolized and fatty acids, which are hydroperoxidized.[5,6]

The reactions are shown in Figs. 3 and 4. The extent to which the reactions are not dependant solely upon storage conditions since the material has a very strong influence. For instance, the tissues from which animal fats are prepared cannot be stored for a long time and have to be processed promptly. A similar situation is present in most of the oleaginous fruit. In seeds the phenomena are relatively weak but sufficient transformation takes place to make refining mandatory.

Extraction comprises several steps, each of which affects to a different extent the integrity of the oleaginous cell. As the result of that action, the oil comes into contact with air and more intimate contact between the oil and enzymes occurs. The two reactions shown previously may develop more rapidly at least in the first part of those mechanical treaments where the concentration of enzymes is still high. Another reaction is made possible by the presence of an unlimited amount of oxygen, caused by the increasing subdivision of the material and spontaneous heating caused by friction of the material on itself and on the iron parts of the machinery.

As FFA are present, they react with small particles of metal to enhance chiefly the concentration of iron, the catalytic action of which on auto-oxidation is well known. Fig. 5 shows, in a simplified form, the reactions involved in this step but we have to consider that most of the reaction occurs on the triglycerides.[7]

The speed of reactions A and B and of the enzyme-catalized reactions presented in Fig. 5 is of course proportional respectively to oxygen, iron and enzyme concentration.

The presence of water is always considerable but in fruit extraction technology, the high quantity of water present and the morphology of the material cause a more intimate mixture and enhanced possibility of reactions occurring.

The transformation of the hydroperoxides, which are formed by enzymatic action during storage, commences at this point. Generally speaking, peroxides transformation consists of the breaking down of the fatty acids molecules or in an oxidizing action on other molecules. The general result of the foregoing is the formation of polar triglyceride molecules and of relatively short chain compounds (Fig. 5). The latter form the rancid flavour of the oil.

We do not know anything about interaction of the new molecules with the substrate but, speculatively, they must exist.

The rapidity of the reactions described thus far, with the exception of enzymatic reactions, is counteracted by the inhibiting presence of antioxidant, at least in the raw materials in which they exist; for animal fats, the low content of those compounds is minimal and no prevention is to be expected. Effectively, the processing of animal fats requires special care technology.[8,9,10]

The situation before extraction, may be described as shown in Fig. 6 and we have to observe that the amount of damaged fat in a correct technology is very low. In practise, the higher figures are not attained under normal conditions and statistically the figures are very close to the low values.

To continue the exposition on the incidence of present technology on the quality of the fatty part of the processed material, we shall now enter more deeply into the extraction procedure.

When the material has been prepared, we extract the fats by applying pressure or solvents or by a combination of the two.

Technology has at this point several branches, described in Fig. 6.

When line 1 is applied the material is sent to a coocker where it is heated and brought to the correct moisture content.

The cooker is virtually a reactor where several reactions, most of which are related to enzyme deactivation and the tranformation of several fatty acids derivatives take place. Hydroperoxides are broken down and the reaction of fragments with protein probably takes place but no literature relating to those matters exists. It is likely that sugars present in the matrices react and thus also give rise to Maillard reactions.[11,12]

The extent to which this last group of reactions takes place is proportional to the hydroperoxides content and therefore is of low entity.

Data are given with the aim of quantifying the transformations in Table II. The computations are based upon the few data available in the literature.

188

```
A          RH ---> R* +H*
           R* +O2 ---> ROO*
           ROO* +RH ---> ROOH+R*

                (Fe)n+
B          ROOH --------> RO* +OH*


+-OCO-()-CHO      -=-CHO        Aldehides
+-OCOR2
+-OCOR3          R4-NH=CH-=-    Nitrogen deriv.

Aldoglycerides
```

Fig.5 Autoxydation and related reactions

TABLE I   By-products(ppm) present before extraction*

COMPOUND	SOURCE	AMOUNT SEDDS	AMOUNT FRUITS	AMOUNT ANIMAL
Hydroperoxides	ENZYM.	100	20	20
	AUTOXID.	50	10	30
By-Products	BREAKING	18	4	7
Interaction				
compounds	REACTION	2	1	1
FFA	ENZYM.	10000	3000	3000

*Mean values from literature

```
---------      -----------      -----------      ---------
CONDITIO  +  CONTIN.    +  RECONDITIO  +  SOLVENT
NING   A     PRESSES       NING            EXTRACT.
---------      -----------      -----------      ---------

---------      -----------
CONDITIO  +  SOLVENT
NING   B     EXTRACTION
---------      -----------

---------      -----------
PREPARA   +  DISCONTIN.
TION   A     PRESSES
---------      -----------                -------------------------
                              OCCASIONAL       SOLVENT
                           +  EXTRACTION.THE FATS ARE
---------      -----------     NOT CLASSIFIED AS   THE
PREPARA   +  HORIZONTAL    PRODUCT  FROM  PREVIOUS
TION   B     CENTRIFUGE              OPERATION
---------      -----------                -------------------------
```

Fig.6 Combination of technologies in the extraction
      of fats and oils

The "Final Results" figures take into account the amount remaining and formed after the various reactions.

Treatment in continuous presses has an impact on fats and oils because of the relatively high temperatures attained as a function of the heat generated by friction; temperature is kept at a reasonable value by partial recirculation of the separated oil.

For similar reasons and owing to the presence of FFA, the iron content of the oil increases; owing to the presence of air, most of the reactions [13,1.] are to be associated with autoxidation even when hydrolysis can take place.

Reconditioning and flaking of the cake are still present as a major cause of residual oil transformations and autoxidation.

The impact of solvent extraction on the fatty substance is heavily dependant upon the solvent utilized. Hexane is widely utilized on account of its low polarity which enables components of polarity very similar to that of the solvent itself to be extracted selectively.[15,16]

For instance, the oil extracted with the aid of trichloroethane from olive residue, in a few plants that are still operative, has a far higher content of oxidized derivatives of fatty acids than hexane-extracted oil.[17]

Table III shows transformations caused by the technology utilized, in the extraction of olive oil and other oils from fruits; the absence of the products of hydroperoxides reactions and of the other product is partially due to the mild treatment during preparation and extraction, partly to the presence of vegetation water and partially to the absence of strong concentrations of protein and phospholipids. Virtually, the presence of water enhances hydrolytic actions but prevents esterification reactions.

When solvent extraction procedures are utilized, one more operation has to be executed on the extracted fats in order that the solvent be removed. Solvent removal necesitates high temperature distillation in the presence of water vapour or, in the final stage, in vacuo.[18]
This action causes further transformations as shown in Table V and VI.

The balance for the oil extraction technologies considered is shown in Table VII : the amount of by-products in the oils derived from seed when the proper technologies are used is higher than those from other products processed by other technologies.

Most of the differences are to be attributed to the technology.

Other components are affected by the treatments; autoxidation and the transformation of the hydroperoxides also takes place in proportion to the contents of those components, which are very low in comparison with fatty acids or glycerine derivatives. If that is true for technologies where water is a minor constituent, in the case of fruits we may observe a partition of some of the components between the oil and acqueous phases.

TABLE II By-products formed during an extraction
procedure(ppm)§

REACTION	OPERATION				
	1	2	3	4	5
FFA	13000	18000	16000	16500	17250
HYDROPEROXIDES	200	350	60	50	200
Hydroperoxides decomp.product	100	240	100	110	175
Hydroperoxides reaction prod.	100	250	200	200	225
Proteins-Sugars interact.prod.	50	50	25	10	30
Phospholipids hydrolysis	30	20	10	5	12
Phospholipids reactions	20	10	10	2	6
Lipoproteins reactions	10	10	5	0	5
FFA-Metals interaction	0	10	5	0	5

1:CONDITIONING   2:CONTINUOUS EXTRACTION(PRESSES)
3:RECONDITIONING 4:HEXANE EXTRACTION
5:FINAL RESULTS
4:   " B+CENTRIF.

§Mean values from literature·

TABLE III By-products formed in olive oil extraction(ppm)§
Discontinuos presses

REACTION	OPERATION	
	Prepa ration	Extraction
FFA	3500	6000
HYDROPEROXIDES	800	1000
Hydroperoxides decomp.products	80	100
Hydroperoxides reaction prod.	−	−
Proteins-Sugars interact.prod.	−	−
Phospholipids hydrolysis	−	−
Phospholipids reactions	−	−
Lipoproteins reactions	−	−
FFA-Metals interaction	−	−

§Mean values from literature

TABLE IV By-products formed in olive oil(OO) & pork fat(PF).
extraction(ppm) §Centrifuges

REACTION	OPERATION			
	Prepa ration		Extraction	
	OO	PF	OO	PF
FFA	3500	3000	4000	4000
HYDROPEROXIDES	800	1000	800	1500
Hydroperoxides decomp.products	80	100	80	150
Hydroperoxides reaction prod.	–	–	–	–
Proteins-Sugars interact.prod.	–	–	–	–
Phospholipids hydrolysis	–	–	–	–
Phospholipids reactions	–	–	–	–
Lipoproteins reactions	–	–	–	–
FFA-Metals interaction	–	–	–	–

§Mean values from literature

TABLE V By-products formed in solvent extraction(ppm)§
of seeds oils

REACTION	OPERATION	
	Conditi tioning	Extraction
FFA	13000	13500
HYDROPEROXIDES	200	250
Hydroperoxides decomp.product	100	110
Hydroperoxides reaction prod.	100	200
Proteins-Sugars interact.prod.	50	60
Phospholipids hydrolysis	30	35
Phospholipids reactions	20	22
Lipoproteins reactions	10	10
FFA-Metals interaction	–	1

§Mean values from literature

TABLE VI By-products formed in desolventizing of miscellas from seeds oils(ppm)§

REACTION	OPERATION	
	Hexane solution	Desolv.
FFA	16500	16500
HYDROPEROXIDES	200	100
Hydroperoxides decomp.product	175	245
Hydroperoxides reaction prod.	225	255
Proteins-Sugars interact.prod.	30	30
Phospholipids hydrolysis	12	12
Phospholipids reactions	6	6
Lipoproteins reactions	5	5
FFA-Metals interaction	6	6
Residual hexane		30

§Mean values from literature

TABLE VII Balance of by-products formed by various technologies(ppm)§

REACTION	TECHNOLOGY				
	1	2	3	4	4 (Anim.fat)
FFA	17250	13500	6000	4000	4000
HYDROPEROXIDES	200	250	1000	800	1500
Hydroperoxides decomp.product	175	110	100	80	150
Hydroperoxides reaction prod.	225	200	–	––	–
Proteins-Sugars interact.prod.	30	60	–	––	–
Phospholipids hydrolysis	12	35	–	–	–
Phospholipids reactions	6	22	–	–	–
Lipoproteins reactions	5	10	–	–	–
FFA-Metals interaction	6	6	–	–	–
Residual hexane	15	30	–	–	–

1:COND.A+CONT.PRESS.+RECOND.+SOLV.EXTR.
2: " B+   "   "
3:PREP.A+DISC.PRESS.
4: " B+CENTRIF.
§Mean values from literature

We have demonstrated, for instance, that some of the phenolic components are extracted by the aqueous phase and that some hydrolysis intervenes at the level of complex phenolic consituents; small losses of flavouring compounds are due to the same causes. A summary of the results of computations is given in Table VIII.

Extracted oils, with one notable partial exception, have to be submitted to refining before consumption. Most of the refining is accomplished in accordance with the conventional scheme shown in Fig. 7 but increasing quantities of oils and fats are refined by a relatively new process which is also shown in Fig. 7. The dwell time of oil in tanks, and transportation before refining have been omitted but they have a considerable influence on the autoxidative deterioration of the oil.

Degumming is the operation by which phosphatides and similar undesiderable products are removed from oils. It consists essentially of the hydration of those substances to render them insoluble in oil; the complete separation of the two phases is accomplished by centrifugation.
The procedures vary according to phosphatide content, high phoshatide oils being treated with water and low phosphatide oils with mineral or organic acids.[19]
When water is used, as in soyabean oil extraction, no problems relating to contamination arise but even when solubility in the oil and the subsquent alkaline treatment eliminate all the added reagent and any impurities present in it.

When no alkaline deacidification is utilized, no contamination is possible when an efficient washing system is applied.

Mineral acids, in contrast, decompose the hydroperoxides present and react with epoxides previously formed to open the ring. Effectively, we observe a decrease in the peroxide value after that operation. In addition, during the operations, some autoxidation takes place because of mixing and centrifuging taking place in the presence of air. Table IX describes the situation after degumming.

Alkaline refining is a very beneficial operation for the purification of edible oils and effectively, besides reduction of the FFA content to a very low level, side reactions have a positive effect.[20,21]

For instance, the phospholipid content is still reduced and likewise associations of carbo-hydrates-proteins, acidic derivatives from autoxidation, metals, specific and non-specific contaminants like gossypol and chlorophyll.

Hydroperoxides also react in an alkaline medium while in the different neutralization phases, new peroxides are formed as the result of mixing in the presence of air. The situation is shown in Table X.

Bleaching implies treatment of the product with diatomaceous earth and/or carbon black under a low vacuum. In addition to eliminating several products responsible for oil colour, bleaching has a profound effect on po-

194

TABLE VIII Minor components lost or transformed in
                extraction technologies*

COMPOUND TRANSFOR MED OR NOT EXTRAC TED	TECHNOLOGY			
	1	2	3	4
SQUALENE (Oxyd.)	10	10	2	2
CAROTENES (")	10	10	2	2
TOCOPHEROLS (NE)	-	-	10	10
ESTERS OF ALCOHOLS (Oxyd.)	10	10	1	1
TRITERPENICS . COMP. (NE)	-	-	20	20
STEROLS (NE)	-	-	5	5
OTHERS (NE+oxyd.)	10	10	10	10

* % TRANSFORMED OR NOT EXTRACTED ON THE TOTAL
PRESENT IN THE OIL.
Oxyd.=OXYDATION AND RELATED REACTIONS
NE=NOT EXTRACTED
§Mean values from literature

CLASSICAL

#	NAME	M	C	F	REAGENTS	T(dC)	P(Hg,mm)
					+		
1	Degumming	+	+	-	H /H2O	70	760
2	Neutralizing	+	+	-	NaOH/H2O	80	760
3	Bleaching	+	+	+	Earth/vac.	110	300
						110	760
4	Deodorizing	+	-	-	-	260	5
5	Dewaxing	+	-	+	-	15	760
6	Canning	+	-	-	-	25	760

MODIFIED

#	NAME	M	C	F	REAGENTS	T(dC)	P(Hg,mm)
					+		
1	Degumming	+	+	-	H /H2O	70	760
2	Bleaching	+	+	+	Earth/vac.	120	300
						110	760
3	Deodorizing	+	-	-	-	270	2
4	Dewaxing	+	-	+	-	15	760
5	Canning	+	-	-	-	25	760

M=mixing;C=centrifuging;F=filtering;dC=Celsius

Fig.7 Current refining procedures for oils & fats

TABLE IX Variation of by-products concentrations
during degumming(ppm)§

COMPOUNDS	A	B	B1
FFA	17250	17250	17250
HYDROPEROXIDES	200	150	150
Hydroperoxides decomp.product	175	250	205
Hydroperoxides reaction prod.	225	320	255
Proteins-Sugars interact.prod.	30	20	25
Phospholipids hydrolysis	12	1	5
Phospholipids reactions	6	.6	2
Lipoproteins reactions	5	.5	3
FFA-Metals interaction	6	10	6
Residual hexane	15	15	15
Specific natural contaminants	600	550	600
Phospholipids	10000	500	1500

A:oil to be refined
B:acid treated oil
B1:Water treated oil
§Mean values from literature

TABLE X Variation of by-products concentrations
during alkaline deacidification(ppm)§

COMPOUNDS	Before	After
FFA	17250	100
HYDROPEROXIDES	150	140
Hydroperoxides decomp.product	230	100
Reaction prod.	320	120
Proteins-Sugars interact.prod.	20	0
Phospholipids hydrolysis	1	0
Phospholipids reactions	.6	0
Lipoproteins reactions	.5	0
FFA-Metals interaction	10	6
Residual hexane	15	15
Specific natural contaminants	550	150
Phospholipids	500	150

§Mean values from literature

lar compounds formed by the conversion of hydroperoxides and eliminates some of the metals present, residual phospholipids and other contaminants.[22]

Bleaching virtually has to be seen as an adsorption process where the
polarity of the eluent is determined by the strong mass of the triglycerides. The new situation is shown in Table XI.

Finally, deodorization brings the oil to high temperatures under a high
vacuum during violent mixing caused by the injected water vapour.[23,24]

Deodorizers can be considered as reactors in which chemical action is
also performed in addition to physical action. The most obvious physical
result is the elimination of all volatile substances and also of non-vola-
tile compounds of molecular weight up to 400.

The chemical actions are chiefly to be found in hydrolysis and the
transformation of heat-labile compounds like peroxides (Table XIII).

They are transformed partly into volatile products, eliminated and
partly into polymeric compounds which remain in the oil.

I have not spoken of pesticides and other agricultural contaminants
during the lecture because they are included in Dr. Prevot's lecture but
if they were present, they would be eliminated chiefly at this stage.

Oxygen is stryctly excluded in deodorizing so that there are no actions
resulting from its presence.

After deodorization, no further changes with reference to the chemical composition of the oil are possible. Some autoxidation, mostly yielding
peroxides is possible during dewaxing and canning. In this latter stage, if
the oil is not kept under inert gas, oxygen, which was eliminated completely during deodorization, will dissolve in the oil, to approach saturation,
the limit of which is 35 ppm. That oxygen will react rapidly in the canned
oil to give hydroperoxides which will be transformed slowly into decomposition products, reverting the flavour and enhancing the content of decomposition products, although in completely closed storage media, no further
action will take place.[25]

The parallel situation when the modified refining line is utilized is
shown in Table XIV.

To return to the equation set out in Fig. 2, leaving aside ϕ (FA)
ϕ (MC) which have nothing to do with me, it is possible to define ϕ
(CN), the function related to contaminants as a 0-tending function. A statement should be made about ϕ (SP) but the intriguing action of oxygen
has a tendency to send such a function toward a small but definite value.

When quality 1 is kept high ϕ (K1) becomes high if the quality of
the oil refining process and storage is high.
In the present situation, we have seen that a high value can be attained,
provided that a good storage situation is assured.

TABLE XI Variation of by-products concentrations during bleaching(ppm)§

COMPOUNDS	Before	After
FFA	100	100
HYDROPEROXIDES	140	100
Hydroperoxides decomp.product	230	120
Reaction prod.	320	120
Proteins-Sugars interact.prod.	0	0
FFA-Metals interaction	6	8
Residual hexane	15	10
Specific natural contaminants	250	10
Phospholipids	150	120

§Mean values from literature

TABLE XII Variation of by-products concentrations during deodorization(ppm)§

COMPOUNDS	Before	After
FFA	100	50
HYDROPEROXIDES	100	0
Hydroperoxides decomp.product	120	90
Reaction prod.	120	60
FFA-Metals interaction	8	8
Residual hexane	10	5
Specific natural contaminants	10	5
Phospholipids	120	100

§Mean values from literature

TABLE XIII Minor components lost during
            refining technologies*§

COMPOUND	%
SQUALENE	50
CAROTENES	80
TOCOPHEROLS	50
TERPENICS	20
ALCOHOLS	30
OTHERS	50

* % ON THE TOTAL
PRESENT IN THE OIL.
§Mean values from literature

TABLE XIV Variation of by-products concentrations
            during physical refining(ppm)§

COMPOUNDS	SP	B	D
FFA	17250	17250	20
HYDROPEROXIDES	150	70	O
Hydroperoxides decomp.product	230	370	180
Reaction prod.	320	600	320
Proteins-Sugars interact.prod.	20	O	O
Phospholipids hydrolysis	1	O	O
Phospholipids reactions	.6	O	O
FFA-Metals interaction	10	12	12
Residual hexane	15	10	2
Specific natural contaminants	550	25	10
Phospholipids	200	160	120

SP:degummed oil
B:bleached oil
D:deodorized neutralized oil
§Mean values from literature

The keeping quality 2 ϕ (K2), relating to the cooking of the product, is a function mainly of the fatty acids composition of refining and storage: the higher the insaturation of the oil, the lower is ϕ (K2) but as far as refining and storage are concerned, the advice given for ϕ (K1) is applicable.

In conclusion, we may state that present extraction and refining situations tend to bring Q towards high values but work has to be done mostly to try to extract and to preserve nutritionally useful comonents present in the natural material of departure as well as to try to preserve the oil from damage resulting from autoxidation.

REFERENCES

1. E. Fedeli, "RFCG" 30, 51 (1983)
2. E. Fedeli, "JAOCS" 60, 404 (1983)
3. N. Cortesi and E. Fedeli, "RISGAD" 60, 341 (1983)
4. E. Fedeli, A. Lanzani, P. Capella and G. Jacini "JAOCS" 43, 254 (1966)
5. H.N. Moore, "JAOCS" 60, 189 (1983)
6. G. Florin and H.R. Bartesh "JAOCS" 60, 193 (1983)
7. W.G. Merteus, C.E. Swindels and B.F. Teasdale "JAOCS" 48, 544 (1971)
8. B.N. Stuckey "Handbook of food additives" T.E. Furia Ed., Chemical Rubber Co, Cleveland, 209-245 (1968)
9. W.W. Nawar, "Food Chemistry" O.R. Fennema Ed., M. DEKKER Inc., Chapt 4, 176-205 (1985)
10. E. Fedeli, A. Brillo and G. Jacini "RISGAD" 50, 102 (1973)
11. E. Fedeli, "RISGAD" 59, 185 (1982)
12. W. Heimann, "Fundamental of food chemistry", Chapt. 5, Ellis Horwood Publ., AVI Publ. Co. West Port, USA
13. D.K. Bredeson, "JAOCS" 55, 762 (1978)
14. D.K. Bredeson, "JAOCS" 60, 211 (1983)
15. N.H. Witte, "JAOCS" 57, 854A (1980)
16. P.L. Christensen, "JAOCS" 60, (1983)
17. E. Fedeli, "RISGAD" 61, 355 (1984)
18. N.W. Myers, "JAOCS" 60, 224 (1983)
19. G. Haraldson, "JAOCS" 60, 251 (1983)
20. M. Kock, "JAOCS" 60, 198 (1983)
21. T.K. Mag , "JAOCS" 60, 380 (1983)
22. D.R. Taylor and C.B. Ungermann, "JAOCS" 61, 1372 (1984)
23. E.G. Latondress, "JAOCS" 60, 257 (1983)
24. F.A. Dudrow, "JAOCS" 60, 272 (1983)
25. A. Forster and A.J. Harper "JAOCS" 60, 265 (1983)

HYDROGENATION OF EDIBLE OILS

Timothy L. Mounts

Northern Regional Research Center, ARS-USDA
1815 N. University Street
Peoria, Illinois 61604

INTRODUCTION

Dietary fat consists of a mixture of vegetable oils and animal fats that are generally characterized by their fatty acid composition. As shown in Table 1, liquid oils are high in mono- and polyunsaturated fatty acids, while solid fats consist mainly of saturated fatty acids. Partial

Table 1 (Anon. 1982)

FATTY ACID COMPOSITION OF DIETARY FATS

Fat or Oil	Saturates	Monoun- Saturates	Polyun- Saturates
Sunflower oil	10.8	22.0	67.2
Soybean oil	16.0	25.0	58.0
Corn oil	14.5	29.0	56.5
Cottonseed oil	26.7	18.0	55.3
Peanut oil	21.0	52.0	27.0
Chicken fat	34.0	47.5	18.5
Pork fat	41.4	46.3	12.3
Palm oil	50.1	40.2	9.7
Tallow	50.4	44.3	5.3
Butter fat	65.5	30.6	3.9
Coconut oil	91.7	6.4	1.9

and selective hydrogenation of liquid oils is practiced to change the physical properties for solid fat products or to improve the stability of highly unsaturated oils during high temperature use. Hydrogenated vegetable oils are used in the formulation of shortenings, margarines, salad/cooking oils, and other foods. In the United States, the total consumption of visible and non-visible fat is 169 g per person per day (Table 2).   Of this total, 72 g (42.6%) is visible fat, of which about 39.6% (28.5 g) is estimated to be hydrogenated vegetable oils, mainly hydrogenated soybean oil.   Non-visible fat contains about 11.8% (11.4 g) hydrogenated vegetable oils. Combined, these products constitute the second most important source of fat calories and dietary fat. Thus, the hydrogenation process is a technology having significant nutritional impact.

Table 2 (Rizek, et al, 1983)

DIETARY SOURCES OF ENERGY FOR THE U.S. POPULATION

	Total Calories	Visible Fat	Non-visible Fat	Total Fat	
	(%)	(%)	(%)	Grams[a]	(%)
Butter and animal	24.5	29.4	79.8	98.6	58.3
Soybean oil	3.3	13.3	3.8	13.3	7.9
Hydrogenated oils	9.8	39.6	11.8	39.5	23.4
Vegetable oils (other)	4.4	17.7	5.1	17.6	10.4
Total fat	42.0	42.6	57.4	169	100
Total carbohydrate	46.0				
Total protein	12.0				

[a]Per person per day.

HYDROGENATION PRACTICE

Most hydrogenations are performed in batch autoclaves, such as shown in Figure 1. In general, the batch hydrogenator is a cylindrical pressure vessel having a capacity of 5-20 tons (section A, Fig. 1).

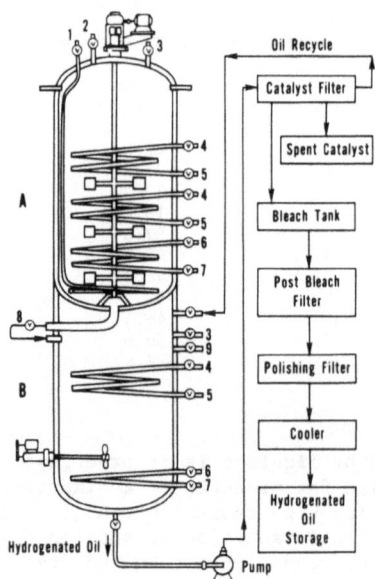

Figure 1 (Mounts, 1980)

The oil going to the hydrogenator must be refined, bleached and dry. The hydrogenation of the oil is accomplished with a nickel catalyst that has been chemically reduced with hydrogen to an active state. Commercially, two types of hydrogenation are performed: nonselective hydrogenation under conditions such as 50 psig, 0.05% catalyst at 121 C; and selective hydrogenation under conditions such as 5-14 psig, 0.05% catalyst, at 177 C. Two or more agitators are fitted to a vertical axial shaft. Hydrogen gas, high purity and as dry as possible, is sucked from the headspace and dispersed into the oil as fine bubbles. The interior of the vessel contains pipe coils for heating and cooling the charge. Circulating water through the cooling coil is used to control the heat reaction during hydrogenation, and after hydrogenation to cool the charge to filtration temperature. To increase utilization of the hydrogenator, the charge can also be cooled by draining it into a drop tank (section B, Fig. 1), which is also equipped with agitators and heating and cooling coils. A fresh charge can be delivered to the hydrogenator and reaction initiated while the first charge is cooling. The cooled oil is then filtered to remove catalyst. The filtered oil is mixed with citric acid in the bleach tank to facilitate removal of any "colloidal" nickel and then the batch is passed through a post bleach filter, a polishing filter, cooled and stored.

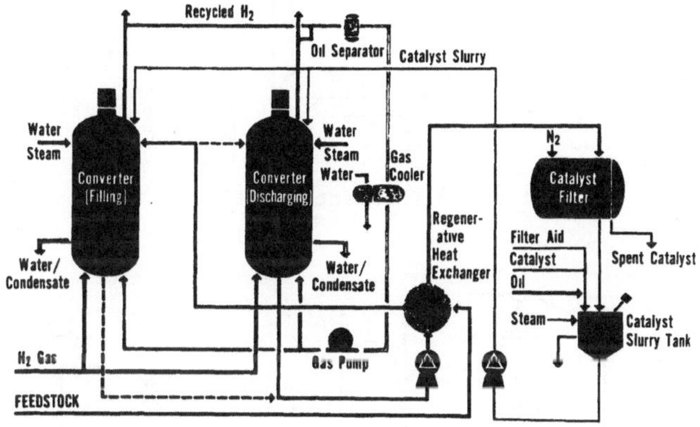

Figure 2 (Courtesy of the Sullivan Systems, Inc.)

A semi-continuous hydrogenation system, shown in Figure 2, utilizes two alternating reactors employing high-shear mechanical agitation aided by a moderate amount of hydrogen recycle from the headspace back into the oil. The recycle of hydrogen back into the oil increases the hydrogen available at the catalyst surface, thus increasing the rate of reaction. While a batch of oil is being hydrogenated in one reactor, the batch in the other reactor is being filtered. The flow of oil from reactor, to filter, to post treatment, and to storage tank is continuous.

The loop reactor shown in Figure 3, is enjoying some commercial success. It employs an injector type mixing nozzle based on the Venturi principle. The injector mixing nozzle produces a jet of liquid that picks up and accelerates a gas from its surroundings. An intimate

mixture of the oil/catalyst slurry and the hydrogen gas results, which
increases the hydrogen concentration at the catalyst surface and thus the
rate of reaction.

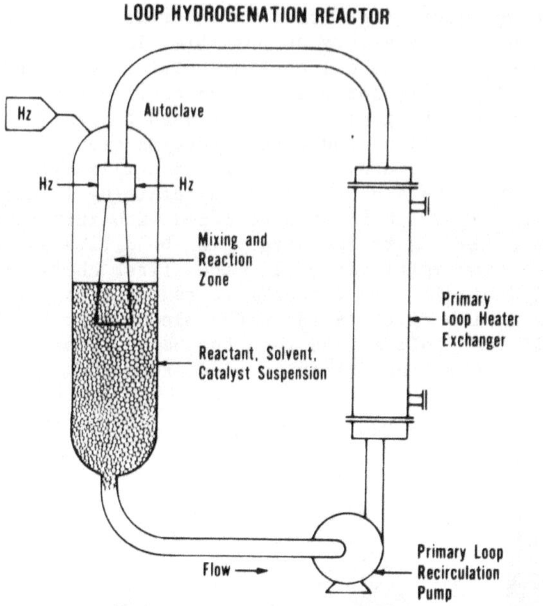

LOOP HYDROGENATION REACTOR

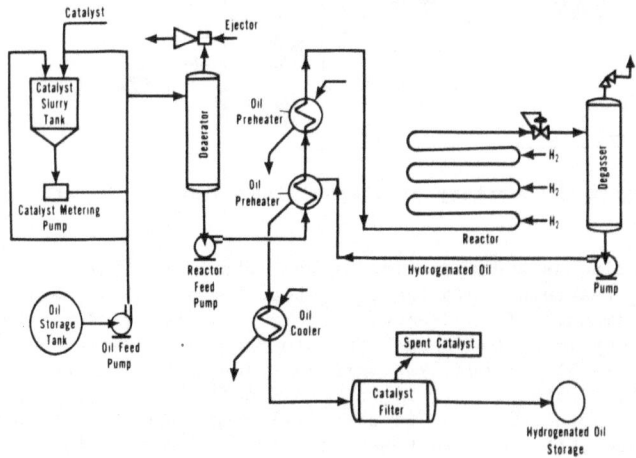

Figure 3 (Hastert, 1981)

Figure 4 (Mounts, 1980)

There is much interest in the development of truly continuous
processing systems. One such system, diagrammed in Figure 4, involves
the concurrent flow of oil, catalyst, and hydrogen gas in a horizontal
pipe reactor. The basic features of this system are: stoichiometric
consumption of hydrogen; an essentially liquid-filled reactor; minimized

back-mixing; and agitation provided by jet addition of hydrogen along the reactor length. The kinetic reaction curves (Fig. 5) generated for the hydrogenation of soybean oil in both commercial and laboratory-scale versions of such an open tube reactor are quite similar to those for batch hydrogenations of the same oil.

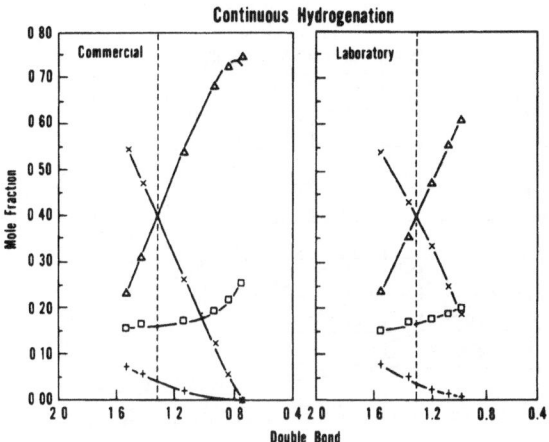

Figure 5 (Mounts, 1980)

HYDROGENATION THEORY

Selectivity is an important concept to be considered for hydrogenation of vegetable oils. Only the unsaturated fatty acids of an oil participate in the hydrogenation reaction. A simplified reaction sequence is shown in Figure 6. As the hydrogenation proceeds the fatty

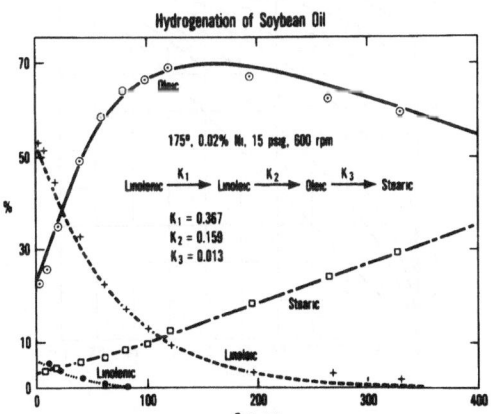

Figure 6 (Allen, 1981)

acid composition of the oil changes. The reaction rate of the hydrogenation of linolenic acid to linoleic is k_1. k_2 is the hydrogenation rate of linoleic to oleic and k_3 is the hydrogenation rate of oleic to stearic acid. Catalyst selectivity is based on the rates of reaction during hydrogenation, that is, the linolenic selectivty is defined as the ratio of k_1/k_2, while the linoleic selectivity is the ratio of k_2/k_3. For the reaction shown, linoleic

205

selectivity is 0.159/0.013 = 12.2 (ie. linoleic acid is being
hydrogenated 12.2 times as fast as is oleic acid). To determine the
selectivity ratio for each acid, one must know the fatty acid composition
of the beginning oil and of the hydrogenated product, as determined by
gas chromatographic analysis. The graph shown in Figure 7, used to

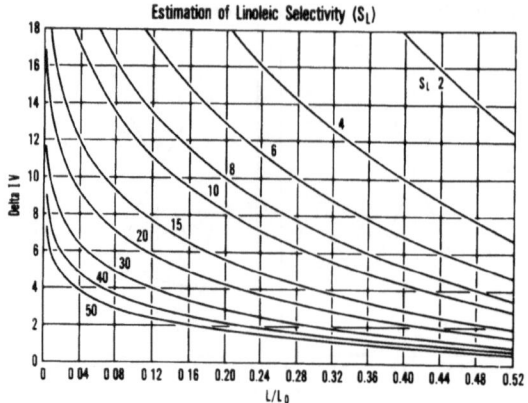

Figure 7 (Allen, 1978)

estimate linoleic selectivity requires that the ratio of the starting to
final linoleic acid concentration and the change in iodine value be
known. The graph shown in figure 8 has been developed to estimate the

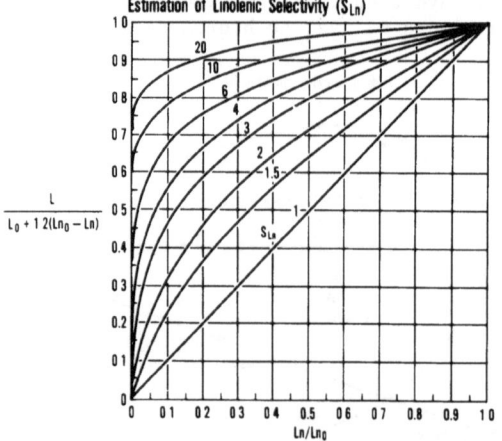

Figure 8 (Allen, 1978)

linolenic selectivity and requires that the initial and final contents of
linoleic and linolenic acids be determined. The selectivity is defined on
those lines of concurrence of the analytical data. In the past the term
selectivity was used in the sense that the hydrogenated product should be
as soft as possible at a given iodine value. By defining selectivity as
the ratio of reaction rates, as described here, it becomes a more
quantitative term. Selectivity values have been used to classify

catalysts. Since linoleic selectivity is a measure of the amount of
stearic acid formed as the linoleic acid is reduced, the catalysts that
show the highest selectivities should be used. That is, more linoleic
may be reduced to increase stability before much saturates are formed, so
that the product will not be too hard. Linolenic selectivity is used to
select catalysts to hydrogenate linolenic acid with little or no
hydrogenation of linoleic acid. This finds particular application in
the partial hydrogenation of soybean oil or rapeseed oil to a stable
liquid oil. In commercial practice, nickel catalysts show a linolenic
acid selectivity of about 2.

INFLUENCE OF PROCESS CONDITIONS

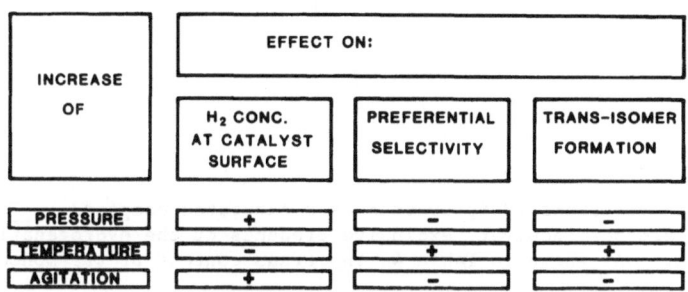

INCREASE OF	EFFECT ON:		
	H₂ CONC. AT CATALYST SURFACE	PREFERENTIAL SELECTIVITY	TRANS-ISOMER FORMATION
PRESSURE	+	−	−
TEMPERATURE	−	+	+
AGITATION	+	−	−

Table 3 (Hastert, 1986)

Changes in process conditions produce observable and measurable
effects on activity, selectivity, and isomerization, which have been
attributed to a change in hydrogen concentration at the catalyst
surface. As shown in Table 3, increasing the pressure, temperature, or
agitation, affects the hydrogen concentration at the catalyst surface
and results in the indicated effect on selectivity and isomerization.
The rate of reaction is increased by an increase in each of the process
parameters.

ISOMERIZATION

Hydrogenation of vegetable oils causes structural modifications of
the double bond configuration from cis to trans; and movement of double
bonds to new positions in the fatty acid chain. The cis and trans
isomers of monounsaturated 18-carbon fatty acid are diagrammed in Figure
9. Oleic acid, the cis isomer is bent at the double bond, while the
carbon atoms of elaidic acid, the trans isomer, lie along a straight
line. In this respect, elaidic acid is quite similar to the saturated
fatty acid, stearic, which is a straight-chain fatty acid. The trans
isomer has a higher melting point than the cis isomer and in many

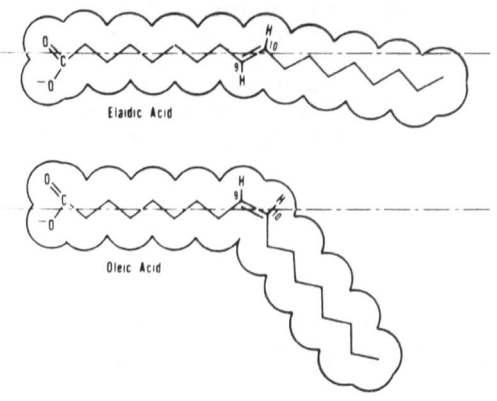

Orthogonal Projections of Oleic and Elaidic Acids

Elaidic Acid

Oleic Acid

Figure 9

enzymatic reactions it is treated as a saturated compound.   Trans
isomers are not, however, exclusive products of the hydrogenation
reaction.   As shown in Table 4, the trans content of the monoene

Table 4 (Emken, 1983)

trans Isomer Content in Butter and
Partially Hydrogenated Vegetable Oils

Sample	Percent *trans* in	
	Total 18:1	Total Sample
Butter (summer)	44.7	7.6
Butter (winter)	17.3	4.3
Salad Oil (typical)	25.0	15.0
Shortening (typical)	32.0	16.0

fraction from butter can amount to as much as 44.7%, substantially
greater than the monoene trans content found in typical hydrogenated
products such as salad oils and shortenings.   The dietary trans isomer
contribution of various fats and oil products has recently been updated
to 1984 (Table 5).   In the twenty-year period indicated, the total trans

Table 5 (Hunter, 1985)

Trans Acid Levels in the Diet, (g/person/day)

Year	Salad and Cooking Oils	Household Shortenings	Margarines	Food Service Fats & Oils	Industrial Fats & Oils	Meat & Dairy Products	Totals
1963	0.36	1.01	2.48	NA	1	1.54	-
1970	0.38	0.76	2.73	1.35	1	1.48	7.70
1975	0.49	0.75	2.73	NA	1	1.41	-
1980	0.35	0.6C	2.73	1.54	1	1.33	7.55
1984	0.31	0.55	2.85	NA	1	1.38	-

content has remained relatively constant. However, in recent years the trans content of salad/cooking oils and shortenings has tended to decline. In 1984, margarines showed a slight increase in trans content relative to previous year values.

As mentioned, movement of double bonds to new positions in the fatty acid chain is also an effect of hydrogenation. Cis-monoenes in unhydrogenated oils have the double bond in the 9-position of the chain. Analysis of a commercial salad oil, presented in Figure 10, shows the double bond position distribution in the cis and trans

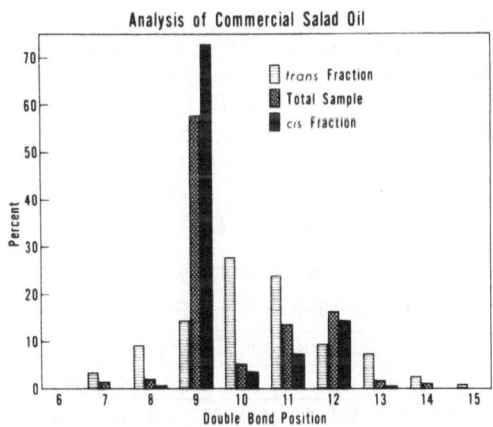

Figure 10 (Scholfield, et al, 1967)

fractions isolated from the monoenoic fatty acid. The cis fraction has double bonds in positions 8 thru 13, with the greatest amount remaining in the 9 position; while the trans fraction has the double bonds spread from position 7 thru 15 with positions 10 and 11 being the most predominent. Four vegetable oil shortenings and two liquid oils described as hydrogenated soybean oil, were also analyzed for monoene double bond position. Results of the analysis of the <u>trans</u> fraction are presented in figure 11. All of the double bonds are widely scattered. Again the 10- and 11-monoenes are present in the greatest amount and are thought to be formed by the hydrogenation of linoleic acid.

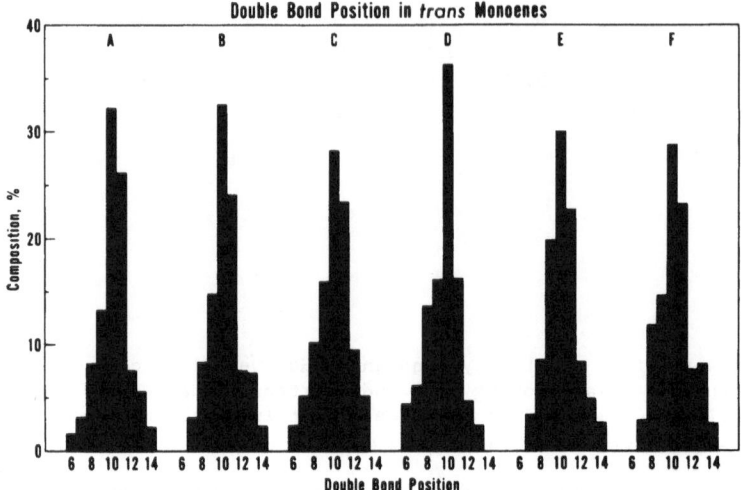

Figure 11 (Scholfield, et al., 1967)

NUTRITIONAL IMPLICATIONS

It is these structural modifications that have raised concerns about the nutritional impact of hydrogenation and have prompted research at our Center into human metabolism of geometric and positional isomers of hydrogenated oils. The estimated total daily consumption of <u>trans</u> and <u>cis</u> positional isomers is shown in Table 6. On this basis the total daily

Table 6 (Emken, 1985)

Estimated Daily Consumption of Positional Octadecenoic Acid Isomers

	Positional Octadecenoic Acid Isomer (g)									
	6	7	8	9	10	11	12	13	14	Total
c-18:1	0.01	0.14	0.24	6.41	0.37	0.54	0.68	0.14	0.03	8.6
t-18:1	0.01	0.27	0.65	1.56	1.53	1.26	0.82	0.48	0.29	6.8
Total isomer content minus $9c$-18:1 = 9.0 g										

Based on an estimated content of 25% c-18:1 and 20% t-18:1 in HSBO and on a daily per capita consumption of 34 g hydrogenated vegetable oil.

consumption of isomers generated during the hydrogenation reaction amounts to 9.0 grams. The experimental approach to evaluating the human metabolism of these positional isomers is visualized in Figure 12. Isomers were labeled with deuterium and a multiple deuterium isotope technique was developed to follow the incorporation of a specific isomer into various lipid classes. The experimental design consisted of feeding mixtures of triglycerides containing deuterated fatty acids, to young adult male subjects and drawing blood samples at various times. The blood was fractionated (Fig. 13), and the plasma and lipoprotein lipid classes were isolated and analyzed by gas chromatography-mass spectrometry. In this series of experiments, completed recently, the metabolism of all the <u>cis</u>- and <u>trans</u>-isomers having positions 8 thru 13

Human Metabolism of Hydrogenated Soybean Oil Isomers

Blood samples drawn at 0, 2, 4, 6, 8, 12, 16, 24 & 48 hrs

Figure 12 (Anon., 1978)

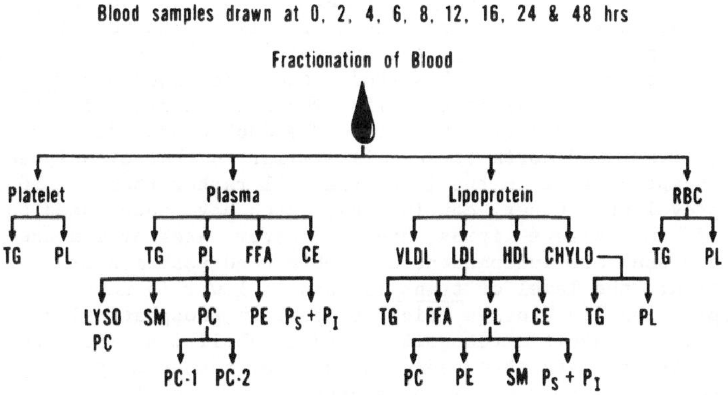

Blood samples drawn at 0, 2, 4, 6, 8, 12, 16, 24 & 48 hrs

Fractionation of Blood

Figure 13

have been evaluated. These are the predominent isomers generated during hydrogenation. The findings of this research are: monoene isomers are well absorbed; numerous examples indicate that the monoene isomers are preferentially metabolized; differences in the utilization of the positional isomers are dependent on the experimental model; and none of the isomers accumulate in human tissues.

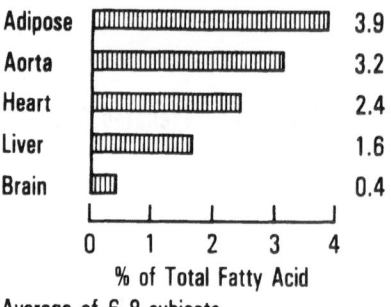

Weight Percent of *trans*-18:1 Isomers
in Human Tissue Total Fatty Acids

Figure 14 (Ohlrogge, 1981)

This latter conclusion is based on analyses of human tissues obtained from routine autopsies. As shown in Figure 14 the weight percent of <u>trans</u> monoene isomers ranged from 0.4% of the total fatty acid in brain tissue to 3.9% in adipose tissue. The fatty acid profile of adipose tissue has been suggested to partially reflect the fatty acid composition of the diet. The <u>trans-</u> double bond distribution in diet and in human tissue is compared in Figure 15. The distribution of double bond positions in adipose tissue is remarkably similar to the pattern observed in dietary hydrogenated vegetable oils but rather different from the pattern reported for butter and ruminant fat. This suggests that, for the subjects analyzed in this study, the major source of <u>trans</u> isomers is hydrogenated vegetable oils, whereas contributions from dairy and ruminent fat are relatively minor. The double bond distribution in the heart, liver, and aorta tissues is similar to that of adipose tissue, with the exception that in the liver the 11 rather than the 10 isomer was most prevalent. Other conclusions, which have been based on the analysis of human tissue lipids, are: turnover rates of monoene isomers in adipose tissue are independent of double bond position or configuration; the level of <u>trans</u> isomers is lower in membrane phospholipids than in storage triglycerides; in phosphatidyl choline of liver and heart, specific recognition and metabolism of individual positional isomers are observed; and rates of turnover of the positional isomers are evidently adequate to prevent major accumulation of any unusual isomer.

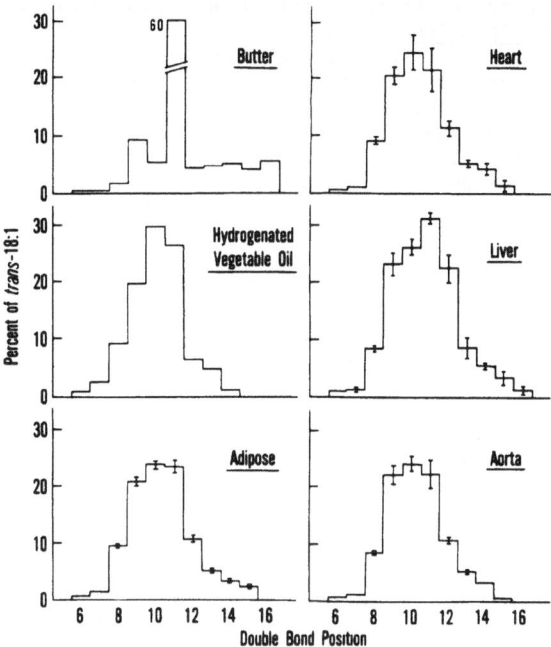

trans- Double Bond Distribution in Diet and in Human Tissues

Figure 15 (Ohlrogge, 1981)

REFEREFNCES

Allen, R.R., 1978, Principles and catalysts for hydrogenation of fats
    and oils. J. Am. Oil Chem. Soc., 55:792.
Allen, R.R., 1981, Hydrogenation. Ibid. 58:166.
Anon., 1978, NRRC: Heart of soybean research, Ibid, 55: 193A.
Anon., 1982, "Food fats and oils," 5th Edition, W.H. Meyer, ed.
    Institute of Shortening and Edible Oils, Washington, D.C.
Emken, E.A., 1983, Human studies with deuterium labeled octacecenoic
    acid isomers, in "Dietary Fats and Health," E.G. Perkins and W.J.
    Visek, eds., American Oil Chemists' Society, Champaign.
Emken, E.A., 1985, Nutritional considerations in soybean oil usage, in
    "Proceedings of World Soybean Research Conference III," R.
    Shibles, ed., Westview Press, Boulder and London.
Hastert, R.C., 1981, Practical aspects of hydrogenation and soybean
    salad oil manufacture, J. Am. Oil Chem. Soc., 58:169.
Hastert, R.C., 1986, Hydrogenation, in "Proceedings of World Conference
    on Emerging Technologies in the Fats and Oils Industry," A. R.
    Baldwin, ed., American Oil Chemists' Society, Champaign.
Hunter, J.E., 1985, Current, previous, and predicted levels of trans
    fatty acids in the U.S. diet, in "Health Aspects of Dietary Trans
    Fatty Acids," F.R. Senti, ed., Federation of American Societies
    for Experimental Biology. Bethesda.
Mounts, T.L., 1980, Hydrogenation practices, in "Handbook of Soy Oil
    Processing and Technology," D.R. Erickson, E.H. Pryde, O.L.
    Brekke, T.L. Mounts, and R.A. Falb, eds., American Oil Chemists'
    Society, Champaign.

Ohlrogge, J.B., Emken, E.A., and Gulley, R.M., 1981, Human tissue
    lipids: occurrance of fatty acid isomer from dietary hydrogenated
    oils, J. of Lipid Res. 22:955.
Rizek, R.L., Welsh, S.o., Marston, R.M., and Jackson, E.M., 1983,
    Levels and sources of fat in the U.S. food supply and in diets of
    individuals, in "Dietary Fats and Health," E. G. Perkins and W.J.
    Visek, eds., American oil Chemists' Society, Champaign.
Scholfield, C.R., Davison, V.l., and Dutton, H.J., 1967, Analysis for
    geometrical and positional isomers of fatty acids in partially
    hydrogenated fats, J. Am. Oil Chem. Soc. 44:648.

MARGARINE PRODUCTION IN THE U.S.A. AND ITS NUTRITIONAL CONTRIBUTION

TO MAN'S DIET

Ahmad Moustafa

Consultant to the Fats and Oil Industry
2476 Little Dry Run Road
Cincinnati, Ohio 45243

INTRODUCTION

Margarine is an engineered product invented in 1869 at a time when demands were high for oils and fats because of a butter shortage in Europe. Its evolution to the present day nutritious spread is a prime example of the technological advancement made in the food industry through the efforts of food technologists, oil chemists, nutritionists, and chemical engineers. It has taken its place world wide as an excellent nutritive food because of its concentrated source of food energy, its uniform supplement of vitamins A and D, its high content of polyunsaturated essential fatty acids, its satiety value and its appetizing flavor and complementary effect it has on other foods.

In the progress of science a means had been discovered by which a new article of food could be produced which was equally as nutritious and less expensive than butter and for which it could be used as a substitute. The original process has been told many times (1), (2). It consisted essentially of the four steps; (a) mincing and washing of fresh fat; (b) digestion of the minced fat with artificial gastric juice; (c) expression of the softer portion of the fat from the harder; (d) digestion and agitation of the soft fat with milk and mammary-gland tissue extract.

From our present day knowledge, step (b) established the principle of low-temperature rendering for the production of a bland fat free of odors and flavor. The rendered oil was crystallized and fractionated. The softer fraction, called oleo oil was the raw material for margarine manufacture. We also know that beef fat (3) and heated pork fat (4) both have a certain content of delta lactones. The occurrence of delta lactone in butterfat and animal fat is due generally to the fact that a portion of the fatty acids in the triglycerides of these fats consists of delta hydroxy fatty acids. When heated or stored at room temperature, the triglycerides may liberate delta hydroxy fatty acids from which free lactones are formed spontaneously (5).

215

TECHNOLOGICAL ADVANCES

From its inception to date, the margarine industry has seen a lot of technological changes. Digestion of the minced fat with gastric juice and oleo oil with mammary-gland tissue was eliminated. Prevention of graininess by quick chilling and the use of soured milk was described by Mott (6). The mixing of liquid cottonseed or rapeseed oils with tallow to reduce the melting point of the mixture was followed by the use of lard in the tallow-cottonseed blend. The use of other plant base oils such as peanut, seasame, coconut and nut oils which did not need a high degree of refining followed.

Development of refining technology followed by the development of the hydrogenation process opened the door for the use of refined and hydrogenated vegetable oil stocks in the manufacture of margarine. It was estimated that in 1948 cottonseed oil accounted for 62% of the oils used in margarine in the USA, while soybean oil accounted for 35%. In 1956, soybean oil accounted for 68% of the total oil used in margarine, and cottonseed oil for 25%. Today soybean oil constitutes almost 90% of the oil phase in any margarine blend. These developments enabled the formulation of oil blends of varying melting points and plasticity ranges that were acceptable to the consumer.

Numerous product improvements brought margarine to its present high standard of acceptance. Not long after its invention, the use of surface-active agents was used to prevent spatter-ing. Egg yolk was the first agent used. However, soybean lecithin is utilized almost universally as the sole emulsifying agent or in conjunction with mono- and diglycerides. Equally important was the engineering development in the USA of the brine cooled revolving drum for the rapid chilling of the fat emulsion (7) and the ultimate development of the closed continuous internal chiller and plasticizer which became known as the Votator (8), (9), (10).

The early digestibility experiments on the common cooking oils, cottonseed, peanut, and olive oils, as well as butter and lard, concluded that the digestibility coefficient obtained for the oleomargarine was found to be about 95% (11). These experiments were confirmed by Deuel at USC in 1946 (12). As a result of these findings the digestibility of margarine was no longer questioned.

The discovery that the growth promoting factors A & D were lacking in margarine prompted the addition of cod liver oil and cod liver oil concentrates both of which were known to be rich in both factors. The original Standard of Identity for Margarine, promulgated in 1941 specified the addition of the above two additives. However, when synthetic vitamin A became available in the late 1940 it was included in the amended Standard in 1952 (13). Improving the vitamin A content of margarine by the addition of a carotene preparation dates back to 1936 (14), and the successful industrial trials using synthetic B-carotene was first reported in 1955 (15), (16).

Because of the nature as a butter substitute, there has been much deception in the sale of margarine. Margarine manufacturers resorted to the addition of naturally colored oils mainly to avoid the sales tax, or to sell it as butter. This practice gave way to the use of a combination of a mixture of food approved coal tar dyes. FD & C yellow 3 and 4 were dissolved in liquid oil and encapsolated in gelatin and placed in the Peters bag (17). With the repeal of the tax on colored margarine in 1950, a new era for the margarine industry began.

With the delisting of the above mentioned synthetic dyes, the revised Standard of Identity allowed the use of annatto and turmeric extractives as well as Beta carotene and some of the derivatives, as the only permissible colorant for margarine.

Packaging is another factor of the process that had to undergo considerable amount of mechanical improvement. Margarine for household consumption has for many years been marketed in individually wrapped one pound or four 1/4 pound prints. Present day automatic combination printing, wrapping and cartoning machines were adapted from the butter industry. It was not until 1963 that margarine appeared in two half pound plastic oblong tubs in a linerless, zip open, folding carton (18).

MARGARINE LAWS

This presentation will not be complete without a few words about the legislative battle margarine as an industry had to endure in the USA. Because of its nature as a butter substitute, margarine has been subject to many federal and state legislative laws. The first federal law taxing margarine was passed in 1886, and amended in 1902 and 1930. These laws were aimed at protecting the butter industry from the sale of the low price spread and, by taxing colored margarine at such a high rate, practically prohibited its manufacture. Without these laws the demand for margarine would have been greater than the record shows (2 pg. 59). In addition, the practice of incorporating as much moisture as the product would bear gave the spread a bad image. With the establishment of the Federal Standard of Identity in 1941 and the repeal of the tax laws on colored margarine in 1950, the per capita consumption took a sharp rise. By the years 1956-1957 the balance started to shift from butter to margarine, Table 1. That coincided with the removal of the restrictive laws by all the States. It would appear that despite these laws, outside forces beyond these legislative bodies seemed to promote public demand for margarine. The food shortages during World Wars I and II, the development of the Peters bag, as well as the sharp rise in butter prices beyond $1.00 per pound on the east coast in 1947, all attributed to the increase demand for the lower priced spread.

In the thirty five years since the repeal of all restrictions on margarine production and sales, its volume has tripled and the per capita consumption amounted to two and half times that of 1949, as shown in Table II.

Early margarines were hard to spread at ordinary refrigerator temperatures of 6-7 degrees Celcius, Table III. When cottonseed and soybean oils became available, the common practice was to hydrogenate the individual oils separately or as a final blend under selective conditions promoting the formation of maximum quantity of iso-oleic acid and a minimum of stearic acid. The margarine made with this blend would be printable in the open hopper type packaging machinery at about 14-16 degree Celcius and have a sharp melt-down and a fast get-away in the mouth. When the same packaging machines were totally enclosed in the early sixties, oil blends with

Table I

PER CAPITA CONSUMPTION OF MARGARINE AND BUTTER IN THE USA
Pounds per Year*

Year	Margarine	Butter
1900	1.4	19.6
1905	0.5	18.6
1910	1.6	18.3
1915	1.4	17.2
1920	3.4	14.9
1925	2.0	18.1
1930	2.6	17.6
1935	3.0	17.6
1940	2.4	17.0
1945	4.1	10.9
1950	6.1	10.7
1955	8.2	9.0
1956	8.2	8.7
1957	8.6	8.3
1960	9.4	7.5
1965	9.9	6.5
1970**	11.0	5.3
1975**	11.2	4.4
1980**	11.3	4.5
1985**	10.3	5.0

(*) Historical Statistics of the USA. Colonial times to 1970. US Dept. of Commerce. Bureau of Census 1975. p.330
(**) NAMM Private Communications.

Table II
MARGARINE AND BUTTER PRODUCTION
million pounds

Year	Margarine	Butter
1970	2277	1143
1971	2333	1144
1972	2350	1102
1973	2434	919
1974	2346	962
1975	2709	984
1976	2532	979
1977	2535	1086
1978	2520	994
1979	2553	985
1980	2589	1145
1981	2583	1228
1982	2596	1259
1983	2451	1299
1984	2481	1120
1985*	2480	1125

Source: USDA
(*) NAMM Estimated

Table III

SPREADABILITY COMPARISON OF MARGARINE OIL BLENDS

Temperature °C	S O L I D S F A T I N D E X			
	1	2	3	4
10.0	41.0	32.0	27.0	23.0
15.6	33.0	26.0	22.0	17.0
21.1	23.0	18.5	16.0	15.0
26.7	15.5	13.5	10.0	8.0
33.3	4.5	4.5	2.0	4.5

(1) Animal vegetable oil blend
(2) Cottonseed soybean oil blend
(3) Partially hydrogenated soybean oil blend
(4) Partially hydrogenated and liquid soybean oil blend

Table IV

FATTY ACID PROFILE

	M A R G A R I N E			
	S T I C K 1*	2**	S O F T TUB	L I Q U I D SQUEEZABLE
	%	%	%	%
cis,cis EFA	3–10	20–40	30–60	40–60
Monounsaturate	50–70	20–50	15–42	20–33
Saturates	16–25	13–23	10–20	10–16
SOLID CONTENT INDEX Temperature °C				
10.0	25–30	16–24	8–14	1.5–4.0
21.1	14–18	10–15	5–8	1.5– 4.0
33.3	2–4	1.5–4.0	0.5–2.5	1.0–3.0

(*) Regular margarine
(**) High polyunsaturated margarine

much lower solids at the 10 degree Celcius were developed. With close control over temperature at the different critical control points, printability was excellent and machine efficiency increased.

The physical attributes such as spreadability, printability, stand-up properties, and meltdown or get away properties were not sacrificed. These are governed by the slope of the solids curve as determined either by dilatometry or NMR. In the long plastic-range fat, the melting point can be somewhat higher than body temperature if the solids content at that temperature is not more than two and one half per cent. For a wide plastic range fat the slope will be relatively flat, and the solids content curve can extend somewhat beyond body temperature even though the solids content at that temperature is relatively low. For a sharp-melting fat the slope will be relatively steep, and therefore, the opposite will result, Table II and Table III (19).

With all federal and state restrictions abolished, margarine established itself as important and indispensable food item competing for the consumer dollars. A spark ignited the imagination of oil chemists and food technologists and the fifties and sixties saw a great deal of technological innovations. The more important have been the introduction of the first colored quarter pound stick margarine in aluminum foil wrap (1950), soft spreadable stick margarine at refrigerator temperatures (1952), use of butter in a margarine formula (1956), whipped stick margarine or six sticks to a pound (1957), soft corn oil margarine in a one pound can with the first high P/S ratio (Emdee 1958), followed by the first high P/S ratio soft consumer pack margarine (Chiffon 1963), safflower oil margarine (1961), spice flavored spread (1961), liquid margarine in a squeezable bottle (1963), diet or 40% fat margarine (1964), fruit flavored spreads (1971, and 1975 saw the introduction of the 60% fat spreads, (1 p.61), (20), (21).

NUTRITIONAL CONTRIBUTION OF MARGARINE

Since 1952, the plasma cholesterol lowering effect of polyunsaturated fat in the human have been demonstrated by many investigators, (22), (23). It was not simply the amount of fat in the diet, but the type of fat as was clearly demonstrated by feeding subjects coconut oil and corn oil which led to vastly different plasma cholesterol levels, (24). In 1958 the special premium type margarine EMDEE (25) appeared in drug stores. Made with liquid corn and hardened coconut oils, it contained as much as 40% essential fatty acids (EFA) and a P/S ratio of 2.1 to 1.0. This resulted in an increased consumer demand for other all-vegetable oil margarines high in the polyunsaturated fatty acid linoleic acid. Corn, safflower, soybean and sunflower oils base stick and soft tub margarines soon appeared with P/S ratios ranging from 2 : 1 and up 7 : 1. Table IV shows the fatty acid profiles of these special high linoleic acid margarines versus the regular margarines (26).

Many research reports published in the 1970s have comprehensively reviewed the relationship between diet and serum lipids. They showed the profound effect diet has on serum cholesterol and LDL. It is now well established that the former is lowered by 1) reduction in total caloric intake, 2) reduced intake of saturated fat and cholesterol, 3) polyunsaturated fat, 4) dietary fiber, and 5) possibly vegetable protein. Alteration of polyunsaturated and saturated fat and cholesterol content of the diet produces predictable changes in the serum total cholesterol. Two units of dietary polyunsaturated fat counteract the cholesterol raising effect of one unit of saturated fatty acid.

The current American diet contains about 40% of total calories as fat, of which about 15-17% are derived from saturated fats and 4-7% as linoleic acid (27). A fundamental goal of The American Heart Association (AHA) has been to educate the American public about the relationship between diet and reduced incidence of Coronary Heart Diseases (CHD). In 1968 the intake of fat was set at 30-35% of calories as fat with a distribution of one third saturates, one third monoun-saturates and one third polyunsaturates (28). Included in the last group members of the omega-6 and omega-3 essential fatty acids linoleic and linolenic acids. The first will reverse all the symptons of EFA deficiency and cannot be substituted by linolenic acid. On the other hand, Linolenic acid play a role in prostaglandin synthesis and acts as a precursor for eicosapentaenoic acid (EPA) (C20:5 W-3) (Table V & VI)

It is well known that these two essential polyunsaturated fatty acids(EFA) cannot be made from other dietary constituents. Also, it appears that the body tends to avoid the accumulation of PUFA rather than to store them. Even though it has been proven that diet modifi-cation by increasing the comsumption of PUFA can prevent atherosclerosis, there may still be risks associated with it (29). Hence, it would appear that while replacement of saturated fats by polyunsaturates seems safe, it may be prudent not to exceed 10% of the total calories.

What role can margarine play in human nutrition? Awareness of the nutritional aspects of this product by the margarine manufacturer has resulted in the production of the high polyunsaturated fatty acid formulations, the low fat spread and the different butter blends.

It has been shown in adults (30), infants (31) and animals (32) that the dietary requirement for vitamin E increases when the intake of polyunsaturated fatty acids increases. If we believed that a food should be self sufficient nutritionally, enough vitamin E should be added to maintain the E/PUFA ratio within the recommended RDA. In this manner, a consumer who obtains most of his dietary fat from that type of product in addition to getting his vitamin E daily needs is protected against the impact of the free radicals formed during the oxidation of these PUFA. Margarine's total disappearance from the visible portion of the fat is shown in Table VII.

Table V

MAJOR FAMILIES OF POLYUNSATURATED FATTY ACIDS

Family Designation	Parent Fatty acid	Major Metabolites	Characteristic Structure	Principal Source
n–6	C18:2 n–6 Linoleic Acid	C20:4 n–6 Arachidonic Acid	H₃C-C-C-C-C-C=C-R'COOH (over C: 6)	Many Vegetable Oils
n–3	C18:3 n–3 Linolenic Acid	C20:5 n–3 Eicosapentaenoic Acid	H₃C-C-C=C-R'COOH (over C: 3)	Some Vegetable Oils
		C22:6 n–3 Docosahexaenoic Acid		Marine Oils (20:5 , 22:6)

For the characteristic structures:

$$H_3C\text{-}C\text{-}C\text{-}C\text{-}C\text{-}\overset{6}{C}\text{=}C\text{-}R'COOH$$

$$H_3C\text{-}C\text{-}\overset{3}{C}\text{=}C\text{-}R'COOH$$

Table VI

PROSTAGLANDIN SYNTHESIS FROM POLYUNSATURATED FATTY ACIDS

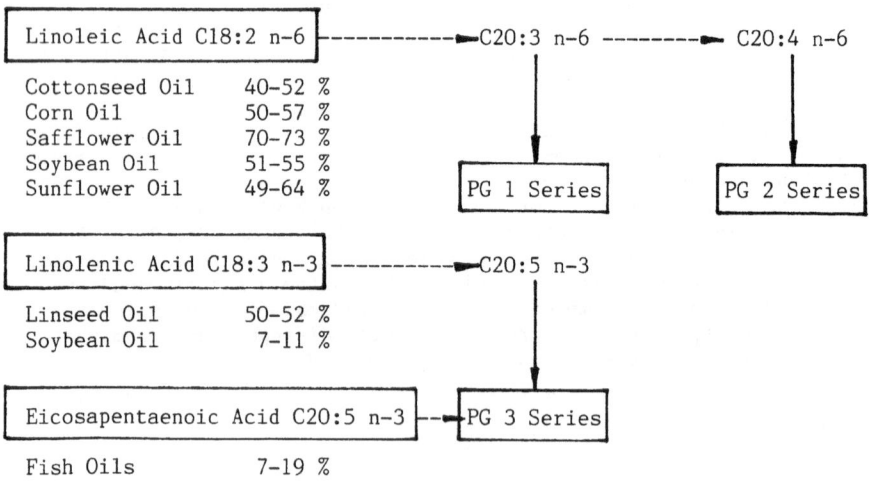

| Linoleic Acid C18:2 n–6 | ------------► C20:3 n–6 --------► C20:4 n–6 |

Cottonseed Oil   40–52 %
Corn Oil   50–57 %
Safflower Oil   70–73 %
Soybean Oil   51–55 %
Sunflower Oil   49–64 %

PG 1 Series   PG 2 Series

| Linolenic Acid C18:3 n–3 | ---------► C20:5 n–3 |

Linseed Oil   50–52 %
Soybean Oil   7–11 %

PG 3 Series

| Eicosapentaenoic Acid C20:5 n–3 | --► PG 3 Series |

Fish Oils   7–19 %

Table VII

SOURCES OF NUTRIENT FAT IN THE AMERICAN DIET
FROM THE "VISIBLE" FAT
Pounds per capita

Year	Total	Margarine	Butter	M+B % of Total
1931	44.1	1.5	14.6	36.36
1935	44.1	2.4	14.1	37.41
1940	46.4	1.9	13.6	33.41
1945	39.1	3.2	8.7	30.43
1950	45.9	4.9	8.6	19.41
1955	45.9	6.6	7.2	30.07
1960	45.1	7.5	6.0	29.93
1965	47.7	7.9	5.1	27.25
1970	52.6	8.6	4.3	24.53
1975	52.3	8.8	3.8	22.95
1980	55.8	9.0	3.6	22.58

Source: USDA

Table VIII

MARGARINE LABELLING IN THE USA

NUTRITION INFORMATION PER SERVING

Serving size.........................14 g
Serving per container*..............32
Calories............................100
Protein............................. 0 g
Carbohydrates....................... 0 g
Fat (100 % of calories)............. 11 g
Polyunsaturates**................... 3 g
Saturates**......................... 3 g
Cholesterol (80 mg/100g)**.......... 8 mg
Sodium (79 mg/100g)**...............110 mg

Percentage of U.S. Recommended Daily
Allowance (U.S.RDA) of Vitamin A.... 10 %

Contains less than 2.0% of U.S.RDA of
protein, vitamin C, thiamin, riboflavin,
niacin, calcium and iron.

(**) This information on fat and cholesterol
content is provided for individuals who on
the advise of a physichian, are modifying
their total dietary intake of fat and
cholesterol

PREPARED FROM SWEET CREAM BUTTER, LIQUID CORN
OIL, PARTIALLY HYDROGENATED CORN OIL, SKIM
MILK AND/OR WATER AND NON FAT DRY MILK SOLIDS,
SALT, LECITHIN, MONO- AND DIGLYCERIDES,
ARTIFICIALLY FLAVORED, COLORED WITH BETA
CAROTENE, VITAMIN A PALMITATE ADDED.

(*) one pound container

The fact that margarine is an excellent source of calories per unit weight suggests a complementary role with high protein foods. This is true when formulating special diets high in protein content or fortified with certain amino acids such as lysine, cysteine and methionine for geriatrics and the underfed. In this type of product, energy will be supplied by the margarine and both protein or amino acids will be utilized efficiently. Thus margarine may be a good vehicle to get the limiting amino acids to people with fixed food habits (33).

In the last three decades enormous progress has been made by scientists in the fields of epidemiology, physiological chemistry, and animal experimentation, as well as in human metabolism and nutrition. Where we are today has been arrived at through a combination of educational, scientific, and economic forces, the sum total of which has been considered to represent progress (34). Obviously, we have not solved the relationship of diet to heart diseases since the problem is further complicated by the heterogeniety of the human species and to the likelihood that multiple causes are differently expressed in different individuals.

Unlike the high price spread, margarine is a product of COMBINATION. Because of this entity, it was used in 1962 for the National Heart Institute dietary studies to supply the high polyunsaturated fats in the diet. It can be used again to formulate a spread with the proper ratio of saturate:mono: polyunsaturated fats derived from either natural sources or synthetic fats (e.g. fractionated oils and/or synthetic products).

SUMMARY

In terms of time, Margarine is only one hundred and fifteen years old (115 years). It has become an integral part of man's diet since it was first introduced as a butter substitute in France in 1869. However, it has become the spread of choice for the consuming public.

National trends in amounts and types of fat consumed have not alleviated our problems in nutrition and health. The role of polyunsaturated fatty acids (PUFA) and their ratio to saturated fatty acids (P/S ratio) in an oil is still questionable even after more than thirty five years of intensive clinical investigations and studies.

In attempting to focus attention on nutrition, the Food and Drug Administration (FDA), has been promoting labeling of foods (Table VIII). However, it is only through education that the public will be able to make intelligent choices in attempting to balance and control its food intake with respect to fats, proteins, and carbohydrates in order to ensure good nutrition and health.

As an Engineered or Combined foods, Margarine can assist investigative teams in all scientific disciplines in supplying the type of fat, P/S ratio, natural or synthetic blends, etc., needed by the different investigators.

Then it becomes the responsibility of the margarine manufacturer to periodically review and evaluate his product, both when new nutritional findings are made or when compositional changes occur.

REFERENCES

1. Riepma, S. F., The Story of Margarine pp.iii, Washington, D. C., Public Affairs Press (1970)
2. Snodgrass, K., Margarine as a Substitute. Food Research Institute, Stanford University pp. 121, (1930)
3. Watanabe, K., and Sato, Y., Studies on the Changes of Meat fats by Various Processings. Part II. Gas Chromatographic Identification of Aliphatic (8) and (6) Lactones Obtained from Beef Fats. Agr. Biol. Chem. 32:(2),191 (1968)
4. Ibid. Lactones in the Flavor of Heater Pork Fat. Agr. Biol. Chem. 33:(2),242-249 (1969)
5. Als, Gert, Delta-lactones in flavoring. Food manufacture January 1973.
6. Crump, G. B. The Technology of Margarine Manufacture. "Progress in the Chemistry of fats", Vol. 5, pp.285-321 (1957)
7. Snodgrass, K., Margarine as a Substitute. Food Research Institute, Stanford University pp. 149 (1930)
8. Slaughter, J. E., and McMicheal, C. E., Plasticizing and Packaging. J. Amer. Oil Chem. Soc. 26:632 (1949)
9. Joyner, N. T. The Plasticizing of Edible Fats. J. Amer. Oil Chem. Soc. 30:526 (1953)
10. Mattil, K. F., Norris, F. A. and Stirton,, A. J., Bailey's Industrial Oil and Fat Products. Daniel Swern ed. 3 rd. edition. Solidification, Homogenization, and Emulsification, pp. 1063 (1964)
11. Holmes,A.D., Digestibility of Oleomargarine. Boston Med. Surg. J., 192,1210-1212 (1925)
12. Deuel,H.J., Studies on The Comparative Nutritive Value of Fats. J. Nutrition 32:69 (1946)
13. U.S. Margarine Standard of Identity (1952)
14. Matzko, S. N. Z. Unters. Lebeusm. 72:143-148 (1936)
15. Schuchardt, W. Du B-Carotene Synthetique de Son Eemploi Ceomme Colored Alimentaire. Oleagineaux, 10:259-264 (1955)
16. Klaui, H. and Bauernfeind, J. C., Carotenoids as Food Colors. Carotenoids as Colorants and Vitamin A Precursors J. C. Bauernfeind, ed. Academic Press, New York, pp.201, (1981)
17. Peters, L., Methods and Means of Packaging and Mixing Plastics. U.S.Pat.#2,347,640 (1944)
18. Moustafa, A., Soft Margarine in the U.S.A. Rev. Fran. des Corps Gras. 26:485 (1979)
19. Struble, C. H., The Physical Requirements of a margarine fat. J.Amer. Oil Chem. Soc. 31:(5)34 (1954)
20. Miami Margarine Company Files. Cincinnati, Ohio.
21. Miksta, S.C., Margarine:100 Years of Technological and Legal Progress. J. Amer. Oil Chem. Soc., 48:169A (1971)
22. Kinsell,L. W., Partridge, J., Boling, 1., Margen, S., Michaels, G.P., Dietary Modification of Serum Cholesterol and Phospholipid Levels. J. Clin. Endocrinol. 12:909-913 (1952)
23. Ahrens, E. H., Blankenhorn, D.H., Tsaltas, T. T., Effect on Human serum Lipids of Substituting Plant for Animal Fat in the Diet. Proc. Soc. Exp. Biol. Med. 86:872-878 (1954)
24. Ahrens, E. H., Hirsch, J., Insull, W. et al. The Influence of Dietary Fats on Serum Lipid Levels in Man. Lancet, 1:943-953 (1957)

25. Phillips, R. A., Emulsified Oleaginous Spread Containing Essential Fatty Acids and Process of Making Same. U.S. Pat. #2,890.959 (1959)
26. Massiello, F. J., Changing Trends in Consumer Margarines. J. Amer. Oil Chem. Soc. 55:262 (1978)
27. Rizek, R. L., Freind, B., and Page, L. Fat in Today's Food Supply - Level of use and Sources. J. Amer. Oil Chem. Soc. 51:244-250 (1974)
28. American Heart Association. Diet and Heart Disease. American Heart Association (1968)
29. Kabara, J. J. Polyunsaturated Fatty Acids. To Eat Or Not to Eat. Bul. European Org. for the Control of Circulatory Diseases. 8:57-77 (1984)
30. Horwitt, M. K., Interrelations Between Vitamin E and Polyunsaturated Fatty Acids in Adult Men. Vit. Horm. 20:541-558 (1962)
31. Hassan, H., Hashim, S. H., Van Itallie, T. B. and Sebrell, W. H., Syndrome in Premature Infants Associated with Low Plasma Vitamin E Levels and High Polyunsaturated Fatty Acid Diet. Amer. J. Clin. Nutr. 19:147-157 (1966)
32. Dam, H., Interrelations Between Vitamin E and Polyunsaturated Fatty Acids in Animals. Vit. Horm. 20:527-540 (1962)
33. Bonnell, R. H., Bornstein, B. and Schutt, G. W., Future Considerations in Margarine Fortification. J. Amer. Oil Chem. Soc. 48:175A (1971)
34. Ahrens, E. H., Diet and Heart Diseases. Arteriosclerosis 2:85-86 (1962)

NON-CONVENTIONAL HYDROGENATION OF FAT

Michele Rossi

Centro di Studio sulle Metodologie Innovative
di Sintesi Organiche
Dipartimento di Chimica, Università di Bari, Italy

INTRODUCTION

In recent years non conventional hydrogenation techniques have been adopted or proposed for industrial application. These techniques differ from the conventional way of hardening fats in relation to either hydrogenation method used or the chemical modification aims. Some of these will be illustrated and a new technique for improving vegetable oil stability toward oxygen will be presented.

ALTERNATIVE METHODS TO HYDROGENATE FATS

Different hydrogen sources have been explored as substitute for molecular hydrogen. Among these alcohols have proven very effective hydrogen transfer reagents, with aldehydes or ketones as their by-products[1]. In this context methanol may be a very attractive reagent because of its low cost, safety in handling and easy removal of gaseous by product (CO). Carbon monoxide and water can be also considered low cost hydrogen sources on account of the water gas shift reaction

$$CO + H_2O = CO_2 + H_2$$

which can be coupled with different hydrogenation reactions[2]. Beside the forementioned, other hydrogen donors have been studied (hydrazine, cyclohexene) but no application of the so called "conjugated hydrogenation" is known.

During the last decades many noble-metal based catalysts have been studied for oil hydrogenation. It has been established that palladium is a very active catalyst in soybeen oil hydrogenation[3]. Moreover, its selectivity performance and its accessible cost make this metal a candidate for industrial application. On the other hand soluble metal complexes should be considered only of scientific interest. The importance of copper in selective hydrogenation of linolenate will be presented below.

DIFFERENT CHEMICAL MODIFICATIONS

Ever since copper chromite was found to be a catalyst in hydrogenating esters to alcohols[4], many attempts have been made to carry out this reaction under milder conditions. In view of the results reported about the reduction of the free –COOH group in various substrates[5] the application of ruthenium-based catalysts for the hydrogenation of fatty acids to fatty alcohols should be taken in consideration. Furthermore, selective double bonds hydrogenation of vegetable oils has been extensively studied in recent years[6]. Partial-selective hydrogenation technique aims at reducing linolenate content in order to stabilize the product. Copper chromite catalyst at 170-200° C and at 2-6 atm. of hydrogen pressure has been commonly used for this process. In particular, a recent development in this field led to the ultra-selective hydrogenation[7,8] which will be discussed in detail.

COPPER CATALYST SELECTIVITY

Among noble and non-noble metals, copper is known to have peculiar properties of selectivity in hydrogenation reactions. Consequently, it is used as catalyst for many industrial chemicals. The following are aspects of selectivity.

Synthesis of Different Products

The reaction of CO with H_2 leads either to CH_3OH or to CH_4:

$$CO + 2H_2 = CH_3OH$$

$$CO + 3H_2 = CH_4 + H_2O$$

Partial hydrogenation of CO can be obtained on copper catalyst, whereas nickel is the favoured catalyst for total hydrogenation to CH_4[9].

Chemio-Selectivity

Furfural can be reduced to furfuryl alcohol through selective hydrogenation of the –CHO group on copper, whereas this substance undergoes indiscriminate hydrogenation of aldehyde and of olefinic double bonds on nickel[10].

Alkyne: Alkene Selectivity and Stereoselectivity

Copper behaves better than nickel during partial hydrogenation of alkyne to alkenes[11] and like palladium it yields a high quantity of cis alkenes[2].

Polyene: Monoene Selectivity

Eterogeneous copper catalysts show a very high linolenate: linoleate and linoleate: oleate selectivity. In the form of copper chromite,it is used to reduce the linolenic acid in soybean oil.

"COPPER CHROMITE " CATALYST

The chemical and spectroscopic characterization of this catalyst shows that the original copper oxide-copper chromite bifasic system undergoes a partial reduction during soybean oil hydrogenation. The two components produce initially Cu(I) and Cu(0) species, which are both present in the "active" catalyst. The "spent" catalyst contains only the Cu(0) species[12].

$$\overset{I}{CuO} \cdot \overset{II}{CuCr_2O_4} \xrightarrow{H_2} \overset{0}{Cu} \cdot \overset{I}{Cu_2Cr_2O_4} \xrightarrow{H_2} \overset{0}{Cu} \cdot Cr_2O_3$$

In the 1,4 -diene hydrogenation, isomerization to conjugated double bonds is an essential step prior to hydrogen addition[13]. Comparison of the activity of the different copper species, present on the catalyst surface, suggests that Cu(I) is active in promoting, 1,4 → 1,3 diene isomerization, whereas Cu(0) is active in hydrogen addition. The behaviour of different catalysts originated from a "copper chromite" Girdler G 89 will be compared.

PROCEDURE

Starting from G 89 we prepared the "active" catalyst, G 89 U, by washing a sample of the catalyst used in soybean oil hydrogenation for 0.5 h at 200° C and 6 atm. G 89 R was prepared through dry, careful reduction of G 89 under hydrogen at 270° C and 1 atm. By means of the E.S.C.A. technique comparable amounts of Cu(0) and Cu(I) were found on the surface of the G 89 U, whereas only Cu(0) was detected on G 89 R. Table I shows the results obtained in the hydrogenation of soybean oil by using 0.3 % of different catalysts at 200° C and 6 atm.

The different behaviour of G 89 and G 89 U with respect to G 89 R is quite evident. In fact both G 89 and G 89 U are similarly active whereas G 89 R is inactive in the hydrogenation of isolated dienes. Another important aspect is that the active catalysts allow formation of conjugated polyenes up to a maximum value of ca 1.4% (Fig. 1).

The activity of G 89 and G 89 R was compared also in the hydrogenation of polyenes containing conjugated double-bonds. By operating at 140° C and 1 atm., selective hydrogenation to monoenes with both catalysts was obtained by using different substrates, i.e., preconjugated soybean methylesters and cis,cis-cicloocta-1,3-diene. However, only G 89 catalyst requires a long induction time to form the active species whereas the reduced catalyst is immediately active as shown in figs. 2 and 3.

Experiments on soybean oil and on preconjugated soybean methyl esters confirm that Cu(0) can hydrogenate only the conjugated polyenes thus suggesting that Cu(I) plays a determining role in the isomerization of isolated to conjugated isomers.

Table 1. Hydrogenation of Soybean Oil.

Cat.	t(h)	C18:0	C18:1	C18:2	C18:3
		3.4	23.3	54.4	7.2
G89	1	3.4	38.0	45.3	1.4
G89U	1	3.6	41.1	43.1	1.5
G89R	1	3.4	23.8	54.1	7.0

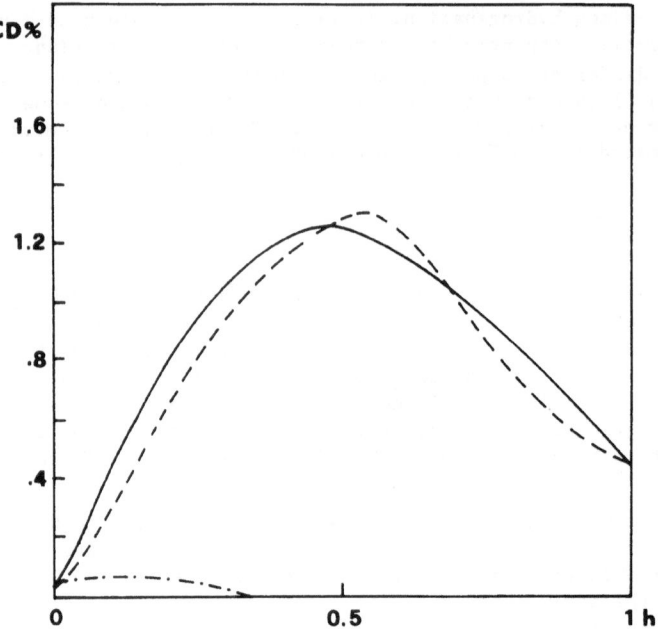

Fig. 1. Accumulation of Conjugated Dienes during Soybean
Oil Hydrogenation.
—— G89; ––– G89U; – · – G89R.

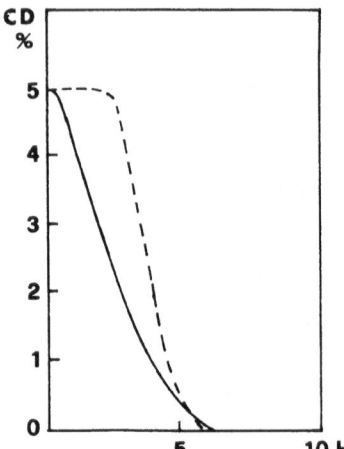

Fig. 2. Hydrogenation of Conjugated Soybean Oil Me-
thylesters. ––– G89;
—— G89R.

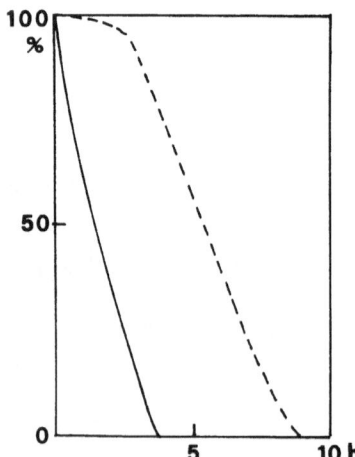

Fig. 3. Hydrogenation of 1,3-
Cyclooctadiene. ––– G89;
—— G89R.

The presence of trance-amount of conjugated dienes in oils is easily detectable by U.V. spectroscopy . As shown in fig. 4, an aged sample of soybean oil exibits a characteristic three-band pattern centered at 268 nm. because of the conjugated trienes . It also shows a shoulder at 232 nm. due to conjugated dienes along with oxidized compounds[14]. After hydrogenation on the G 89 catalyst, the bands at 268 nm. disappear, but the accumulation of conjugated dienes causes a strong absorbance at 232 nm. . In the case of G 89 R only the disappearence of the 268 nm. band without any increase at 232 nm. can be observed. Therefore, G 89 R is an effective catalyst for reducing the trace-amount of conjugated trienes. The reduction of conjugated dienes is hard to prove owing to the absorbance complexity at 232 nm. . Nonetheless, previous experiments suggest such a possibility.

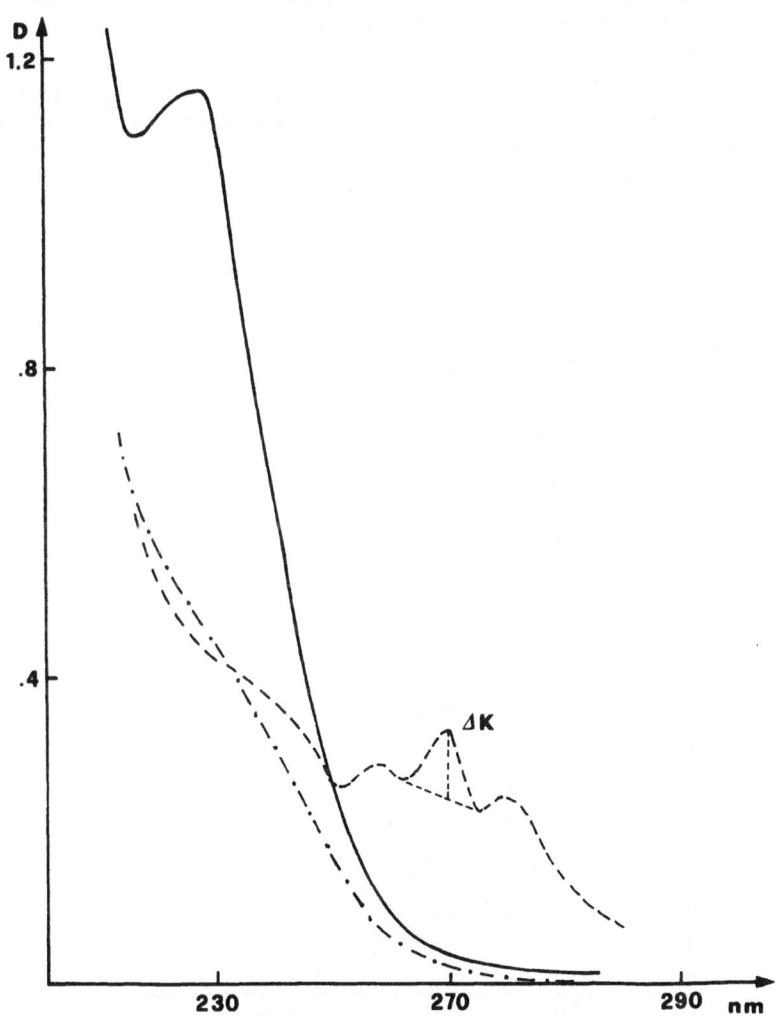

Fig. 4. Removal of Trace-Amount of Conjugated Trienes in Soybean Oil.
--- Soybean Oil; ——— Soybean Oil on G89; –·– Soybean Oil on G89R.

This technique aims at bettering the oil quality by improving its resistance to ageing without introducing dangerous chemical modifications. The method bases on a very mild hydrogenation of the oils (PH_2 = 1 atm.; T= 140°C) in the presence of prereduced copper catalyst in order to reduce the trace amount of conjugated polyenes. In particular, a process for aged olive oils will be illustrated.

Oxidized olive oil shows a U.V. pattern similar to the one already reported for soybean oil (fig. 5). Blaching treatment on activated earths can reduce the 232 nm. band almost completely. However, in the meanwhile it causes a strong increase of the 268 nm. absorbancies. By reacting bleached and unbleached oils with hydrogen at 140° and 1 atm. we observed an immediate disappearence of the bands at 268 nm. only in the presence of the G 89 R catalyst. Under these conditions, G 89 was in fact much less effective as can be seen from the results of fig. 6.

During treatment with G 89 R catalyst we measured a very low hydrogen consumption (1 – 3 mmols H_2 per Kg. oil). G. 1. c. analysis reveled no modifications in total acidic composition, and I.R. spectroscopy showed no increase in trans isomers.

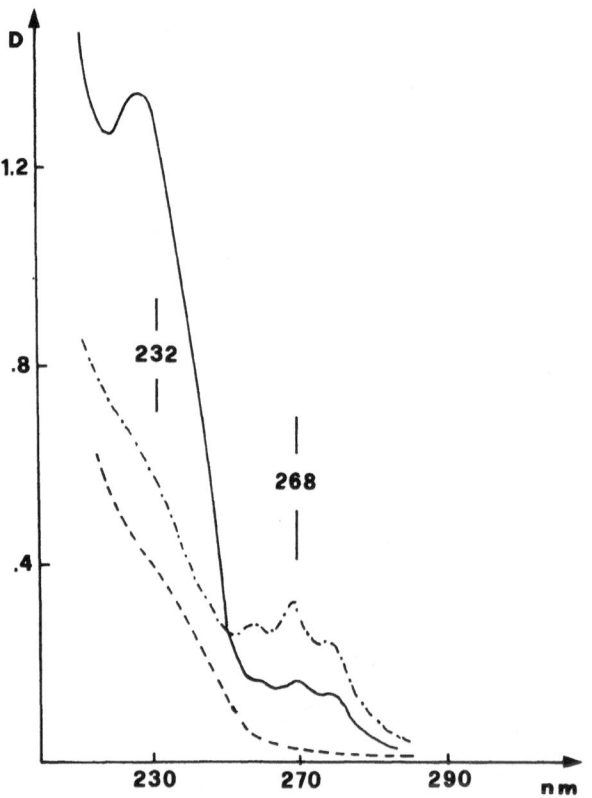

Fig. 5. U.V. Spectra and Olive Oil "Quality".
- - - Extra-Vergin Olive Oil; ——— Aged Olive Oil;
- . - Bleached-Aged Olive Oil.

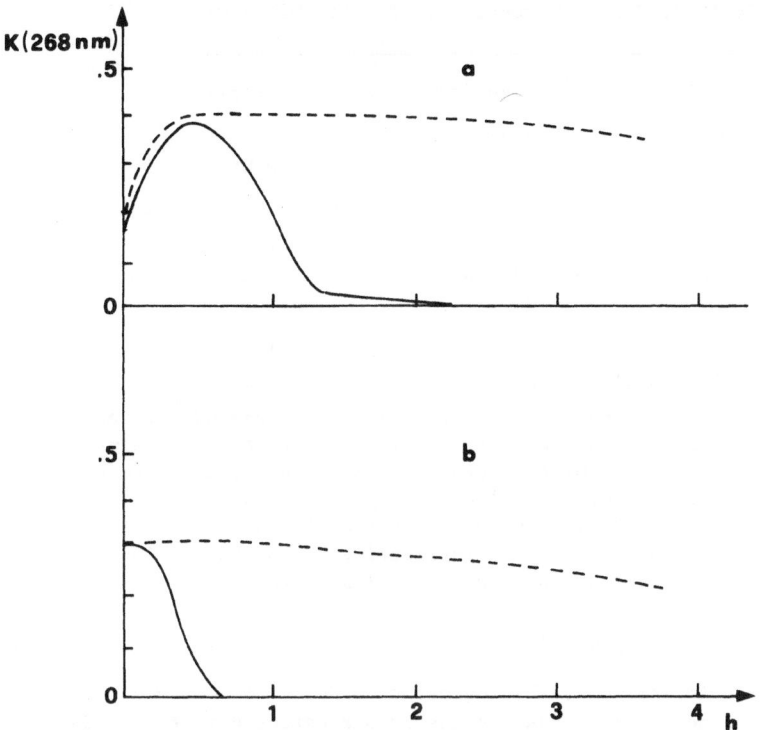

Fig. 6. Ultraselective Hydrogenation of Olive Oil. a: Aged Olive Oil; b: Bleached-Aged Olive Oil.
—— G 89 R; --- G 89.

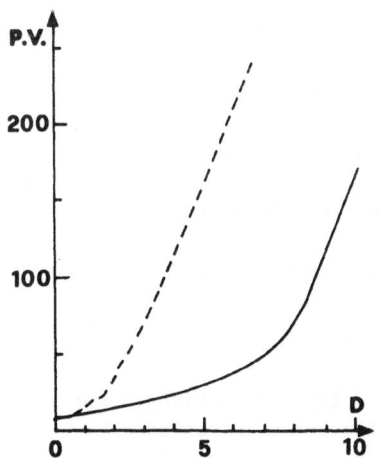

Fig. 7. Peroxide Value during
Ageing Test for Bleached
Olive Oil. --- Untreated
Oil; —— G89R Treated Oil.

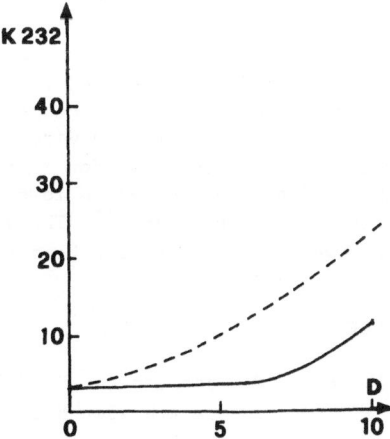

Fig. 8. K232 nm. Absorbance during
Ageing Test for Bleached
Olive Oil. --- Untreated
Oil; —— G89R Treated Oil.

233

Table 2. Copper Removal from Treated Olive Oil

Catalyst	T(°C)	PH$_2$(atm)	t(h)	Cu(ppm)
				0.06
G 89	140	1	0.5	9.4
"	"	"	1.5	6.5
"	"	"	7	1.7

The comparative ageing test in air at 25°C for treated and untreated samples of olive oil was followed by determining the peroxide value and by monitoring the U.V. spectrum. As shown in figs. 7 and 8, the bleached olive oil, which has an induction time of ca. 2 days, becomes much more resistant to oxygen after treatment with the G 89 R catalyst. In fact the new induction time is now ca. 7 days. Clearly, ultraselective hydrogenation improved the stabilization of oil against autoxidation quite markedly.

Copper Removal from Treated Oils.

During the ultraselective hydrogenation of olive oil with G 89 R catalyst part of the metal dissolves because it was detected in the product after centrifugation at 9000 rpm. . No investigations have been made to distinguish between small colloidal metal particles and soluble copper compounds. However the copper amount was found to be dependent by the contact time as illustrated in table 2 [15] .

The observed decrement of metal with time suggests that copper in contact with oxidized oil produces a soluble metallorganic complex. Moreover, this latter can be slowly reduced to unsoluble metal by hydrogen at 140°C.

The removal of copper was achived by a standard procedure. In fact, activated clays of the type used for bleaching edible oil were found able to cut down the metal. In particular, an oil containing 9.4 ppm. of copper was purified by treatment with Girdler Tonsil C activated earth (1%) under nitrogen at room temperature for 15 min. . The resulting product contained less than 0.5 ppm. of copper. A lower residue (0.02 ppm) can be obtained with other simple treatments[16]. These concentration values indicate that ultraselective hydrogenation can be performed without hazard due to presence of heavy metals in oil.

REFERENCES

1. R. A. W. Johnston, A. H. Wilby and I. D. Entwistle, Heterogeneous Catalytic transfer Hydrogenation and Its Relation to Other Methods for Reduction of Organic Compounds, Chem. Rev. 85:129(1985).
2. C. Fragale, M. Gargano and M. Rossi, Selective Hydrogenation of 1,3-Cyclooctadiene and Diphenylacetylene on Copper Using the Water-Gas Shift Reaction as a Hydrogen Source, J. Catal. 80:460 (1983).

3. J. D. Ray, Hydrogenation of Soybean Oil with Palladium, J.A.O.C.S. 62:1213(1985).
4. W. A. Lazier, U.S. Pats. 1,964,000-01(1934).
5. G. Pez, Hydrido Ruthenate. Catalytic analogs of Lithium Aluminium Hydride, Abstracts 3rd Int. Symp. Homog. Catal. 15(1982).
6. S. Koritala, K. J. Moulton, J. P. Friedrich, E. N. Frankel and W. F. Kwolak, Continuous Slurry Hydrogenation of Soybean Oil with Copper-Chromite Catalyst at High Pressure, J.A.O.C.S.,61:909(84).
7. C. Fragale, M. Gargano and M. Rossi, Catalytic Hydrogenation of Vegetable Oils: II.The Activity of Prereduced Copper Chromite Catalyst, J.A.O.C.S. 59:465(1982).
8. C. Fragale, M. Gargano and M. Rossi, Improvement of Vegetable Oils Quality by Ultraselective Hydrogenation, Riv. Ital. Sost. Grasse, LXII:187(1985).
9. G. Henrici Olivé and S. Olivé in "The Chemistry of the Catalyzed Hydrogenation of Carbon Monoxide" Springer Verlag, Berlin (1984).
10. G. Seo and H. Chon, Hydrogenation of Furfural over Copper-Containing Catalysts, J. Catal. 67/424(1981).
11. R. S. Mann and K. C. Khulbe, The Reaction of Methylacetylene with Hydrogen Catalyzed by Nickel, Copper and their Alloys, Can. J. of Chem., 46:623(1968).
12. L. E. Johansson and S.T. Lundin, Copper Catalysts in the Selective Hydrogenation of Soybean and Rapeseed Oils: I. The Activity of Copper Chromite Catalyst, J.A.O.C.S., 56:974(1979).
13. S. Koritala, R.O. Butterfield and H. J. Dutton, Selective Hydrogenation with Copper Catalysts: II, Kinetics, J.A.O.C.S., 47:266 (1970).
14. R. T. O'Connor, Ultraviolet Absorption Spectroscopy, J.A.O.C.S., 32:616(1955).
15. R. Capelli, C. Fragale, M. Gargano and M. Rossi, unpublished results.
16. O. Popescu, S. Koritala and H. J. Dutton, High Oleic Oils by Selective Hydrogenation of Soybean Oil, J.A.O.C.S., 46:97(1969).

FATS OF TROPICAL ORIGIN

J. Graille

Division of Fats and Oils Chemistry
CIRAD-IRHO BP 5035, 34032, Montpellier

First of all I would like to give an idea of the world production of the main tropical fats and oils in comparison with the total world production.

Thus this lecture will deal with palm, palm kernel, coconut and peanut oils.

The world production reaches almost 70,000,000 metric tons (T) to day-about 68,155,000 T - but this figure is not yet definitive (1).

- Oils from palmacea are as follows :

. Palm	6,270,000 T
. Palm kernel	870,000 T
. Coconut	2,640,000 T

- Peanut oil 3,195,000 T

That is to say a total of 12,975,000 T or 19 % of the world production, about a 1/5 th !

Nowadays everybody hears about the implication of fats and oils in human health :

In the scientific world this impulsion provides a lot
of work for searchers in the fields of Essential Fatty Acids
(EFA), Hydrogenated Fats (HF), Long Chain Fatty Acids (LCFA),
Satured Fatty Acids (SFA), New Chemical Species (NCS) arising
from thermal treatment or catalytic hydrogenation and
oxidation and also in the field of New Chemical Species
arising from Minor Compounds (NCSMC).

Then I will try to give an oversiew about those different topics in relation with oil milling, refining,
transformation process and food utilization.

On the table I, typical fatty acids contents are given :

Table I

	Palm	Kernel	Coconut	Peanut
C 6	-	0,1	0,6	-
C 8	-	3,3	7,7	-
C 10	-	3,4	5,7	-
C 12	0,1	47,0	48,0	-
C 14	1,0	17,3	17,8	-
C 16	46,9	9,1	9,0	10,5
C 16:1	0,2	-	-	0,2
C 18	5,1	3,5	2,9	2,5
C 18:1	36:2	12,5	7,0	58,5
C 18:2	12,3	3,8	1,3	21,8
C 18:3	0,2	-	-	-
C 20	-	-	-	1,3
C 20:1	-	-	-	1,3
C 22	-	-	-	2,6
C 24	-	-	-	1,5
I.V.	50,0	14,0	6,0	100,0

At a glance, it is very easy to see that the table I
determines three types of oils according to the iodine value:

- Two very saturated oils (I.V. 8 and 14) which are
palm kernel and coconut oils (lauric oils).

- One, with a medium I.V. (50) represented by palm oil
(a palmitic oil).
- One, with a high I.V. (100) represented by peanut oil
(an oleic oil).

Most of our considerations here after will be governed
by fatty acids composition of each oil.

Palm oil

Oil milling does not really affect fatty chains,
however one must be carefull to avoid acidification of oil.
To day everybody knows how to obtain a good quality crude
oil.

During refining however, some problems can arise depending on the type of refining.

High temperature during deodorization, both in the
alcaline process or the physical process (240 - 260 C) can
induce interesterification (2) but this reaction has only
technological consequences.

On the other hand carotenoïds are affected ; as a
matter of fact the typical red colour disappears completely
during this stage of refining.

A polemic exists around this (3) :

Some people pretend that carotene breacks down to give
toxic aromatic hydrocarbons, others pretend that carotene
breaks down to give only volatiles which are then stripped
off from the oil (4).

In fact, nothing has really been clearly demonstrated.

Palm   oil   is   traditionnally   fractionated   by
crystallization   into   stearin   and   olein.   Olein is enriched
with   Unsaturated Fatty Acids (UFA) and is suitable for salad
dressing   and   frying. Stearin is used in margarine industry.

Although olein contains more UFA, this oil is still poor in EFA.

Two ways are studied to day to enrich oil with EFA :

- One is hybridation of elais guineensis (the common oil palm) with elais melanococca, an american ecotype from Amazonia.

Table II shows fatty composition improvement of the new palm oil.

Table II

	E.G.			E.M.			G.M.		
16:0	46,9	)		23,6	)		37,7	)	
18:0	5,1	)	52,0	1,0	)	24,6	2,8	)	37,5
18:1	34,2	)		59,9	)		45,4	)	
18:2	12,3	)	46,5	16,3	)	73,2	14,5	)	59,9
I.V.	48,5			80,5			65,5		

This work is still in progress at IRHO. Tissue culture seems to be a good way for good specimen propagation (5).

Another way is the interesterification with unsaturated oils like soya-bean, rape seed or rice bran oils which could be good counter oils.

Enzymatic interesterification with 1-3 specific lipase could become a new industrial technique in a very near future. This type of interesterification will keep orginal UFA on the 2 position of glycerol (6and 7).

Palm kernel and coconut oils

Owing to their particular compositions, Lauric oils are rich in medium chain triglycerides (MCT) which balance the fact that those oils are very poor in EFA.

As a matter of fact MCT are very easily hydrolyzed during digestion and short fatty acids are very well absorbed by intestine and by mitochondria in the absence of carnitine.

Moreover, MCT are recommanded both for people having problems for digestion and utilization of lipids (8).

For healthy people lauric oils are not criticized.

Processing has no harmfull nutritional effects.

Peanut oils

Here also processing has no harmfull effect on human nutrition. In fact industry has now a very good mastery of oils and fats extraction and refining.

However according to new trends in the field of crude ecological food products, crude peanut oil is now distributed to consumers. Oil is obtained from high quality seeds by a simple expression at room temperature and filtration. Some samples have been found to contain aflatoxin which is very toxic. However I do think that this is an exception, but this type of extraction does not guarantee aflatoxin free oils. Everybody knows that aflatoxin is eliminated in the course of refining. Peanut oils are criticized for their rather high content of LCFA, 7 to 10 % depending on genotypes, see table I.

Here also breeders try to select good strain to produce low LCFA and high linoleic acid containing oils (9).

Here also the toxicity of LCFA in peanut oils has to be proved.

Hydrogenation and New Chemical Species

Obviously, those topics are common to numerous fats and oils and scientists have a lot of work to do.

I think that it is wise to question oneself because those treatments do deeply modify fats and oils which lose their natural structure.

Tropical oils are well scored for deep frying fats, they give low NCS. Moreover they could play a very important role for fats tayloring. As a matter of fact, consumers can buy now margarines quaranted without hydrogenated fats (10) and sometimes quaranteed sithout both hydrogenated and interesterified fats in dietetic stores.

Conclusion

I think that we can say that tropical oils have a very good future and that critics in the field of nutrition are up to now much more relevant to the economic war than relevant to scientific evidences.

REFERENCES

1. BROCHE, G., : OLEAGINEUX (1985), 40, 339-349.

2. PAIRAUD,D., : REVUE FRANCAISE DES CORPS GRAS (1979), 26, 5-8.

3. DAVIS, J.B., ROBINSON, J.M., SILVA, N.K., and BARRANCO A.: J. FOOD TECHNOL. (1979), 14, 253-264.

4. STAGE, H., : J.A.O.C.S., (1985), 62, 299-308.

5. NOIRET, J.M., and WUIDART, W. : OLEAGINEUX, (1976), 31, 464-474.

6. HANSEN, T.T., and EIGTVED, P., : AOCS World Conference on Emerging Technologies in the Fats and Oils Industry, Nov. 3-8, 1985, Cannes, France, Submitted for publication in J.A.O.C.S.

7. MUDERHWA, J.M., RATOMAHENINA, R., PINA, M., GALZY, P., and GRAILLE, J. : J.A.O.C.S., (1985), 62, 1031-1036.

8. HASHIN, S.A., GERGEN, S.S., KRELL, K., and VAN ITALLIE, T.B. : CLIN. J. INVEST., (1964), 43, 1238-1241.

9. BOCKELEE-MORVAN, IRHO, 11 Square Pétrarque, 75016 PARIS: Results not yet published.

10 . MAJUMDAR, S., and BHATTACHARYYA, D.K., : OLEAGINEUX, (1986), under press.

THE IMPACT OF NON-CONVENTIONAL TECHNOLOGIES ON THE CHEMICAL COMPOSITION OF FATS

Enzo Fedeli

Stazione Sperimentale per le Industrie degli Oli e dei Grassi
Milano, Italy

The following conclusions were drawn from the previous exposition on conventional technologies:

1) The impact of the storage of raw materials on the quality of the final product varies according to the source material.

2) The quality of the extraction technology is very important for the obtention of raw oils for refining and even more important for oils, such as virgin olive oil, which are utilized without subsequent refining.

3) Present refining technology is really apt to eliminate impurities originating in the substance of departure and in the extraction technology.

4) The refining technologies in use at present also remove nutritionally important components from oils.

5) Autoxidation and the transformation of hydroperoxides are among the contaminants the presence of which most greatly influences product quality since the current technology only partially eliminates the products of oxidation.

6) Storage in small vessels influences the quality of the product for consumption.

Several proposals for new technologies have been put forward during the last few years; some of them will be considered in this lecture and compared with the statement given previously.

An attempt will be made to propose approaches for future technologies; no changes in the technologies utilized for the storage of raw materials have been proposed and I consider that it be necessary to suggest new systems

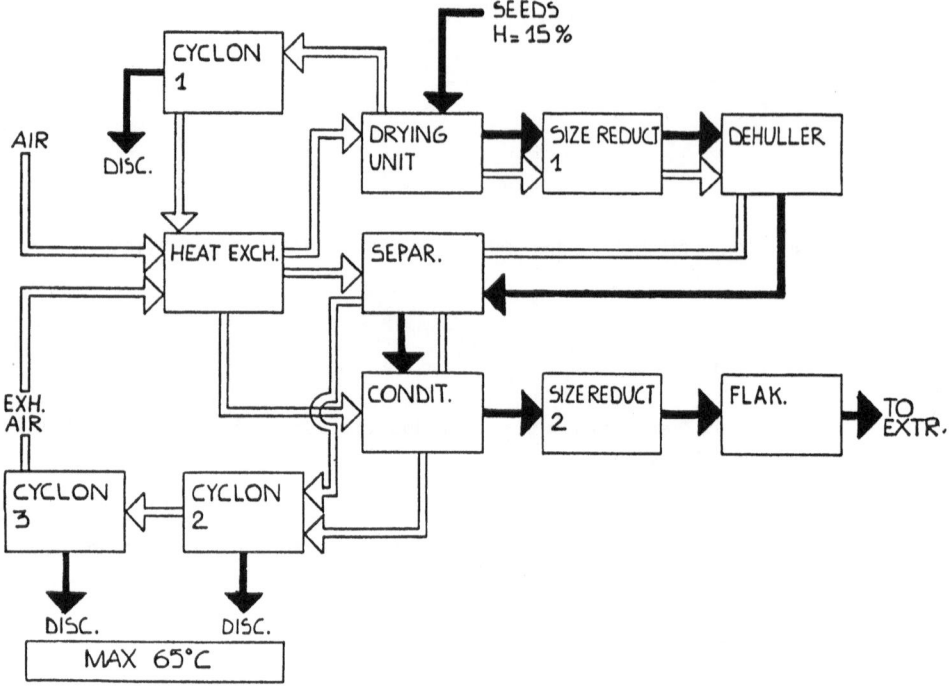

Fig. 1.   Seed preparation before extraction.

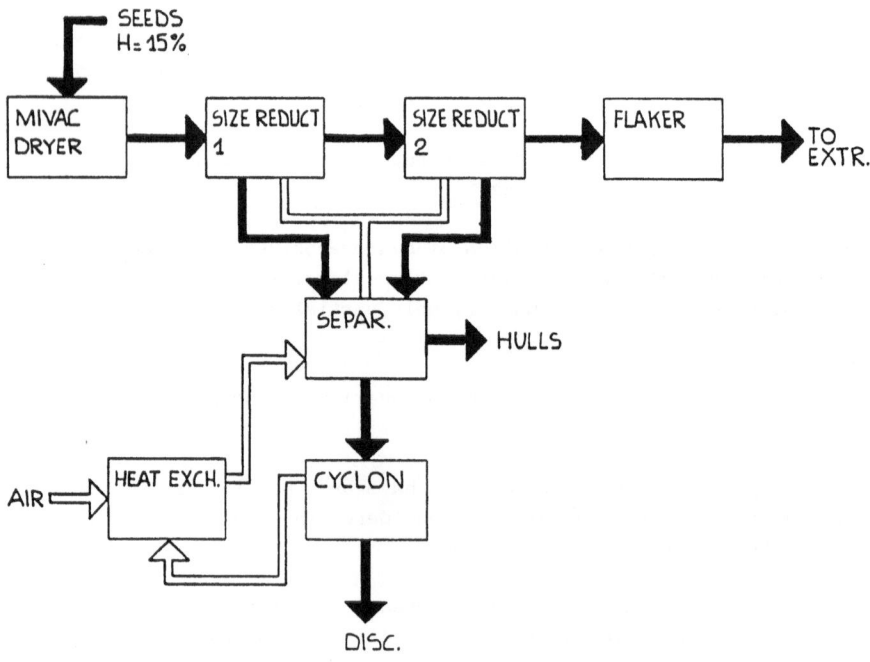

Fig. 2.   Seed preparation before extraction.

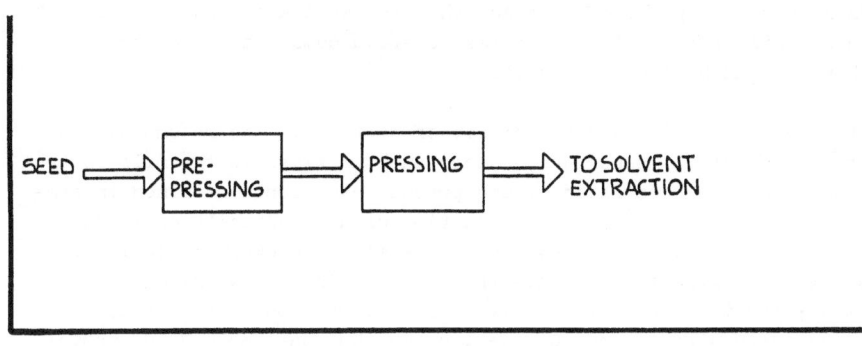

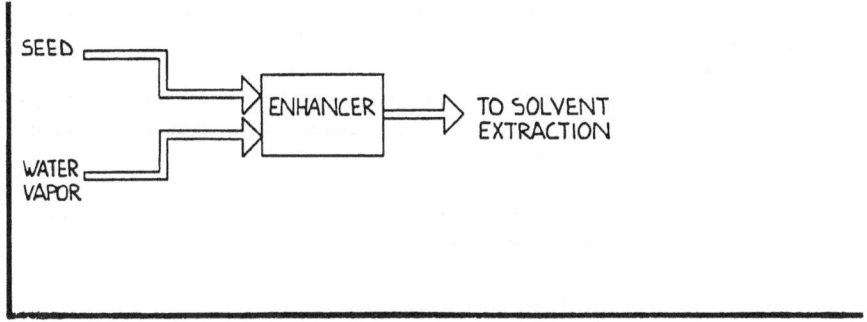

Fig. 3. Two schemes for improvement in mechanical extraction.

STAGE	% FFA	P (ppm)	
ROW CORN OIL	5.2	330	
DEGUMMED		50	
BLEACHED	5.3	12	
DEWAXED		11	
REFINED	0.03	10	DISTILL. FFA 93.5% PEROXIDE VALUE 0

Table I. Quality improvements in corn oil refining by distillation procedure

for seeds storage and propose, instead, that the well-known rules be obeyed in design and utilization of facilitites to avoid damage to seeds resulting from mechanical and biological causes.[1]

In the case of fruit, and especially of olive oil, the recommendation to work the materials of departure as soon as it arrives in the mill is truly proficuous as is demonstrated by the product excellence attained in Italy in recent years following the introduction of high capacity equipment apt to process the material in few hours. A similar recommendation is also applicable to the extraction of fats from animal tissues. Where oils are to be consumed without prior refining, it is also very important that contaminants not be introduced during cultivation and that safe methods of protecting crops from pests be devised.[2]

Two new technologies for the preparation of seeds before extraction have been proposed recently.[1,3,4,5]

The first, which is illustrated in Fig. 1, is a very interesting improvement with respect to conventional technology because it avoids overheating, and p eserves the seeds before extraction.

The aim is achieved by the reutilization of heat from the circulating air so that only minor quantities of heat are used to compensate losses.

A similar purpose is contained in the second proposal (Fig. 2). The low temperatures and the deep drying attained guarantee oil from autoxidation and hydrolysis, and minimal side-reactions.

Improvements to mechanical extraction methods have also been proposed as shown in Fig. 3; advantages correlated with the quality of the oil are increased rapidity which would mean less exposure to oxidation for the first scheme and protection exercised by water vapour in the second.[6]

Proposals have been made and research performed into the utilization of solvents but at present no successful applications have been devised. One of the most attractive possibilities is related to the utilization of carbon dioxide as an extracting medium. The advantages in terms of oil quality would be considerable owing to the selectivity attainable with the aid of that solvent and its protective action. Technological complications involved in the high pressures required in the process raise operational costs to very high figures.[7,8,9,10]

Many improvements have been proposed or utilized in relation to the refining of edible oils and fats in recent years.

Some of them are linked with what is termed physical refining which imposes more thorough purification of oils before submitting them to the action of water vapour and heat.[11,12,13]

Data for the results obtained during preparation are shown in Table I.

Most of the progress in refining technology concerns the combination

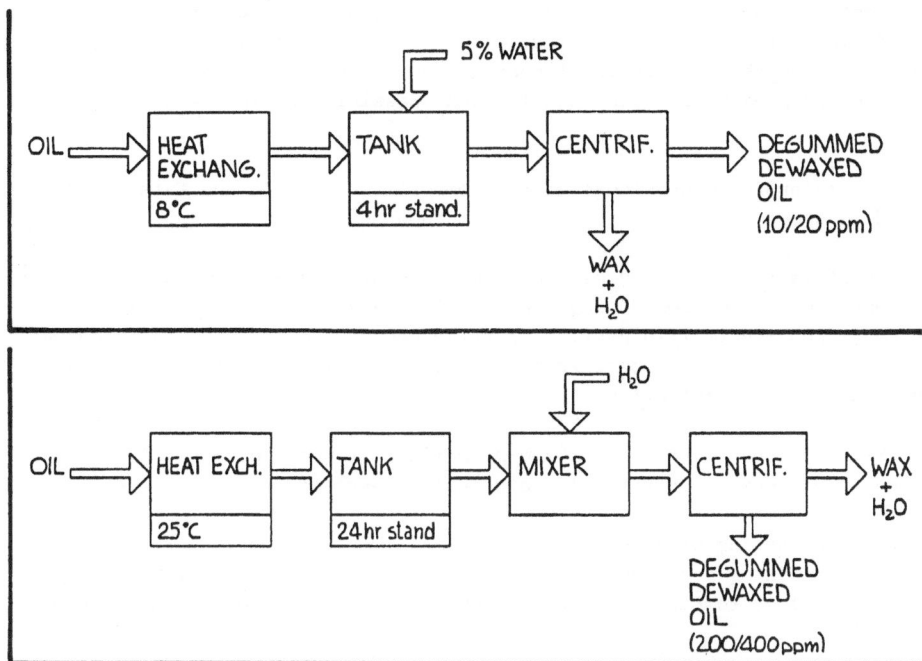

Fig. 4.   Dewaxing - Degumming procedures.

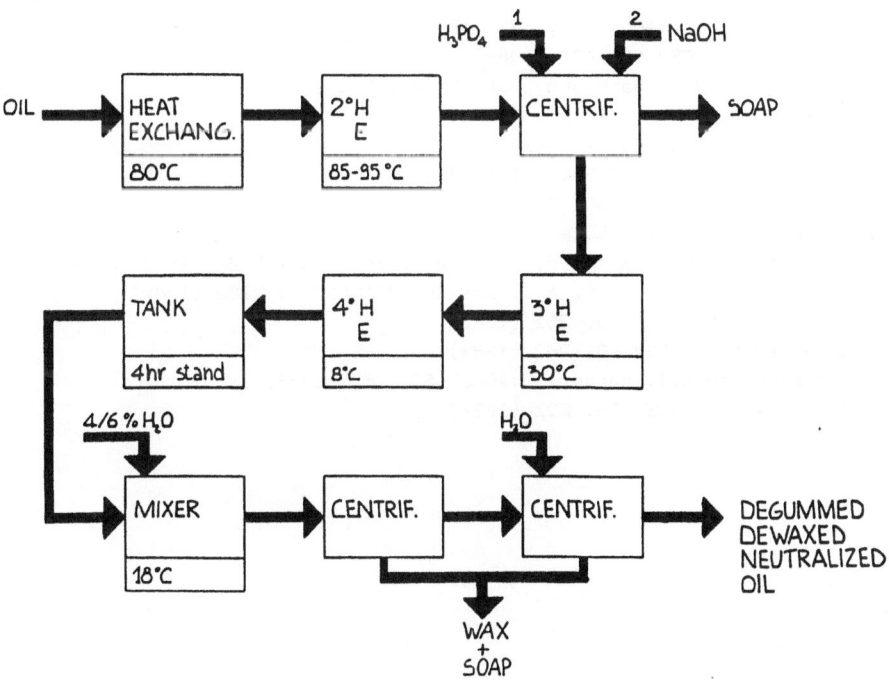

Fig. 5.   Degumming, Dewaxing refining procedures.

of a conventional operation with dewaxing; schemes are shown in figures 5 and 6.

Most of the improvements presented have been proposed for the purpose of reducing extraction or refining costs instead of aiming for oil quality improvement which is an indirect consequence of the necessity to send to the deodorizer-netralizer the least amount possible of impurities in the oil.

A more profound knowledge of the composition of oils and fats is required in order that specific refining methods be developed.

If it is true that the general knowledge of fats and oils has increased in geometric progression during the last 30 years, it is mostly related to the major components with particular reference to the fatty acids in natural and processed fats. Apart from very few exceptions, minor components have received less attention.

Still some of them are not known, neither is their possible influence on the nutritional properties of fats fully understood.

The furthering of research in that field would be very useful for the design of refining equipment trying either to avoid the formation of oxidation products or to eliminate them, provided that they are harmful components.[14]

REFERENCES

1. N. Hunt Mooreé, "JAOCS" 60, 189 (1983)
2. E. Fedeli, "JAOCS" 60, 404 (1983)
3. G. Florin and H.R. Bartesch, "JAOCS" 60, 193 (1983)
4. M. Koch, "JAOCS" 60, 193 (1983)
5. W. Fetzer, "JAOCS" 60, 203 (1983)
6. D.K. Bredeson, "JAOCS" 60, 211 (1983)
7. M.K. Mangold, "JAOCS" 60, 226 (1983)
8. L.A. Johnson and E.W. Lusas, "JAOCS" 60, 229 (1983)
9. W. Schultz and M.K. Mangold, "J. Agric. Food Chem." 28, 1153 (1980)
10. J.P.Friedrich, G.R. List, A.J. Heakin, "JAOCS" 59, 288 (1982)
11. G. Haralson, "JAOCS" 60, 251 (1983)
12. J.G. Segers, "JAOCS" 60, 262 (1983)
13. A. Forster and A.J. Harper, "JAOCS" 60, 265 (1983)
14. A.R. Macrae, "JAOCS" 60, 291 (1983)

RECENT DEVELOPMENTS IN THE TECHNOLOGIES OF

PLANT BREEDING FOR OILSEED IMPROVEMENT

Gerhard Röbbelen

Institute of Agronomy and Plant Breeding
University of Göttingen, Germany F.R.

Within the context of the present symposium the following contribution is meant to indicate the prospects of oilseed production in view of
the recent developments in procedures and trends of genetic plant improvement. Because of the wide scope, the today's activities and discussions
in plant breeding can be compiled in 10 theses only, which will give
sufficient grounds for the subsequent discussion.

1. High specific values of oilseeds have caused
    explosive expansions of production

Since 1935/39 the total oilseed production of the world has grown
2.6 times up today allowing for a worldwide rise of per capita fat consumption from 9.6 to 13.9 kg. A great deal of this increase came from
higher soybean (623%), sunflower seed (582%) and rapeseed (319%) productions (Hatje 1986). Oilseeds rank among the most important agricultural commodities. This high regard is derived from the high value of their
product. Vegetable oils and fats not only exhibit highest energy densities
as well as some essential factors of nutrition, but they also contribute
considerably to the tastiness of human foods and meals.

Many oilcrops in addition are two-product commodities. After oil
extraction they yield meals of high protein content. These meet great
demands as a protein supplement for livestock and poultry. In particular,
soybean oil is usually produced in larger amounts than needed because of
the demand of soybean meal. Other oilseeds have less satisfactory meals.
These are frequently sharply discounted on markets, and at some locations
they are used as a nitrogen manure only. Some vegetable oils are by-products of other productions. Cotton is the most significant example of
this kind and the amount of cotton seed produced is a strict function of
the given requests for cotton fibre. The same is true for hemp seed and
to some extent for linseed, while corn oil is a residue of the starch
industry. Some oils are produced from perennial plants, e.g. olive tree
or oil palm, so that there is no reasonable alternative but to harvest
and market the crop each year once the decision has been made to start
the plantation. For all these reasons the supply of vegetable oils is
relatively inflexible.

2. Progress of breeding and production technologies
has narrowed down the number of profitable oilcrop species

World oil markets are dominated by less than 10 plant species which have attracted the interest of modern crop technologists for one or another reason. Only 15 or 20 are usually listed in the official statistics on vegetable oil markets. However, there are a 100 or more botanical species in the world with reasonable potentials and oil qualities to serve mankind (see e.g. Rehm and Espig 1984).

In small amounts lipids occur in each living cell. But for commercial utilization higher concentrations of lipids are necessary as they are attained particularly in seeds. Here, storage of oils and fats provides for the energy reserves needed for germination and seedling establishment. Within the seed, oils and fats are usually deposited in the nutritive tissue, i.e. the triploid endosperm. Sometimes the diploid perisperm or the embryo itself, particularly its cotyledons (Cruciferae), serve as a storage organ. Rarely are oils and fats stored in the pericarp of fleshy fruits (oilpalm, olive). The most exceptional in this respect is the tiger nut (Cyperus esculentus L.), which contains 20 to 28% oil in its sub-terraneous tubers; these are locally consumed raw, boiled or roasted as a vegetable in North Africa. In this way, many fat producing plants are only of local importance. But even those contributing to human fat supply, in that their intact fruits or seeds are consumed in the form of nuts or vegetables, should not be omitted from total supply estimates.

All together, much more care should be given to the wealth of traditional, indigeneous oilcrops of the various regions. Governmental Developmental Aid authorities should lend more expertise to support conservation and development of other than the five leading oilcrops. For example, rapeseed breeders have achieved fascinating developments e.g. in the European (B.napus) and the Canadian (B.campestris) rapeseed species. But worldwide at least the same acreage is devoted to other Cruciferae species. Only recently breeders and agronomists have discovered major advantages of B.juncea for the more arid regions or of B.carinata with its high yield potentials. All these plant species provide essentially the same product. Indeed, vegetable oils and faty are highly interchangeable with each other and their demand is mainly determined by availability and consumer habits. Different oilcrops, therefore, offer a special chance to contribute to the genetic diversity of our cultivated fields and crop rotations.

3. Efficient analytical techniques set
the stage for effective oilcrop breeding

The effectiveness of plant breeding programmes directed to improve oil yields and qualities of a crop is dependent on the available methods to determine the oil content and fatty acid composition of the respective plant materials.

The classical method for the determination of oil contents has been the tedious oil extraction by n-hexane after Soxhlet from the ground seed meal. Because of its high costs and relatively large seed samples needed for such analysis, the procedure almost exclusively has been used for breeding lines ready for release. In other words, effective selection for oil content has almost entirely been absent until recently. Several proposals of other methods have been made as early as 50 years ago. Von Seng-busch (1942) suggested to press a single lupine seed between cardboards and to measure its oil content by the diameter of the resulting grease spot. By immersion of seed lots in sucrose solutions or mixtures of kerosene and bromobenzene seeds with high oil contents were separated by floating from those descending in the solution because of their low oil content (Tröeng 1955). Both essential principles, the non-destructive and the single seed measurement, were combined for the first time by the technique of low resolution nuclear magnetic resonance (NMR). Only after the introduction of this instrument were plant breeders prepared to include a selection step into their early breeding materials (Conway and Earle 1963). Today similar results are also achieved by another purely physical method, the near infrared reflectance spectrometry (NIR). By measuring several ten thousands of single plant derived samples each season, plant breeders now do stringently screen in their pedigrees for oil content improvements.

Even more so has the selection for oil quality been started only after efficient methods became available. Routine analysis of oil quality has been limited for long to the detection of double bonds within the fatty acid molecule by iodine binding. Downey and Harvey (1963) were the first to apply gas liquid chromatography (GLC) for analysing fatty acid patterns within one cotyledon of a seed using the rest embryo for growing the correspondent plant. This half-seed-technique was the decisive break-through in worldwide rapeseed breeding programmes for the elimination of erucic acid from the rapeseed oil. Later, paper chromatographic methods were developed by Thies (1971) to facilitate a simple, cheap prescreening within series as large as 1000 half-seed samples per man and day.

Both the quick and simple methods for use at the prescreening level within large populations of early breeding materials, where genetic selection is still highly effective, and the accurate, reliable analytical methods for final evaluation of the variety and the commercial seed lots, are indispensable for any successful breeding towards quality improvements of oilseed cultivars.

4. Prerequisite for genetic progress is the availability of appropriate genetic resources

Genetic selection is dependent on preexistent genetic diversity. Therefore, the first phase of every plant breeding programme is directed to create useful variation.

Quality characteristics, e.g. a specific fatty acid pattern, may be controlled by one or only a few genes. If the biogenetic pathway is known, at which an unwanted constituent can be genetically blocked or at which a wanted product can be accumulated prior to such a block, than a corresponding mutant allele can be searched for. In this way, high oleic acid sunflower and safflower (Knowles 1969) mutants have been found as well as the first zero erucic rapeseed. If breeding stocks or natural populations do not contain the desired variant, then physical or chemical mutagenesis will be helpful.

Quantitative characteristics, however, such as seed or oil yield per acre are generally controlled by a great number of different genes. Such polygenic traits are much more difficult to improve. Only by experimental crossing of parental lines and by testing of the resulting progenies can the breeder find out, whether the available material is valuable and contains genetic components, the combination of which in a descendent individual will lead to an improved genotype. There is no hope for genetic gain from crosses, where the parents are not sufficiently different. But if the genetic distance of the parents is beyond a certain optimum, genetic disequilibrium will result in the cross progeny. With the help of in-vitro embryo culture, wide crosses are no longer a problem today.

For example, hundreds of new hybrids have been obtained by crossing cabbage and kale, B.oleracea, with turnip rape and chinese cabbage, B. campestris, to resynthesize rapeseed, B.napus, from its ancestral species. But such primary hybrids usually do not provide more than raw basic materials, which needs many decades of intensive recombinational treatment by plant breeders until any sufficiently adapted cultivar may be expected to arise (Gland 1982).

For such reasons, cultivars with good performance and appropriate yields are much better starting material for new breeding cycles. Therefore, one of the most important articles in the Act of Plant Breeders' Rights is the regulation, that anyone is allowed to freely use a protected variety for his further breeding work. Only because trade profits are exclusively reserved to the original breeder, can the free genetic use of the same variety be rightly admitted. Any other patenting system would cause at least temporary inavailabilities of these most important plant genetic resources and would thereby delay or inhibit breeding progress considerably.

5. Utilization of heterosis requires suitable systems of pollination control

Whereever possible, hybrid breeding has greatly contributed to the admirable increases in crop yields during the last decades. By combining different positive alleles of two parental lines into a heterozygous hybrid, heterosis effects yield additional vitality and productivity. Many oilcrops are crossbreeding. But the construction of a heterotic hybrid variety requires the preselection of two parental lines with high combining ability and an exclusive pollination between these two parents. If the seed parent is male sterile, then 100% hybrid seed can be produced in fields with alternating strips of both parents. Genetic systems causing such male sterility are known in many plant species and new ones are continually discovered in others. In some instances, self-sterility can also provide appropriate pollination control. Both systems, however, have draw-backs in that the responsible genes must be transferred by tedious backcrossings into the parent line to be used for production of the hybrid seed. This procedure not only causes delay in breeding progress but also tends to severely narrow the genetic basis of the breeding material. It is, therefore, highly desirable to intensify research and development of suitable chemical gametocides. By application of these snythetic hormones any selected genotype can be made male sterile and thus available for hybrid seed production. Gametocides are at present investigated and applied preferably for wheat. But in many oilcrops profitability and expected yield increases are much higher, since allogamy is widely spread amongst them and much less heterosis has yet been "fixed" by earlier genetic combinations in many of these still more primitive oilcrop species.

6. The sole goal of plant breeding is a better variety

The official regulation allowing the release of a new variety for agricultural production requires a distinct improvement above all the existing varieties in at least one characteristic. But any released variety is also expected to offer an agricultural value in general, i.e. it shall be sufficiently suited for reliable, high productions. For this end, a great number of canopy characteristics is needed, as well as stalk stiffness and shattering resistance, winter hardiness and drought tolerance, earliness and disease or insect resistance. One single deficiency in anyone of these and more traits can make on otherwise best variety useless.

On the other hand, it may be assumed that by molecular gene technology or the more traditional means of repeated backcrossing the transfer of a single gene into a selected genotype will be possible. But whereever thoroughly studied, such gene has never expressed its desired characteristic independent from the recipient rest genome. Indeed, it is nothing but logical that every gene action depends on the support of other genes and that it also provides effects to these in return. In some cases such interactions have found to be even larger in effect than the action of the individual gene per se. Single gene repair of a failure or single gene addition to a given high-bred line has only rarely been a successful approach in plant improvement programmes. Accordingly, mutagenesis has frequently been very helpful in creating new mono- or oligogenic variation. But even single gene mutants generally exhibit unexpected (mostly negative) pleiotropic effects which impede their direct use as an improved variety.

7. Every quality improvement has its genetical
   or even physiological costs

The general experience says that varieties with specific quality improvements show reduced yields. This may result from two reasons which lead to opposite consequences regarding breeding strategy.

Firstly, any additional trait reduces the chance of improving the others. Just selecting for yield in a rapeseed programme, we pedigreed 100 selected individuals from a F_2 population of 2000 individuals. But after an additional selection for low glucosinolate content, we from the same F_2 population just obtained 22 individuals meeting this requirement of contanining the three responsible loci in a homozygote recessive condition. As was to be expected, the genotypic variation in most of the performance characteristics was lower in this smaller population than in the 5 times larger one without the additional quality selection. The consequence of this cognition is that larger population numbers, higher expenses, or more years of breeding work are needed to arrive at the same level of performance, when additional quality selections must be observed.

Another actual example of our present breeding work in rapeseed may serve for the second instance. In the zero erucic rapeseed oil an improvement of the polyenoic fatty acid pattern presently is the next urgent aim of breeding. Monogenic mutants have been selected after chemical induction, producing as high as 35% of 18:2 acid and as low as 3% of 18:3 acid. The agronomical performance of these genotypes is, however, considerably reduced as compared to the original cultivar. By measuring ethane formation after cellular decompartmentation, thylakoid membranes of the mutants were shown to contain less 18:3 acid than normal (Röbbelen 1984). Luminescence response was also negatively affected, as were some measures of photosynthetic activity of the mutant chloroplasts. In this case, physiological costs of a low linolenic acid content may be unavoidable. Recently, however, definite progress has been achieved in this material by two cycles of recurrent selection (Brunklaus 1985). This confirms that predictions of functional restrictions have been premature and the significance of the complex physiological regulations within a plant organism has been underestimated. Such experience provides optimistic views for further breeding progress in this case as well as in general.

8. In-vitro techniques can considerably support conventional plant breeding programmes

More recently, breeding procedures have received great benefit from modern techniques of in-vitro tissue and organ culture. The rescue of weak hybrid embryos from wide crosses in artificial cultures on sterile media was first successfully applied by Laibach (1925) in interspecific crosses of Linum. Quick in-vitro mass propagation of valuable genotypes has become almost standard, where clone varieties are the final end of breeding, or where parental lines of hybrid varieties urgently require multiplication. Long term storage of tissues and cells are of particular value, where seeds are not readily available. But of specific importance are in-vitro mass propagations for the breeding of perennial crops. The pioneering programme in oil palm cloning by the Unilever group at Colworth House has more recently been accompanied by similar activities in favour of the Mediterranean olive trees. The more complex the desired product is - as is the case with the multitude of aroma constituents of the virgin olive oils - the more important is the utilization of vegetative propagation to keep the quality components together when once they have been combined in a variety. On the other hand does in-vitro propagation strengthen the needs of intensive and reliable genotype identification before-hand in the orchard, since every mistake will be multiplied with the greater speed of the new biotechnologies, too.

Another support from tissue culture techniques is directed to phytosanitary purposes. Vegetatively propagated varieties may become seriously affected with time by various virus infections which are transmitted through the propagule from one clone generation to the next. Meristem regions,however, are usually free of virus particles even in most highly infected individuals. If the apical few cells of a shoot tip are carefully dissected and cultured under sterile conditions, healthy plants may be recovered, with which a new clone propagation can be started.

A third point on this thesis deals with the chance to suspend the complications of Mendelian segregations in cross progenies of higher plants and to select haploid gametes rather than diploid zygotes. Microspores from immature anthers can directly be regenerated to plants and after colchicine treatment entirely homozygous diploids are obtained (e. g. Lichter 1982). With these materials selection response is not biased by heterozygosity, segregations are much simplified at the haploid status, and superior genotypes can immediately be used as an origin of a new variety.

9. Biotechnology offers new tools for efficient oilcrop breeding

By further reducing the propagule size in the in-vitro cultures, one arrives at the single cell state. This is most easily achieved by dissolving the cell walls enzymatically and producing naked single cells, called protoplasts. These have been the key to many unconventional techniques of genetic manipulation. For example, damage by fungal infection may result from pathogenic toxins, by which the fungus subjects the cells's larder at its disposal. After mutagenic treatment of the original plant source, toxin resistant protoplasts may be selected on toxin containing media and these may be regenerated to disease resistant plants. Other metabolic changes have also been proposed for such selection, e.g. cold or salt tolerance, herbicide resistance, photosynthetic efficiency etc.

Protoplasts by chemical or physical means can also be induced to fuse. Thereby hybrids may be obtained, the production of which by sexual means is impossible because of incompatibilities or other reproductional barriers. Since throughout evolution these barriers have usually led to considerable genetic divergence of the respective species, the value of protoplast fusion for plant breeding is generally overestimated. But there are cases, where protoplast fusion is an exclusive means to construct a desired breeding line. For example useful male sterility has been transferred from Raphanus into rapeseed by interspecific crossing. But the Raphanus chloroplasts become cold sensitive under control of the rapeseed nucleus. Cytoplasmic recombination of rapeseed chloroplast with the Raphanus mitochondria, which control male sterility, can only be obtained after fusion of the appropriate rapeseed with Raphanus protoplasts (Pelletier et al. 1983).

10. Gene technology is a future challenge for the achievement of particular goals

Protoplasts have been the first requisites for what is today regarded as high technology, i.e. molecular gene transfer. By culturing protoplasts in a suspension of cloned DNA molecules, these may not only be incorporated but obviously also stably inserted at a proper chromosome site and finally expressed by function. Directional gene transfer has also been possible by using specific vectors such as the TI-plasmids of Agrobacterium or certain virus genomes. But several of the necessaries of an effective gene transfer are still in demand. The desired character must be caused by a single gene which must be known and available as a DNA molecule. This sequence must be activated in the host cell, but it must also be of a type which needs no developmental or tissue specific regulation. Last not least, its activity must not interfere with other essential gene functions of the host cell.

Since the DNA language is universal, it may be anticipated that gene transfer can be effective over wide taxonomic distances. Genes have been isolated from soil bacteria, which detoxify herbicides. Plants with such genes could become herbicide resistant, and since some of these systems are highly specific, their utilization is very appealing from both, an economical and an ecological point of view. Other projects have been proposed, but none has been completed so far.

In any event, for reasons mentioned above, these single gene manipulations most probably will yield little more than basic materials for further breeding work. The expertise of the classical plant breeder will still be required as before. But an array of highly potent tools has already become and will soon even more become applicable from the new developments in biotechnology for the benefit of genetic improvements in oilplant species.

REFERENCES

Brunklaus, E., 1985, Nutzung und Analyse des Leistungspotentials von Poly-
    enfettsäure-Mutanten bei Raps. Diss. F.B. Agrarwiss., Univ. Göttin-
    gen. 78 pp
Conway, T.F., and Earle, F.R., 1963, Nuclear Magnetic Resonance for
    determining oil content of seeds. J.Amer.Oil Chem.Soc. 40:265-268
Downey, R.K., and Harvey, B.L., 1963, Methods of breeding for oil quality
    in rape. Canad.J.Plant Sci. 43:271-275
Gland, A., 1982, Gehalt und Muster der Glucosinolate in Samen von resyn-
    thetisierten Rapsformen. Z.Pflanzenzüchtg. 88:242-254
Hatje, G., 1986, World importance of oilseeds and their products. In: Oil
    Crops of the World - Their Breeding and Utilization, G. Röbbelen,
    R.K. Downey, and A. Ashri (eds.). Macmillan Publ.Comp., New York
Knowles, P.F., 1969, Modification of quantity and quality of safflower
    oil through plant breeding. J.Amer.Oil Chem.Soc. 46:130-132

Laibach, F., 1925, Das Taubwerden der Bastardsamen und die künstliche Auf-
    zucht früh absterbender Bastardembryonen. Z.Bot. 17:417-459
Lichter R., 1982, Induction of haploid plants from isolated pollen from
    Brassica napus. Z.Pflanzenphysiol. 105:427-434
Pelletier, G., Primard, C., Vedel, F., Chetrit, P., Remy, R., Rouselle, P.
    and Renard, M., 1983, Intergeneric cytoplasmic hybridization in
    cruciferae by protoplast fusion. Mol.Gen.Genet. 191:244-250
Pryde, E.H., Princen, L.H., and Mukherjee, K.D., 1981, New Sources of Fats
    and Oils. American Oil Chemists' Society, Champaign, Ill.
Ratledge, C., Dawson, P., and Rattray, J., 1984, Biotechnology for the Oils
    and Fats Industry. American Oil Chemists Society, Monograph 11
Rehm S., and Espig, G., 1984, Die Kulturpflanze der Tropen und Subtropen.
    Ulmer, Stuttgart
Röbbelen, G., 1984, Changes and limitations of breeding for improved poly-
    enoic fatty acids content in rapeseed. In: Biotechnology of the Oils
    and Fats Industry. C. Ratledge, P. Dawson, J. Rattray (eds.). Amer.
    Oil Chem.Soc., 97-105
Röbbelen, G., and Thies, W., 1980, Biosynthesis of seed oil and breeding
    for improved oil quality of rapeseed. Brassica Crops and Wild Allies
    - Biology and Breeding. Japan Sci.Soc.Press, Tokyo, 253-283
Sengbusch, R. von, 1942, Süßlupinen und Öllupinen. Landw.Jahrb. 91:719-880
Thies, W., 1971, Schnelle und einfache Analysen der Fettsäurezusammen-
    setzung in einzelnen Raps-Kotyledonen. I. Gaschromatographische und
    papierchromatographische Methoden. Z.Pflanzenzüchtg. 65:181-202
Tröeng, S., 1955, Oil determination of oil seeds. J.Amer.Oil Chem.Soc.
    32:124-126
Weiss, E.A., 1983 Oilseed Crops. Longman, London - New York

PRODUCTION OF FATS OF PARTICULAR

BIOLOGICAL SIGNIFICANCE

Jean-Paul Helme

Bio-Extraction
Paris, France

INTRODUCTION

In our report we will mainly examine the long-chain essential fatty acids (long chain PUFA or LCP). We know the animals, including men, cannot insert double bonds in the n-6 and n-3 positions and cannot synthesize either of these two EFA.

However, animal can add more double bonds to the parent EFA by introducing them between the original double bonds and the carboxyl group ; at the same time, the carbon chain length is extended at the carboxyl end by further two carbons addition. This metabolic process, produces the LCP having 20 and 22 carbon chain length with 3, 4, 5 and 6 double bonds (Fig. 1).

The result of this metabolic process is the production of LCP which are required for cell structures and prostaglandin synthesis. In practice, the most important, for their biological value, of these LCP are :

For the linoleic acid (n-6 family)
- gamma-linolenic acid (GLA)   18:3 n-6
- di-homo-gamma linolenic acid (DHGLA) 20:3 n-6
- arachidonic acid (ARA)   20:4 n-6

For the linolenic acid (n-3 family)
-   eicosapentaenoic acid (EPA)   20:5 n-3
-   decosahexaenoic acid (DHA)   22:6 n-3.

The sequential process of desaturation and elongation is achieved by Δ 6, Δ 5, Δ 4 desaturation enzymes and the elongation enzymes. The Δ 6 desaturation is a rate limiting step.

We do not study in this report the metabolic pathways of the linoleic, alpha-linolenic acids neither the arachidonic acid metabolism. They have been already reported in this symposium by Pr. GALLI. Same remark for prostaglandins derived from EPA by Drs. DYERBERG and WEBER.

The nutritional aspects of the EFA, the criteria used or suggested for determining EFA requirements, the tentative approach to the quantification of EFA, requirements based on clinical studies (Pr.JACOTOT - France) have been reported in an important literature. These requirements concerning the healthy adult still remain a controversial subject (1) (2) (3) (4) (5) (6).

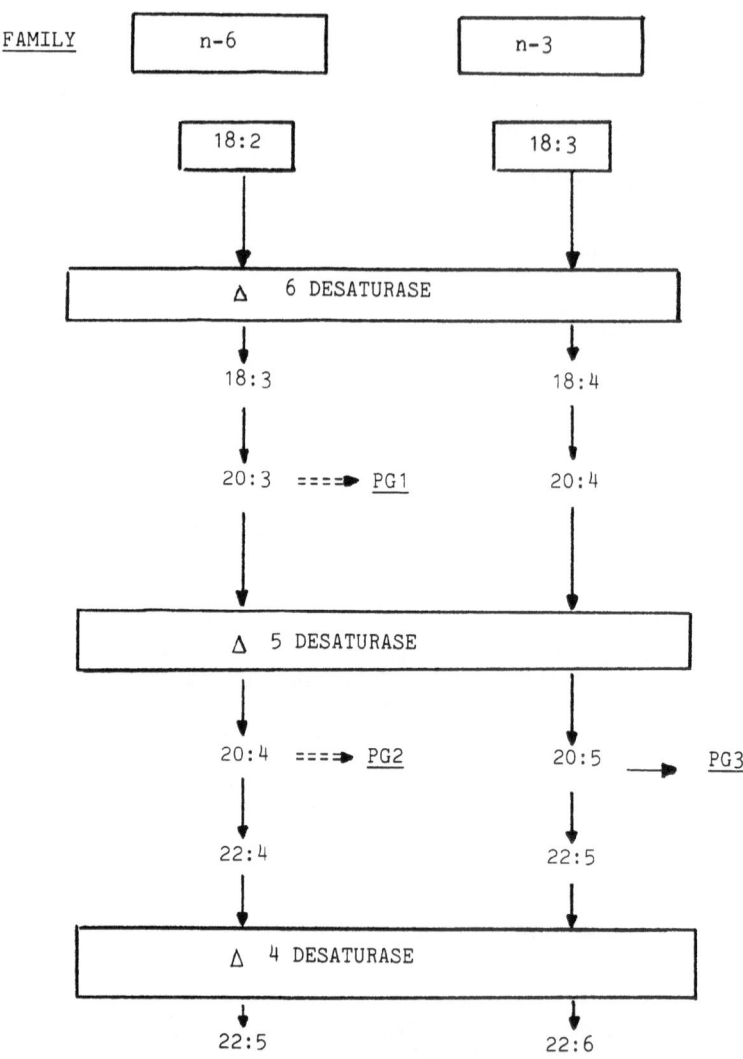

Fig. 1.   E F A Biosynthesis

In addition, these requirements concern the healthy adult able to synthesize in vivo sufficient LCP, owing to desaturases and elongases.

But, in some pathological situations and in aging people (7) (8) in early development (9) a supplementation with the different LCP may be desirable and or necessary.

Our goal in this report is to examine three different natural sources and the fatty acid (FA) composition of these LCP.

NATURAL SOURCES

For the GLA the main vegetable sources are :

- oenagraceae family and oenothera biennis : evening primerose oil
- saxifragaceae family/ribes nigrum and ribes rubrum : blackcurrant oil
- boraginaceae family/borago officinalis :

Table 1.   GLC Ranges of fatty acid composition

	Borage Oil (WT %)	Evening primrose Oil (WT %)	Blackcurrant seed Oil (WT %)
16 :0	15.3	6.6	6 - 7
16 :1	-	0.1	-
18 :0	3.1	1.7	1 - 2
18 :1 n-9	21.1	10.9	9 - 10
18 :2 n-6	37.3	71.5	45 - 60
18 :3 n-6	19.6	8.6	15 - 19
18 :3 n-3	0.1	0.2	12 - 14
18 :4 n-3	-	-	3 - 4
20 :0	2.1	0.3	-
22 :0	0.9	-	-
24 :0	0.4	-	-

We have to note that stearidonic acid (18:4 n-3), precursor of EPA is only present in blackcurrant oil.

For the Di-homo-gamma linolenic acid (DHGLA)

- the human placenta is one of the very rare sources.

We will mainly examine the composition of the lipids extracted from human placenta (10). This study being probably a new one, we will stress it. Let us examine in the following tables the main characteristics of these particular biological lipids extracted from human placenta :

Table 2   Quantitative analysis of lipids extracted from human placenta

Total lipids	3.1 ± 0.2 %
Phospholipids	61.5 ± 3.7 %
Free fatty acids	14.4 ± 1.7 %
Free cholesterol	17.2 ± 2.5 %
Esterified cholesterol	1.7 ± 0.9 %
Triglycerides	1.6 ± 0.2 %

Table 3   GLC ranges of fatty acid composition of
           total lipids extracted from human placenta (WT %)

16 :0	23.6 ± 1.4
16 :1	3.2 ± 0.5
18 :0	11.2 ± 0.7
18 :1 n-9	14.8 ± 1.0
18 :2 n-6	11.2 ± 1.2
20 :1	0.3 ± 0.1
20 :2 n-6	0.6 ± 0.2
20 :3 n-6	4.4 ± 0.4
20 :4 n-6	19.6 ± 1.7
22 :4 n-6	1.5 ± 0.3
22 :5 n-6	1.0 ± 0.3
22 :5 n-3	0.8 ± 0.2
22 :6 n-3	3.2 ± 0.9

Table 4   Quantitative composition of the different
           phospholipids extracted from human placenta (WT %)

Phosphatidylcholine	40.8 ± 3.3
Phospatidylethanolamine	17.6 ± 3.3
Lysophosphatidylcholine	2.1 ± 0.3
Lysophosphatidylethanolamine	7.1 ± 0.8
Phosphatidylenositol	2.9 ± 0.6
Phosphatidylserine	4.4 ± 0.5
Sphingomyelin	25.1 ± 3.1

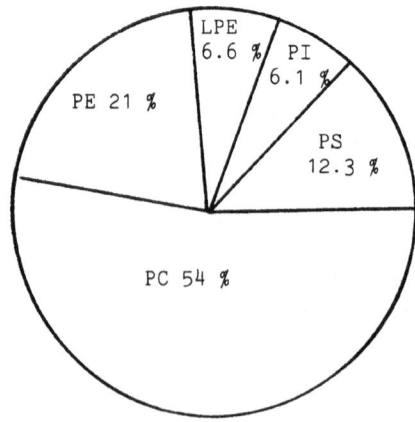

Fig. 2.   Distribution of di-homo-gamma linolenic acid
           in the different phospholipids extracted from
           human placenta.

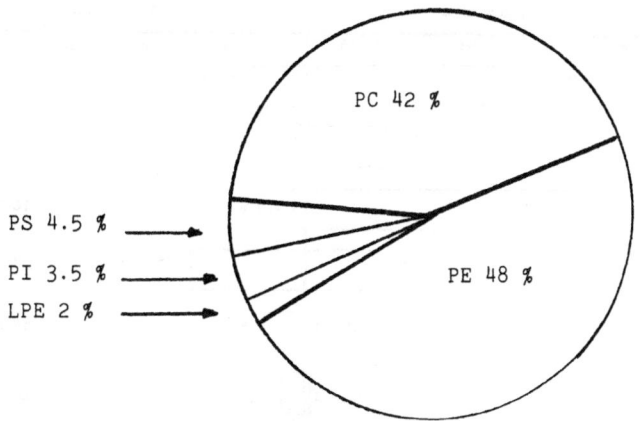

PC 42 %

PS 4.5 % ——→

PI 3.5 % ——→

LPE 2 % ——→

PE 48 %

Fig. 3.   Distribution of arachidonic acid in the different
phospholipids extracted from human placenta.

These particular Fatty acids, extracted from human placenta, contain
three main LCP :

- arachidonic acid precursor of PG2
- di-homo-gamma linolenic acid precursor of PG1
- docosahexaenoic acid (DHA)

The DHA is involved in the organogenesis of myeline, when there is a
high EFA demand for the synthesis of cell structural lipid in early
development. This is true especially for the central nervous system in
which the main period of cell division is prenatal.

The accumulation of LCP in the foetus is achieved by a progressive
increase in chain length and degree of unsaturation in maternal liver,
placenta, foetal liver and foetal brain, resulting in high concentrations
of ARA acid and DHA acid in foetal tissues (according to the publications
Dr. CRAWFORD and al).

Why such a high concentration of arachidonic acid in the human placenta ?

We know that the concentration in free ARA acid in the cells is very
low. We also know that it is only when in the free form that ARA acid is
the precursor of PG2.

The liberation of ARA acid is then the limiting step of the
eicosanoïds synthesis. In fact, equilibrium between deacylation and
transacylation, observed in cells at rest, is disturbed after stimulation
of these cells. Then, there is an increase in the phospholipids hydrolysis
and, consequently an increase of the ARA acid level, in its free form.

The preferential liberation of ARA acid, essentialy esterified on the
2 carbon of the sn-glycerol (as shown in figures 4 and 5), needs the action
of the phospholipase A2. We indicate Figure 6 the sites of hydrolysis
catalyzed by the various phosphalipases.

The first hydrolytic step is applicable to all phospholipids, except
phosphatidylinositol. Several factors - kind of phospholipids, Ca2+ concentration - control arachidonic acid liberation by the phospholipase A2.

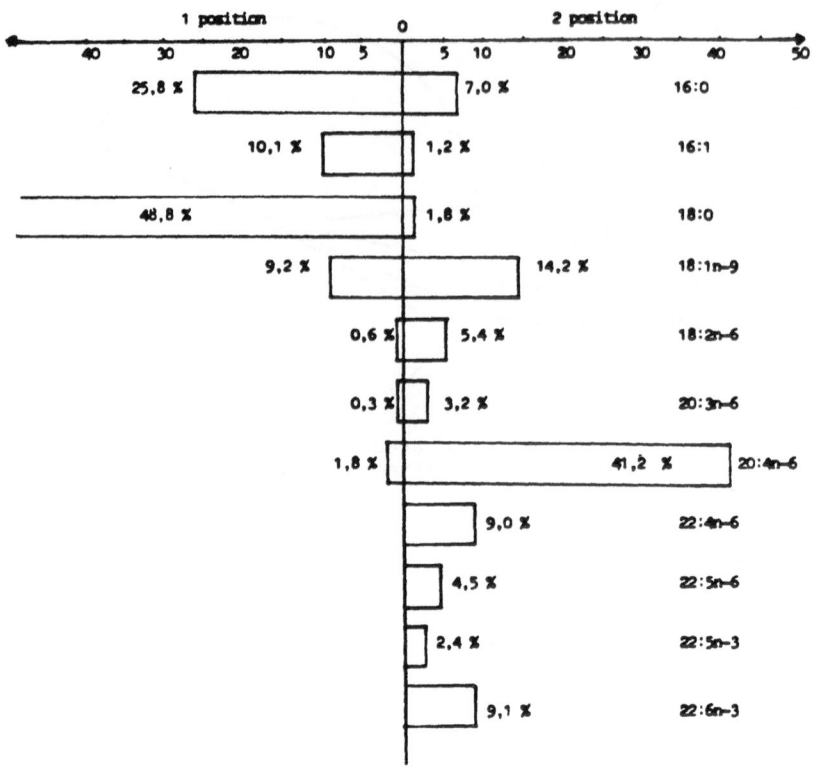

Fig. 4. Positional distribution of fatty acids in phosphatidylethanolamine from human placenta.

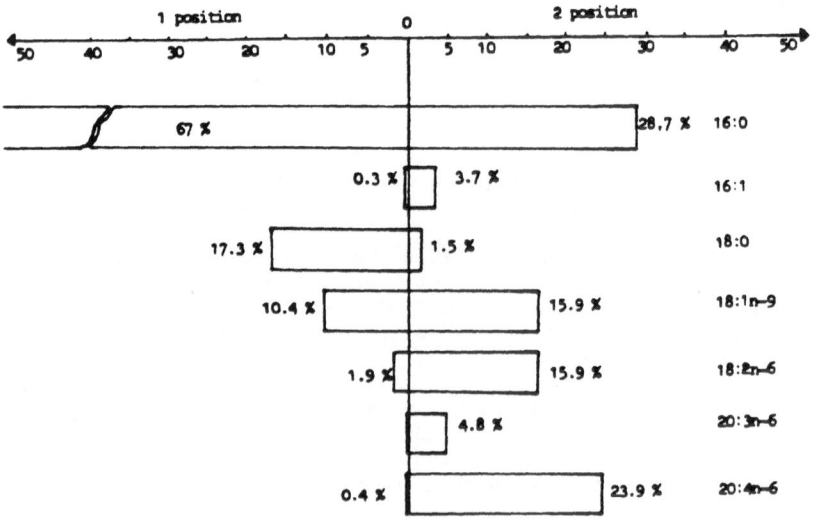

Fig. 5. Positional distribution of fatty acids in phosphatidyl choline from human placenta.

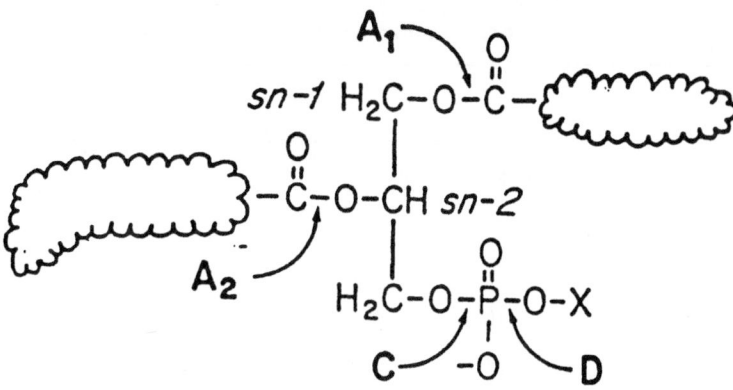

Fig. 6.   The structure of a generalized glycerophospholipid.
The various alcohols that may be found in phosphomono-
ester linkage are depicted as (X). The sites of hydrolysis
catalyzed by the various phospholipases are indicated
by the letters. From Bleasdale et Al, with permission.

The studies of OKITA and JOHNSTON have shown the presence of a specific phospholipase A2 in the human foetal membrane and have explained the
ARA acid mobilization in order to biosynthesize specific prostaglandins
(PG$_s$). These PG$_s$, of the E and F series, increase the intensity of
spontaneous contractions of the uterus during the initiation and
maintenance of human parturition.

The phospholipase A2 activation depends on increase of the CA2+ flow.
This activation is the first step, the second one being the action of a
phospholipase C, specific of the phosphadidyl inositol.

For more details concerning the fondamental enzymatic mechanisms ex-
plaining the ARA acid mobilization, refer to the excellent report of Drs.
J.E. BLEASDALE and J.M. JOHNSTON.

We have analyzed a lot of human placenta coming from different
countries - around the world - and we have found that the variations in the
lipid composition are very weak. In addition, the stability of the lipids
is excellent as shown in table 5 :

The tested lipids have been extracted from human placenta at a semi-
industrial scale. Human milk lipids and problems related to their
replacement have been studied by Dr. U. BRACCO (12). Dr. BRACCO wrote :

"Long chain lipids described as minor components of human milk fat,
possess probably an important physiological significance which is largely
unknown".

The important physiological significance of these lipids explains and
justifies the clinical studies in progress and concerning the utilization
of these particular lipids (as a whole and/or only some fractions).
According to FAO Food and Nutrition paper, confirmed during this
NATO Workshop by Dr. CRAWFORD, "the ideal recommandation for milk
substitutes would be to match the EFA of human milk from well nourished
mothers with respect to both parent and long-chain EFA and the balance of
n-6 and n-3 families of fatty acids".

Table 5   Conservation test
            1 % arachidonic acid fat base

	Aa	Bb	Cc	Dd
Peroxide index (meq/kg)	0.2	2	3.8	3.8
Carbonyl index (micromoles/g)	14	13.5	13	13
E at 232 nm	4.15	4.3	4.7	4.5
E at 270 nm	1.55	1.5	1.6	1.45
E at 303 nm	0.45	0.45	0.5	0.45
E at 318 nm	0.35	0.35	0.35	0.30
% Oleic acidity	5.65	5.5	5.6	5.55
% C20 : 4	0.97	1.01	0.97	1

aA, Beginning
bB, at 0°C/air
cC, After 8 months under nitrogen AT 20°C
dD, at 20°C/air.

Table 6 shows a comparison between human milk (main FA) and the total
fatty acids extracted from human placenta. The LCP - and mainly DHGLA, ARA,
and DHA - are present both in the human milk and in the lipids extracted
from human placenta. But the concentration is higher in the placenta,
comparatively with the human milk : 12.3 x for DHGLA, 27 x for ARA, 17.6 x
for DHA.

Table 6   Comparison between human milk and total fatty acids
              extracted from human placenta

fatty acids	human milk	human placenta	Aa
18 : 2 n-6	13.89	10.50	
18 : 3 n-3	0.65	-	
20 : 2 n-6	0.46	0.45	
20 : 3 n-6	0.39	4.82	$\underline{\frac{12.3}{27}}$
20 : 4 n-6	0.56	15.15	
(20 : 5 n-3	0.27		
(22 : 1			
22 : 4 n-6	0.10	1.15	
22 : 5 n-6	0.05	0.50	
(22 : 5 n-3	0.21		
(24 : 1			
22 : 6 n-3	0.30	5.30	17.6

aA, Ratio $\dfrac{\text{% Fatty acids human placenta}}{\text{% Fatty acids human milk}}$

For EPA and DHA : marine oils

Some marine oils are the third source of LCP. Drs. DYERBERG and WEBER just finished to give their report on "fats from marine animals in human nutrition" and "EPA and eicosanoid system". We only indicate table 7 the FA composition of a marine oil, rich in EPA.

Table 7   Fatty acids composition
marine oil (Japan origin)

	%
10	0.2
12	0.2
14	5.6
16	9.6
16 :1	9.2
17	1.4
18	4.2
18 :1	10.4
18 :2	1.3
18 :3 n-3	1.1
20	4.9
20 :1	3.1
20 :2 n-6	0.15
20 :3 n-6	0.15
20 :4 n-6	1.0
20 :5 n-3	27.8
22 :5 n-6	0.2
22 :5 n-3	3.0
22 :6 n-3	9.2
miscellaneous	7.3

CONCLUSIONS

In conclusion we would like to recall the numerous and different factors influencing the biological significance of the LCP.

Interactions : technologies ◄────────► Nutrition.

. Conditions of storage of raw materials (preservation conditions)
. Extraction processes (conventional and/or non conventional)
. Fractionation and concentration processes
. Characteristics and specifications
. Conditioning (capsules, powder, liquid...)
. Stability conditions (anti-oxydation)
. Biochemical studies on animals
. Clinical studies
. Oral, enteral, parenteral alimentation
. Different "nature" of the lipids : FFA, phospholipids, triglycerides, partial glycerides, esters, salts.

Acknowledgments

We gratefully acknowledge for their research contributions : V. CHIROUZE (Bio-Extraction - Paris) ; Pr. ENTRESSANGLES (University of Bordeaux I - France) ; INSTITUT MÉRIEUX (Lyon - France) and Laboratory LUCCHINI (Geneva - Switzerland) ; SIO SA/Groupe LESIEUR (Arras - France) ; I.T.E.R.G. (Pessac - France).

REFERENCES

(1) Dietary Fats and Oils in Human Nutrition , FAO and Nutrition Paper, Rome 1977

(2) Vles,R.A., Revue Française des Corps Gras, March 1980,No.3

(3) Lundberg,W.Q. On the quantification of Essential Fatty Acids Requirements, Fette Seifen Anstrichm. 81, 9 (1979)

(4) Galli,C. and Socini,A. Dietary Lipids in Pre- and Post-natal Development (Chapter 16) in "Dietary Fats and Health", Perkins,E.G. and Visek,W.J. eds., (1983), AOCS, USA

(5) Holmann,R.J., Johnsohn,S.B., Hatch,T.F., A case of Human Linolenic Deficiency, Am.J. Clin.Nutr. (1982)

(6) Jacotot,Pr., La Famille Linolenique, Cahiers de Nutrition et de Dietetique, vol.XX, April 1985

(7) Spielman,D., Mendy,F., Apfelbaum,M., Hopital Bichat,Paris, Med. Digest. et Nutr. (Q.M.) no. 40, Feb. 14,1985, Q.M., dossier Actualites, Feb. 14,1985

(8) Darcet,Ph., Apports et utilisation des Acides Gras Essentiels chez le Personnes Agees, Centre de Gerontologie, Assistance Publique Hopitaux de Paris

(9) Crawford,M.A., Dietary Fats during Early Development, NATO Advanced Research Workshop, Selvino,Italy, March 1986

(10) Chirouze,V., Obention par Voies Enzymatiques d'un Concentrat d'Acide Arachidonique a partir de Placenta Humain, Laboratoire de Lipochimie Alimentaire, Univ. de Talence, Bordeaux I,France

(11) Bleasdale,J.E., Johnston,J.M., Reviews in Perinatal Medicine, vol. 5, Dallas , USA

(12) Bracco,U. and Bauer,H.,Human Milk Lipids and Problems related to their Replacement, extract from Annals Nestle' no.40 (1978)

AUTO-OXIDATION PROCESS

Giovanni Lercker

Dept. of Prot. e Valorizz
Agro-Alimentare
S. Giacomo 7 - Bologna

INTRODUCTION

Rancidity may be said to be the subjective organoleptic appraisal of the off-flavour quality of food.

Oxidative rancidity is concerned with the changes that result from the reactions between the lipids in foods with atmospheric oxygen.

Oxidation of unsaturated lipids not only produces offensive odors and flavors but can also decrease the nutritional quality and safety due to the formation of secondary reaction products in foods after cooking and processing[1].

The oxidative deterioration of food lipids involves, primarly, the autoxidation reaction, so called because of its autocatalytic mechanism.

The autoxidation is a complex process which proceeds through free ra- dicals by a chain mechanism[2-4].

The Knowledge of the autoxidation process, even if not clear in all the aspects, is mainly due to the study of model systems.

It is generally accepted that the main steps of oxidation characteri- ze the three periods of the process: initiation, propagation and termina- tion (Figure I).

During initiation (induction period) hydroperoxides are accumulated, until their concentration is such to develop the main reaction of the se- cond period (propagation) which produces two radical species from a neu- tral hydroperoxy molecule.

During the last period, the termination, the concentration of radi- cals is so high that bimolecular interactions between radicals take place, allowing the quenching of the process.

PRIMARY OXIDATION PRODUCTS

Methyl oleate hydroperoxides

The most accepted mechanism of autoxidation (Farmer's modified mecha- nism) involves hydrogen abstraction on carbon-8 and carbon-11, with forma- tion of two allylic radicals (Figure II). Double bond rearrangement would

269

produce all four isomeric allylic radicals and after addition of oxygen
we obtain the four isomeric hydroperoxy radicals, which have cis and
trans configuration of double bond.

The evidence of this mechanism was supported by IR, AgNO3-TLC, GLC
and GC-MS analytical techniques[5-6], and the relative percentage of the
four isomeric hydroperoxides of both cis and trans configurations of the
double bond were calculated by mass fragments of the reduction and hydrogenation products of the hydroperoxides. These are isomeric hydroxy stea-
rates (as trimethylsilyl derivatives) obtained from the two fractions
collected by AgNO3-TLC (Table I)[6-7].

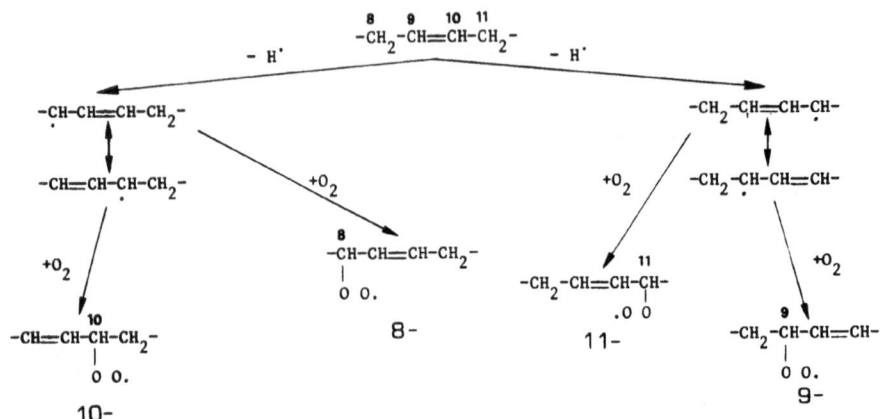

Figure I

INITIATION

$$R-H \quad \xrightarrow{\quad ? \quad \}\{ \quad O_2 \quad} \quad ROO^{\cdot}$$

$$ROO^{\cdot} + R-H \quad \dashrightarrow \quad ROOH + R^{\cdot}$$

PROPAGATION

$$2\ ROOH \quad \dashrightarrow \quad RO^{\cdot} + ROO^{\cdot} + H_2O$$

$$RO^{\cdot} + R-H \quad \dashrightarrow \quad ROH + R^{\cdot}$$

$$ROO^{\cdot} + R-H \quad \dashrightarrow \quad ROOH + R^{\cdot}$$

$$RO^{\cdot} \quad \dashrightarrow \quad \text{decomposition products} + \text{radicals}$$

$$ROO^{\cdot} \quad \dashrightarrow \quad \text{decomposition products} + \text{radicals}$$

$$ROO^{\cdot} + \rangle C = C \langle \quad \dashrightarrow \quad \underset{\text{threo} + \text{erythro}}{\rangle C \overset{O}{-} C \langle} + RO^{\cdot}$$

TERMINATION

$$R^{\cdot} + R^{\cdot} \quad \dashrightarrow \quad R-R$$

$$RO^{\cdot} + R^{\cdot} \quad \dashrightarrow \quad R-O-R$$

$$2\ ROO^{\cdot} \quad \dashrightarrow \quad ROH + R \overset{\pm}{=} O + O_2 \qquad \text{(Russell's mechanism)}$$

$$2\ RO^{\cdot} \quad \dashrightarrow \quad ROH + R \overset{\pm}{=} O$$

Figure II

Table I. Isomeric distribution of oleate hydroperoxides

T°C		IR 8-OOH[9]	AgNO$_3$-TLC 9-OOH[10]	GLC 10-OOH[9]	GC-MS[7] 11-OOH[9]
20	66.9 trans	7.0	11.0	20.9	28.2
	33.1 cis	12.7	1.5	3.6	15.3
40	71.2 trans	6.9	17.4	16.5	30.3
	28.8 cis	10.0	5.0	4.7	12.8
80	76.5 trans	21.5	21.3	22.0	11.8
	23.5 cis	7.4	4.4	3.5	8.2

CNMR/AgNO$_3$-TLC/MS[8]

T°C					
25	70.0 trans	12.3	23.1	21.7	12.9
	30.0 cis	14.1	1.1	1.1	13.7
40	76.0 trans	16.0	22.0	21.7	16.0
	24.0 cis	10.6	1.6	1.7	10.1
75	82.9 trans	19.1	22.5	22.0	19.5
	17.1 cis	6.1	2.7	2.9	5.4

According to the AgNO$_3$-TLC, GLC and GC-MS procedure[7], in our laboratory we determined the isomeric composition of methyl oleate hydroperoxides obtained from methyl oleate autoxidized at 80°C (Table II).

Table II

T°C		8-OOH[9]	9-OOH[10]	10-OOH[8]	11-OOH[9]
80	trans	14.0	24.2	20.9	10.1
	cis	13.8	2.5	1.4	13.1

The higher quantity of 8- and 11-isomers than 9- and 10-isomers found for the cis fraction, allows to hypotheze - as main mechanism - the following mechanism (Figure III).

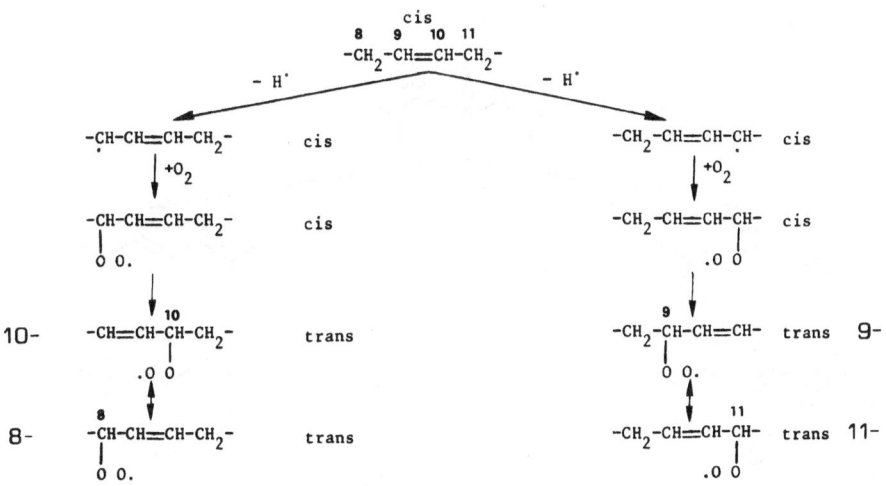

Figure III

The cis 8- and 11- hydroperoxides originate directly from the cis 8- and 11- radicals. The trans 9- and 10- hydroperoxides derive primarly from these cis compounds by hydroperoxy radical isomerisation. In another step the trans 9- and 10- hydroperoxy radicals isomerise to generate the trans 8- and 11-hydroperoxides.

The autoxidizing temperature increases the quantitative difference between the total (cis+trans) isomeric 8- and 11-hydroperoxides and the 9- and 10- corresponding pair: high temperature reduces the difference, normally higher for the 8- and 11-pair[2].

In support of the proposed mechanism are the following arguments:
1) Mercier[9] observed that the autoxidized methyl elaidate (the $\triangle^9$-trans isomer of the methyl oleate) produces the four positional isomers in the trans configuration only: this means that the allylic cis configuration of double bond is not obtained after isomerisation of a trans isomer.
2) Even if the resonance energy of the allylic radicals[10] is only 12 Kcal mole[-1], their rate of isomerisation is very low[11] ($K = 10^2$ 1.mole[-1] sec[-1]) in relation to the rate with which the same radical reacts with oxygen during autoxidation[12] ($K = 10^7$ 1. mole.[-1] sec[-1]). The lack of trans isomers in the unreacted methyl oleate after oxidation[5] and of positional isomeric methyl octadecenoates confirm this hypothesis.
3) The life of the allylic hydroperoxy radicals that are the precursors of hydroperoxides is fairly long. Thus, for the reaction rate constant

$$ROO. + R'H \longrightarrow ROOH + R!$$

a value of 1.19 1.mole[-1]sec[-1] has been calculated[13].

Methyl linoleate hydroperoxides

Methyl linoleate is a substrate with a different behavior than the oleate, from a hydroperoxide formation mechanisms point of view: the presence of two double bonds, methylene interrupted, allows the hydrogen abstraction only from the 11-carbon atom (Figure IV) which is in a very particular condition of reactivity.

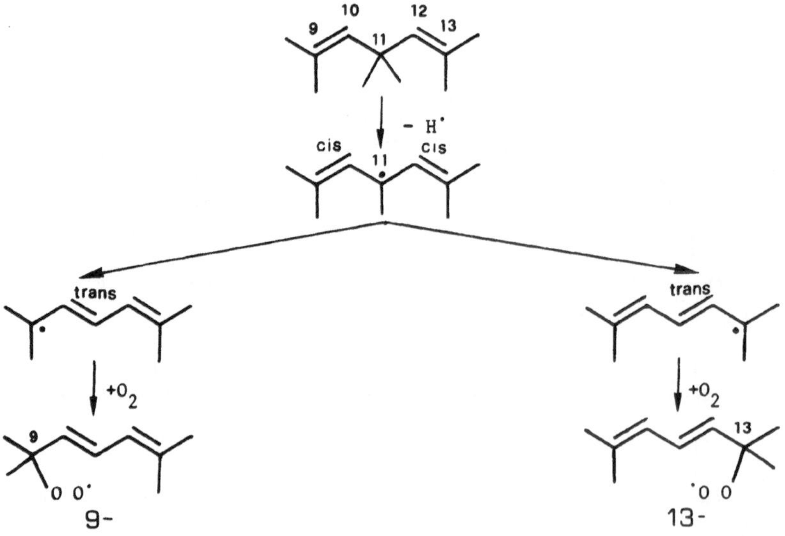

Figure IV

The isomerisation of the bi-allylic radical obtained is favoured by the formation of more stable isomeric conjugated radicals. After reaction with oxygen two hydroperoxy radicals only are produced: the isomers in 9- and 13-carbon atom positions.

Thus, the linoleate gives the isomeric distribution of the hydroperoxides corresponding to that predicted by the free radical mechanism (Table III).

Table III

T°C	P.V.	9-OH	13-OH
40	152	50.2	49.8
	261	51.7	48.3
	686	49.7	50.3
	918	49.6	50.4
60	93	47.3	52.7
	505	51.5	48.5
	1403	49.0	51.0
80	1249	52.5	47.5

Amounts of trans, trans isomers of 9- and 13-hydroperoxides were found in peroxidized linoleate and there is now evidence of the interconversion of 9- and 13- cis, trans-hydroperoxides of linoleic acid (obtained by lipoxygenase) to a mixture of 9- and 13-trans, trans-hydroperoxides[15].

Methyl linolenate hydroperoxides

The mechanism of hydroperoxide production for linolenate is very similar to that of linoleate (Figure V).

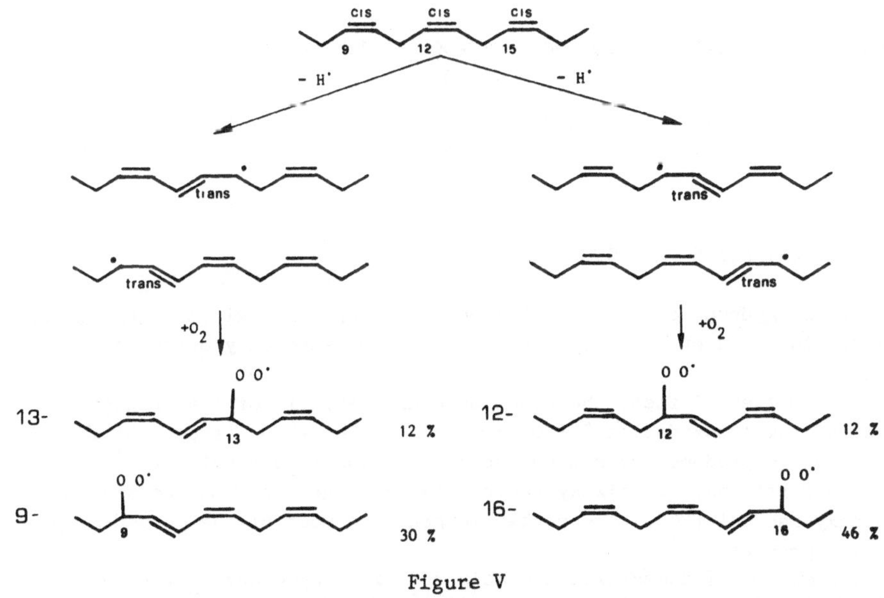

Figure V

Hydrogen abstraction on the doubly allylic methylenes on carbon-11 (left) and carbon-14 (right) produce four different hydroperoxy radicals, after reaction with oxygen. The percentage of the four isomers are indicated in Scheme V[16].

273

Methyl arachidonate hydroperoxide

The mechanism of autoxidation of arachidonate proceeds in the same way of that of linoleate (and linolenate). Hydrogen abstraction at the three doubly allylic carbons[-7], [-10] and [-13] produces 6 different radicals. After the O_2 attack 6 isomeric hydroperoxy radicals are obtained, with a conjugated diene system and 2 methylene-interrupted double bonds[15].

PHOTOSENSITIZED OXIDATION

Another important way that unsaturated lipids can be oxidized involves exposure to light and a sensitizer (chlorophill, erythrosine, riboflavine, etc.).

This oxidation is not a free radical process and the oxygen reacts directly with the double bond of the unsaturated chain of fatty acids when activated to the singlet state ($^1O_2$) by the transfer of energy from the photosensitizer. The singlet oxygen is extremely reactive: linoleate is reported[17] to react at least 1500 times faster with $^1O_2$ than with normal oxygen in the triplet ground state ($^3O_2$).

The isomeric distributions of fatty acid hydroperoxides obtained from the three main model systems by photosensitized oxidation[18] are listed in Table IV.

Table IV

ISOMERIC HYDROPEROXIDES %

Oleate	9-OOH	10-OOH				
	50	50				
Linolenate	9-OOH	10-OOH	12-OOH	13-OOH		
	31	18	18	33		
Linolenate	9-OOH	10-OOH	12-OOH	13-OOH	15-OOH	16-OOH
	21	13	13	14	13	25

SECONDARY OXIDATION PRODUCTS

If the hydroperoxides are considered the primary oxidation products, they are the main precursors responsible of the secondary oxidation products.

As a matter of fact, the reaction characteristic of the propagation period of the autoxidation 2 ROOH $\longrightarrow$ RO. + ROO. +H_2O (step 1 of Figure VI) allows to produce two radical species having good reactivity.

The first one, the alkoxy radical (RO.), together with the second hydroperoxy radical (ROO.) are enable to generate most of the secondary oxidation products.

The Figure VI summarizes the main reactions hypothezing for the methyl oleate secondary oxidation products[19]. The decomposition products of the hydroperoxides take origin from alkoxy and hydroperoxy radicals[18],

even if the major part of these products are produced from alkoxy by carbon–carbon scission[20].

Several "evolution" products are formed from alkoxy and hydroperoxy
radicals by unimolecular reaction, which is the sole active at the vapour
phase (like gas chromatographic injector port).

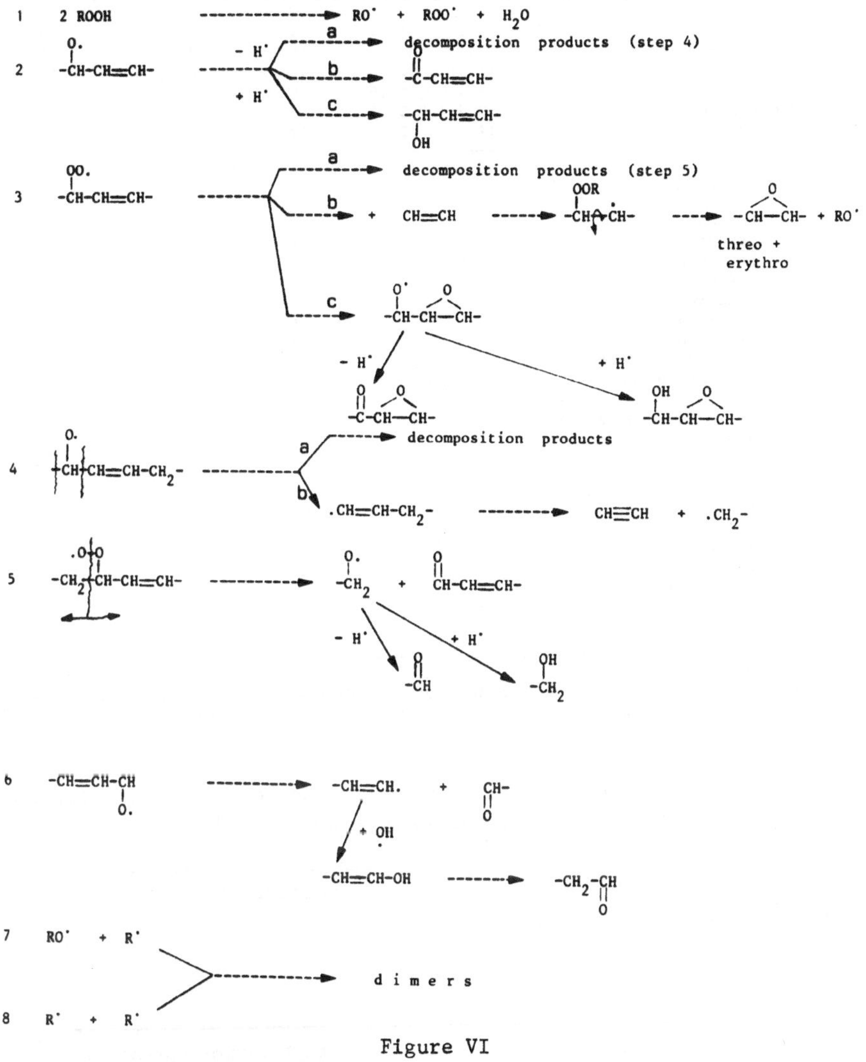

Figure VI

Step 6 was hypotized as to explain the absence of –unsaturated molecules among the decomposition products[21], but step 4 b seems to have
more probability to give account for some identified compounds[18].

The recent experimental determination[22] of the ketone structure corresponding to the step 2 b, showed that only the trans double bond configuration is present.

The presence among the autoxidation products of dimeric components[23],
who's structure may contain (step 7) or not (step 8) one oxygen as bridge
bond between the monomer parts, allows to hypothese a bimolecular formation mechanism. As a matter of facts these dimers originate in the liquid
phase only.

275

Table V

Hydroperoxide source	Volatiles	Me Hydroperoxides		Relative percent Methyl esters			Triolein	
		70	Norm.[a]	105	Norm.[a]	127	188[b]	Norm.[a]
8-OOH	Decanal	3.9	10.5	1	1.7	1	2.8	4.6
	2-Undecenal	1.7	4.6	10	16.9	44	11.1	18.1
9-OOH	Nonanal [c]	(7.5)	20.3	(13)	22.0	(6)	(11.2)	18.3
'	2-Decenal	5.4	14.6	10	16.9	39	16.5	26.9
10-OOH	Nonanal [c]	(7.5)	20.3	(13)	22.0	(6)	(11.2)	18.3
11-OOH	Octanal	11	29.7	12	20.3	4	8.5	13.9
		37	100	59	100	100	61.3	100
	OTHER ALDEHYDS							
	Pentanal			5			0.9	
	Hexanal			6			1.9	
	Heptanal	0.5		11			5.1	
	2-Hexenal			3				
	2-Heptenal			5			0.1	
	2-Octenal			6			0.5	
	2-Nonenal	0.5		4			2.1	
	Me-8-oxooctanoate	3.5						
	Me-9-oxononanoate	15						
	Me-10-oxodecanoate	12						
	Me-10-oxo-8-decenoate	3.4						
	Me-11-oxo-9-undecenoate	5.8						
	OTHER COMPOUNDS							
	Heptane	4.4					8.6	
	Octane	2.7					9.7	
	1-Heptanol	0.4					1.6	
	Me Heptanoate	1.5						
	1-Octanol	0.4					2.5	
	Me Octanoate	5.0						
	Me Nonanoate	1.5						

a Normalized.
b Other minor volatiles included methyl ketones (0.8%), acids (1.2%), and gamma lactones (0.8%).
c Values for nonanal were divided by assuming that they come equally from 9-OOH and 10-OOH.

Table V, shows the methyl oleate hydroperoxide volatile products, identified by means of computerized mass-spectrometry[21-24].

Table VI, reports the results of a similar investigation carried out in our laboratory on the thermal degradation products of methyl oleate hydroperoxides[19].

Table VI

1	n-heptane	10	methyl 7-oxo-heptanoate
2	n-octane	11	2-decenal
3	heptanal	12	methyl 7-hydroxyheptanoate
4	1-heptanol	13	methyl 8-oxo-octanoate
5	octanal	14	2-undecenal
6	methyl heptanoate	15	methyl 8-hydroxyoctanoate
7	1-octanol	16	methyl 9-oxo-nonanoate
8	nonanal	17	methyl 10-oxo-decenoate
9	methyl octanoate	18	methyl 11-oxo-undecenoate

Among the methyl oleate hydroperoxide volatile products which, in our laboratory, were non identified, but found by Frankel and co-workers[21], decanal and methyl nonanoate were the most significant under a quantitative point of view. On the other hand Frankel and co-workers[21] do not report the presence of methyl 7-oxo-heptanoate and methyl 7-hydroxy-heptanoate. Both having the support of an active mechanism to explain many other identified components found.

Decanal, instead, must be justified with a mechanism that needs the collision of the radicals R-CH=CH. and ·OH; which, in the vapour state (as in the GC injector) do not have many chances in finding themselves.

The identification by means of computerized GC-MS can bring to structure hypothesis which, however, should be confirmed through a comparison with standard samples.

Table VII, shows the volatile decomposition products from heated oxidized linoleate and its isomeric hydroperoxides[24].

Table VII

Hydroperoxide source	Volatiles	Relative percent Me Hydroperoxides 70	Norm.[a]	Ethyl eters 76	Norm.[a]	Trinolein 189[b]	Norm.[a]
9-OOH	3-Nonenal[c]	1.4	4.6	6.2	7.1	0.2	0.5
	2,4-Decadienal	14	46.1	72.1	82.9	19	52.2
13-OOH	Hexanal	15	49.3	8.7	10	17.2	47.3
		30.4	100	87.0	100	36.4	100
	OTHER ALDEHYDS						
	Acetaldehyde	0.3					
	Acrolein					4.8	
	Propanal					0.7	
	Butanal					0.4	
	Pentanal					3.6	
	Heptanal					8.2	
	Nonanal			2.9			
	2/3-Hexenal					1.8	
	2-Heptenal	Trace		6.4		15.3	
	2-Octenal	2.7		3.7		4.4	
	2,4-Nonadienal	0.3					
	Me 8-Oxooctanoate	1.3					
	Me 9-Oxononanoate	19					
	Me 10-Oxodecanoate	0.7					
	Me 10-Oxo-8-decenoate	4.9					
	4,5-Epoxy-2-decenal					4.3	
	OTHER COMPOUNDS						
	Pentane	9.9				12.0	
	Hexane (?)					0.7	
	1-Pentanol	1.3				3.4	
	1-Octen-3-ol	Trace				1.7	
	2-Pentyl furan	2.4				0.8	
	Me Heptanoate	1.0					
	Me Octanoate	15					

a   Normalized.
b   Other minor volatiles included acids (1.8%), gamma lactone (0.5%), furan (0.4%), and methyl ketone (Trace).
c   Determined as 2-isomer.

Table VIII, shows the volatile decomposition products from heated oxidized linolenate and isomeric hydroperoxides[24].

The products obtained from the thermal decomposition of hydroperoxides of the model system show the evidence of the cleavage of the hydroperoxide molecule between the carbon atom in α and the one which holds the oxygen of the alkoxy radical. A group of aldehydes and aldoesters, saturated and α, ß-unsaturated, representing the most reactive components among those identified, are therefore obtained. Aldoesters are the type of component which can remain in an oxidized fatty substance even following the refining process, in particular even after deodorization, as glyceridic fragments. Their permanence can represent a threat to the following product preservation, due to the reactivity which this type of substances show. Among the non decomposition products of hydroperoxides, the epoxidic substances stand out for their potential instability and maybe for their toxicity.

Table VIII. Volatile Decomposition Products from Heated-Oxidized Linolenate and Isomeric Hydroperoxides.

Hydroperoxide source	Volatiles	Me Hydroperoxides		Relative percent Methyl esters		Trilinolenin	
		70	Norm.[a]	127	Norm.[a]	187[b]	Norm.[a]
9-OOH	3,6-Nonadenial	0.5	1.5				
	2,4,7-Decatrienal	14	42.6	4(?)	4.8	2	3.8
12-OOH	3-Hexenal[c]	(0.7)	2.1	(0.5)	0.6	(0.5)	0.9
	2,4-Heptadienal	9.3	28.3	30	35.7	24	45.3
13-OOH	3-Hexenal[c]	(0.7)	2.1	(0.5)	0.6	(0.5)	0.9
16-OOH	Propanal/acrolein	7.7	23.4	49	58.3	7/19	13.2/35.8
		32.9	100	84	100	53	100
	OTHER ALDEHYDS						
	Ethanal	0.8		6		2	
	Buthanal	0.1		3			
	2-Butenal	0.5		3		9	
	2-Pentenal	1.6		6		4	
	2-Nonenal			1			
	Me 8-Oxooctanoate	0.6					
	Me 9-Oxononanoate	13					
	Me 10-Oxodecanoate	1					
	Me 10-Oxo-8-decenoate	4.2					
	4,5-Epoxy-2-heptenal	0.2				3	
	OTHER COMPOUNDS						
	Ethane/ethene	10				Trace	
	Ethanol/furan					2.2	
	2-Butyl furan	0.5				1.4	
	Me Heptanoate	1.8					
	Me Octanoate	22					
	Me Nonanoate	0.7					

a Normalized.
b Other minor volatiles included 2-pentene (0.8%), 2-pentenyl furan (0.7%), ethyl furan (1%), and acids (1.5%).
c Values for 3-hexenal were divided by assuming that they come equally from 12-OOH and 13-OOH.

An experimentation to establish the level of toxicity of epoxides, under a nutritional point of view, should be undertaken with modern experimental facilities.

The biological consequences of oxidized lipids that arise from in vivo reactions or from ingested food long have attracted the attention of biochemists and food scientists. Many investigators are now working on several biological systems in which lipid peroxides and free radicals are implicated.

Many lipid oxidation products are known to interact with biological materials to cause cellular demage.

Definitive conclusions on digestion and adsorption of the products of lipid autoxidation have not been reached.

There is considerable indirect evidence that indicates that dietary peroxide may be absorbed[25-28]. But recent studies allow to conclude that the peroxide action must be in the intestinal lumen or the intestinal cells[28] where vitamin A and E antioxidants are destroyed, causing nutritional deficiency symptoms. Furthermore absorbed secondary oxidation products appeared to contribute to worsten the conditions of the liver[28].

REFERENCES

1. S. Tannebaum, Industrial processing, in: "Nutrients in processed foods: proteins", P.L. White and D.C. Fletcher eds., Publ. Sci. Group, Acton, Mass. (1974).
2. N. Uri, Physico-chemical aspects of autoxidation, in: Autoxidation and antioxidants", vol. I, W.O. Lundberg ed., Interscience, New York (1961).
3. W.O. Lundberg, Mechanism and products of lipid oxidation, in: "Lipids and their oxidation", N.W. Schultz, E.A. Day and R.O. Sinnhuber eds., Avi, Westport, Connecticut (1962).
4. R. Hiatt, Hydroperoxides in: "Organic peroxides", vol. II, D. Swern ed., Wiley, New York (1971).
5. M.V. Piretti, C. Cavani and F. Zeli, Mechanism of the formation of hydroperoxides from methyl oleate, Rev. Fr.se Corps Gras 25:73 (1978).
6. E.N. Frankel, Autoxidation, in: "Fatty acids", E.H. Pryde ed., AOCS Monograph 7, American Oil Chemists' Society, Champaign, IL (1979).
7. M.V. Piretti, P. Capella and G. Bonaga, Zusammensetzung des oxydierten ölsäuremethylesters in abhängigkeit von der oxydationstemperatur, J. Chromatogr. 92:196 (1974).
8. R.F. Garwood, B.P.S. Khambay B.C.L., Weedon and E.N. Frankel, Allylic hydroperoxides from the autoxidation of methyl oleate, J. Chem. Soc. Chem. Comm. 364 (1977).
9. J. Mercier, Hydroperoxides formed by autoxidation of methyl oleate and elaidate, and the formation of cis- -ethylenic hydroperoxides, Comp. Rend. Acad. Sci. Paris 269:1002 (1969).
10. S.W. Benson, A.N. Bose and P. Nangia, The kinetics of iodine$_2$-catalyzed positional isomerism of butene-1. The resonance energy of the allyl radical, J. Am. Chem. Soc. 85:1388 (1963).
11. J.K. Koki and P.J. Krusic, Isomerization and electron spin resonance of allylic radicals, J. Am. Chem. Soc. 90:7157 (1968).
12. A.A. Miller and F.R. Mayo, Oxidation of unsaturated compounds. I. The oxidation of styrene, J. Am. Chem. Soc. 78:1017 (1956).
13. J.L. Bolland, Kinetics studies in the chemistry of rubber and related materials. VII. Influence of chemical structure on the -methylenic reactivity of olefins, Trans. Faraday Soc. 46:358 (1959).
14. E.N. Frankel; W.E. Neff, W.K. Rohwedder, B.P.S. Kambay, R.F. Garwood and B.C.L. Weedon, Analysis of autoxidized fats by gas chromatography-mass spectrometry: II. Methyl linoleate, Lipids 12:908 (1977).
15. H.W.S. Chan, C.T. Costaras, F.A.A. Prescott and P.A.T. Swoboda, Specificity of lipoxygenases. Thermal isomerisation of linoleate hydroperoxides, a fenomenon affecting the determination of isomeric ratios, Biochim. Biophys. Acta 398:347 (1975).
16. E.N. Frankel, W.E. Neff, W.K. Rohwedder, B.P.S. Kambay, R.F. Garwood and B.C.L. Weedon, Analysis of autoxidized fats by gas chromatography-mass spectrometry: III. Methyl linolenate, Lipids 12:1055 (1977).
17. H.R. Rawls and P.J. van Sauten, A possible role for singlet oxygen in the initiation of fatty acid autoxidation, J. Am. Oil Chem. Soc. 47:121 (1970).
18. E.N. Frankel, W.E. Neff and T.R. Bessler, Analysis of autoxidized fats by gas chromatography-mass spectrometry: V. Photosensitized oxidation, Lipids 14:961 (1979).
19. G. Lercker, P. Capella and L.S. Conte, Thermal decomposition of methyl

oleate hydroperoxides, Riv. Ital. Sostanze Grasse 61:623 (1984).

20. H.W.S. Chan, F.A. Prescott and P.A.T. Swoboda, Thermal decomposition of individual positional isomers of methyl linoleate hydroperoxide: evidence of carbon-oxygen bond scission, J. Am. Oil Chem. Soc. 53:572 (1976).

21. E. Selke, E.N. Frankel and W.E. Neff, Thermal decomposition of methyl oleate hydroperoxides and identification of volatile components by gas chromatography-mass spectrometry, Lipids 13:511 (1978).

22. G. Lercker, P. Capella and L.S. Conte, Thermo-oxidative degradation products of methyl oleate. Riv. Ital. Sostanze Grasse 61:337 (1984).

23. G. Lercker, Thermo-oxidative degradation of fats: methyl oleate as a model, in: "2. European Conference on Food Chemistry", Euro Food Chem. II, Proceedings, Fecs event n. 45 (1983).

24. E.N. Frankel, Volatile lipid oxidation products, Prog. Lipid Res. 22: 1 (1982).

25. H. Kaunitz, Biological effect of autoxidized lipids, in: "Lipids and their oxidation", H.W. Schultz, E.A. Day and R.O. Sinnhuber eds.,Avi, Westport, Connecticut (1962).

26. F.A. Kummerow, Toxicity of heated fats, in: "Lipids and their oxidation", H.W. Schultz, E.A. Day and R.O. Sinnhuber eds., Avi, Westport, Connecticut (1962).

27. T.A.B. Sanders, Nutritional significance of rancidity, in: "Rancidity in Foods", J.C. Allen and R.J. Hamilton eds., Applied Sci. Publ., London & New York (1983).

28. K. Kanazawa, E. Kanazawa and M. Natake, Uptake of secondary autoxidation products of linoleic acid, Lipids 20:412 (1985).

BY-PRODUCTS FORMED IN HEATED FATS

W.W. Nawar

University of Massachusetts
Department of Food Science and Nutrition
Amherst, MA   01003

Since the beginning of times man heated his food to modify its flavor
or texture, or to keep it from spoiling. With the exception of fresh
fruits and vegetables, most of the food we eat today has been subjected to
heat at least once, often several times, in the course of its preparation,
through for example cooking, baking, broiling, toasting, canning,
concentrating, pateurizing or frying. Depending on the composition of the
food and the conditions of the treatment, heat will produce various
chemical and physical changes in all of the food constituents. Not only
do the different nutrients undergo decomposition reactions, but they also
interact among themselves and form a large number of new compounds.
Indeed, some of these changes are the very goal of the treatment. Other
changes, however, are not so desirable, and may cause problems in both
quality and wholesomeness. In this article, the chemical effects of heat
on the lipids are discussed.

Lipids consist of mixtures of triacylglycerols, diacyl-, and
monoacylglycerols, phospholipids, fatty acids, and fatty acid esters,all
varying widely in their composition. The chemical and physical changes
that occur, when lipids are exposed to heat, depend on the composition of
the lipid and the conditions of the treatment. If oxygen is scarce,
thermolytic reactions will take place. In the presence of air, both
oxidative and nonoxidative reactions will occur simultaneously. In
addition, heat will accelerate the oxidative process and influence the
course of oxidation, both qualitatively and quantitatively. Due to the
obvious difficulties inherent in the study of complicated mixtures, the
bulk of our knowledge regarding thermal oxidation of natural fats has been
gained through investigating model systems of fatty acid esters and
glycerides.

OXIDATIVE REACTIONS

Heat accelerates lipid oxidation probably by enhancing both
hydroperoxide formation and hydroperoxide decomposition, and by destroying
antioxidants. The mechanisms of lipid autoxidation have been extensively
studied, and are discussed in detail elsewhere in the monograph. Only a
very brief discussion of the autoxidative process is given here.

It is generally established that the autoxidation of fatty acids
occurs largely via a free radical chain reaction mechanism. Since direct

281

reaction of unsaturated linkages with oxygen is thermodynamically
difficult, production of the first few radicals necessary to start the
propagation reaction must occur by some other means. It has been proposed
that the initiation step may take place by decomposition of preformed
hydroperoxides (via metal catalysis or heat), by exposure to light, by
direct reaction of metals with oxidizable substrates, or by mechanisms
where singlet oxygen is the active species involved.

Upon the formation of sufficient free radicals, the chain reaction is
propagated by the abstraction of hydrogen atoms at positions alpha to
double bonds, followed by oxygen attack at these locations, and resulting
in the production of peroxy radicals, ROO, which in turn abstract hydrogen
from alpha methylenic groups of other molecules, RH, to form hydroperoxides, ROOH, and yield R' groups which react with oxygen, and so on
(Fig. 1). Due to resonance stabilization of the R' species, the reaction
is usually accompanied by shifting in the position of double bonds
resulting in the formation of isomeric hydroperoxides often containing
conjugated diene groups.

The hydroperoxides, formed as the primary products in lipid
autoxidation, enter into numerous and complex breakdown and interaction
mechanisms and produce a myriad of compounds with significant variation in
molecular weight, flavor threshold, and biological significance. In the
presence of trace metals, scission of the O–O bond in the hydroperoxide
molecule occurs readily yielding the alkoxy radical intermediate

$$R-(CH_2)-\underset{\underset{\cdot}{O}}{C}-(CH_2)-COOH$$

which may decompose by carbon–carbon cleavage on either side of the alkoxy
group to form a variety of aldehydes, hydrocarbons, semi-aldehydes and
acids depending on the original site of oxygen attack. Thus, 2,4–
decadienal, methyl octanoate, methyl 9–oxononanoate, and 3–nonenal arise
from the 9–hydroperoxide of methyl linoleate, while hexanal is the major
aldehyde produced by decomposition of the 13–hydroperoxide isomer.

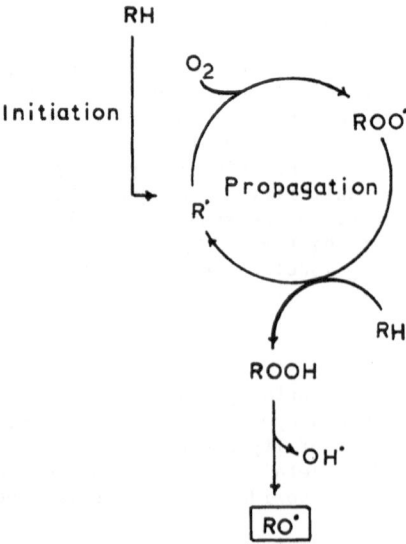

Fig. 1. Free radical autoxidation mechanism

At elevated temperatures the oxidation of unsaturated fatty acids proceeds very rapidly. Although certain specific differences between high and low temperature oxidations have been observed by some investigators, the evidence accumulated to date indicates that in both cases the principal reaction pathways are basically the same. The formation and decomposition of hydroperoxide intermediates, predictable according to location of double bonds (Frankel, et al., 1977a,b,c), appear to occur over a wide temperature range. A large number of decomposition products has been isolated by many investigators from heated fats (Chang et al., 1978; Selke, et al., 1977; Swoboda, et al., 1965). However, the major compounds are typical of those ordinarily produced from room temperature autoxidation.

At higher temperatures the oxidation products themselves readily undergo further oxidation and give rise to more decomposition products. For example aldehydes, the major oxidative products of fat oxidation, may further oxidize to their corresponding fatty acids. In addition, they may undergo autoxidation via the free radical chain reaction mechanism outlined above, giving rise to shorter-chain aldehydes, dialdehydes, hydrocarbons, etc., or epoxidation and cleavage as shown in Fig. 2 for 2,4-decadienal (Matthews, et al., 1971), or condensation to form trioxanes and dioxolanes (Horvat et al., 1966).

In addition to the classic hydroperoxides mentioned above, a variety of cyclic and hydroperoxy cyclic peroxides are commonly formed in the oxidation of polyunsaturated fatty acids. Decomposition of such intermediates gives rise to specific volatile compounds, for example, 3,5-octadiene-2-one from the 12-hydroperoxide of linolenate

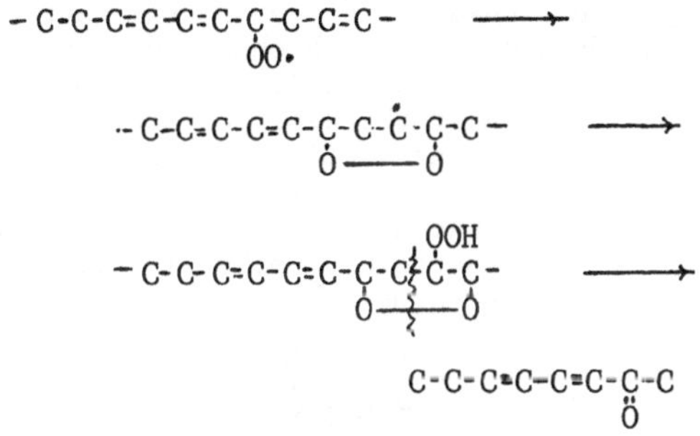

$$- C-C-C=C-C=C-C-C-C=C - \quad \longrightarrow$$
$$\overset{\displaystyle OO\cdot}{}$$

$$-- C-C=C-C=C-C-C-\overset{\displaystyle \cdot}{C}-C-C - \quad \longrightarrow$$
$$\overset{\displaystyle O \longrightarrow O}{}$$

$$\overset{\displaystyle OOH}{- C-C-C=C-C=C-C-C \overset{\displaystyle |}{\underset{\displaystyle \wr}{C}}-C-C - } \quad \longrightarrow$$
$$\overset{\displaystyle O \overset{\displaystyle |}{\wr} O}{}$$

$$C-C-C=C-C=C-C-C$$
$$\overset{\displaystyle O}{\overset{\displaystyle \|}{}}$$

or malonaldehyde from the 13-hydroperoxide isomer

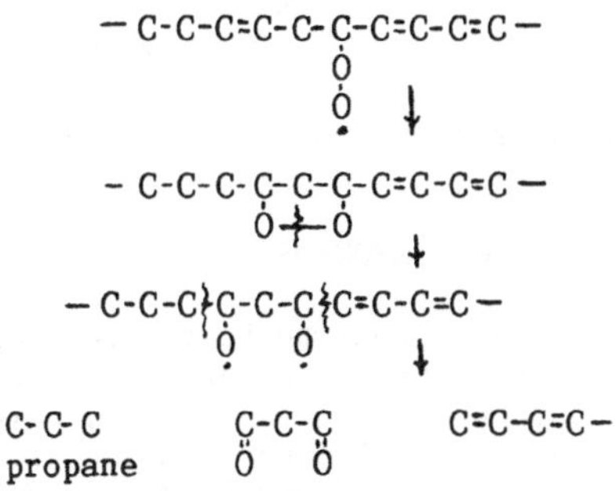

$$- C-C-C=C-C-C-C=C-C=C -$$
$$\overset{\displaystyle O}{}$$
$$\overset{\displaystyle O}{} \quad \downarrow$$

$$- C-C-C-C-C-C-C=C-C=C -$$
$$\overset{\displaystyle O \overset{}{\dashv} O}{} \quad \downarrow$$

$$- C-C-C \overset{|}{} C-C-C \overset{|}{} C=C-C=C -$$
$$\overset{\displaystyle O \quad O}{} \quad \downarrow$$

$$\begin{array}{ccc} C-C-C & C-C-C & C=C-C=C- \\ propane & \overset{\displaystyle \|}{O} \quad \overset{\displaystyle \|}{O} & \end{array}$$

malonaldehyde

Figure 2.

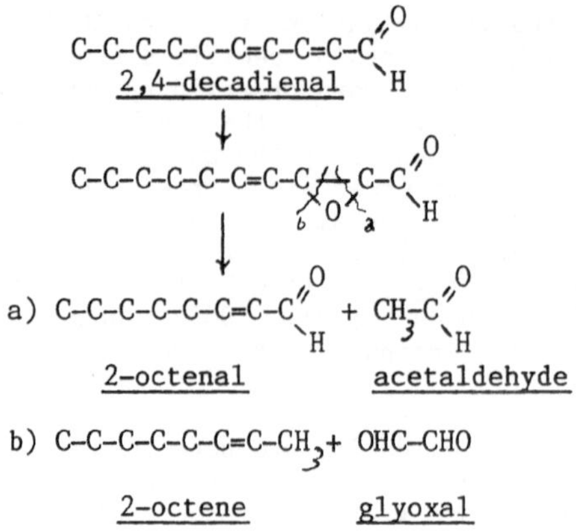

$$C-C-C-C-C=C-C=C-C\overset{\nearrow O}{\underset{\diagdown H}{}}$$
<u>2,4-decadienal</u>

$\downarrow$

$$C-C-C-C-C=C-C\overset{\diagup}{\underset{\overset{|}{O}}{C}}C-C\overset{\nearrow O}{\underset{\diagdown H}{}}$$

$\downarrow$

a) $C-C-C-C-C-C=C-C\overset{\nearrow O}{\underset{\diagdown H}{}}$ + $CH_3-C\overset{\nearrow O}{\underset{\diagdown H}{}}$

    <u>2-octenal</u>   <u>acetaldehyde</u>

b) $C-C-C-C-C-C=C-CH_3$ + OHC–CHO

    <u>2-octene</u>   <u>glyoxal</u>

Figure 2.   Further oxidation of 2,4-decadienal.

Saturated fatty acids and their esters are considerably more stable than their unsaturated analogs. However, when heated in air at temperatures higher than 150° C, they undergo oxidation, giving rise to a complex decomposition pattern. The major oxidative products consist of a homologous series of carboxylic acids, 2-alkanones, n-alkanals, lactones, n-alkanes, and 1-alkenes. It is generally accepted that thermal oxidation of saturated fatty acids involves the formation of monohydroperoxides as a principle mechanism, and that oxygen attack can occur at all the methylene groups of the fatty acid. However, since the dominant oxidative products of saturated fatty acids are those with chain lengths near or equal to those of the parent fatty acids, it is likely that oxidation occurs preferentially at positions near the ester carbonyl group. Oxidative attack at the β-carbon of the fatty acid, for example, results in the formaton of β-keto acids which in turn yield C_{n-1} methyl ketones upon decarboxylation. Cleavage between the α- and β-carbons of the alkoxy radical intermediate gives rise to C_{n-2} alkanal while scission between the β and γ-carbons produces C_{n-3} hydrocarbons

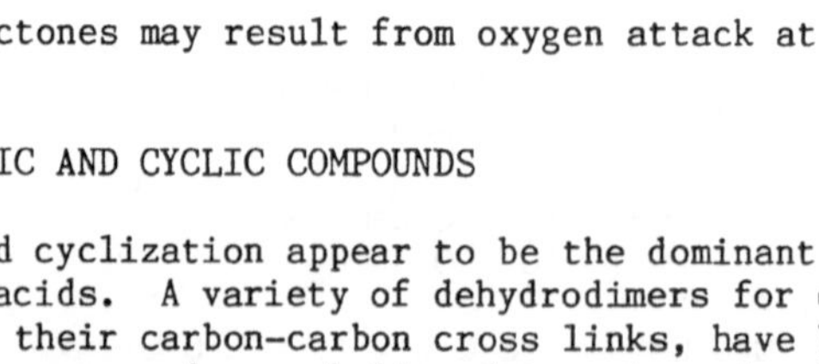

$$\begin{array}{c}
\underset{R_2}{\underset{\overset{|}{C}}{\overset{\overset{|}{C-R_3}}{}}} \quad C-O-\overset{\overset{O}{||}}{C} \big\} C \big\} \overset{\overset{\bullet}{\overset{O}{||}}}{C} \big\} C-C-R
\end{array}$$

             n–3 alkane

             n–2 alkanal

             n–1 Me ketone

Gamma and delta lactones may result from oxygen attack at carbons 4 and 5, respectively.

FORMATION OF DIMERIC AND CYCLIC COMPOUNDS

Dimerization and cyclization appear to be the dominant reactions of unsaturated fatty acids. A variety of dehydrodimers for example, varying in the location of their carbon-carbon cross links, have been reported in heated fats. These compounds are believed to arise via the combination of allyl or pentadienyl radicals (e.g., Types I and II, Fig. 3), saturated dimers with cyclopentane structures (e.g., Type III, Fig. 3), and polycyclic compounds resulting from intramolecular addition to C-C double bonds (Types IV and V, Fig. 3).

284

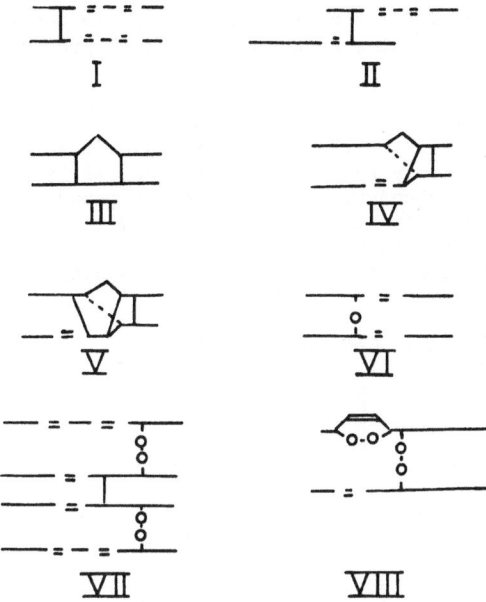

Fig. 3. Dimeric and polymeric products

In the presence of a plentiful supply of oxygen, combinations between alkyl, alkoxy, and peroxy free radicals may result in a variety of dimeric and polymeric acids and glycerides with carbon–oxygen–carbon or carbon–oxygen–oxygen–carbon crosslinks (Type VI and VII, Fig. 3), as well as oxygen-containing dimers, trimers, or cyclic monomers (e.g., VIII, Fig. 3).

Dimerization and polymerization of unsaturated fatty acids can also occur via Diels Alder type reactions. Thus, the conjugated diene from linoleate may react with oleate (or one of the unsaturated linkages in a polyene) to produce a tetra–substituted cyclohexene

If sufficient double bonds are available as in dimers of polyunsaturated fatty acids, a further reaction may take place producing a trimer.

Dimer

Trimer

In the case of glycerides, dimerization can take place between acyl groups in two triacylglycerol molecules or between two acyl groups in the same molecule.

Heat treatment also leads to the formation of various aromatic and other cyclic monomers, as demonstrated with methyl linoleate by Michael (1966a,b), and with partially hydrogenated soybean oil by Artman and Alexander (1968). The mechanisms by which such compounds are formed is believed to involve abstraction of hydrogen at allylic positions and intramolecular free radical to double bond addition. Cyclization occurs more readily in case of the longer chain polyunsaturated acids as shown for arachidonic acid

$$C-(C)_4-C=C-C-C=C-C-C=C-C-C=C-(C)_3-COOH$$

Thermal dimers and cyclic monomers of the types discussed above have been isolated from a variety of natural fats and frying oils (Paulose and Chang, 1973; Perkins, 1976). Dimers and polymers are responsible for the increase in viscosity of the frying oil. They contribute to flavor, texture and appearance of fried foods.

NONOXIDATIVE THERMOLYTIC REACTIONS

The major reactions which unsaturated fatty acids undergo if heated in the absence of oxygen are those of dimerization and cyclization as indicated above. To a much lesser extent, however, they may decompose forming products of lower molecular weight. Thus, no significant decomposition of methyl oleate could be detected at temperatures below 220° C, but when samples were heated at 280°C for 65 hr under argon, hydrocarbons, short- and long-chain fatty acid esters, and straight-chain dicarboxylic dimethyl esters, in addition to the dimers, were produced (Sen Gupta, 1966, 1967).

In general, very high temperatures are required to produce substantial nonoxidative decomposition of saturated fatty acids. However, using very sensitive measurement techniques, thermolytic products can be detected in simple triacylglycerols after they are heated in vacuum for only 1 hr at 180°C. These products consist mainly of series of normal alkanes and 1-alkenes, fatty acids, symmetric ketones, oxopropyl esters, propene- and propane diesters, and diacylglycerols. Acrolein, CO, and CO_2 are also formed (Crnjar, et al., 1981).

THERMAL INTERACTION WITH OTHER FOOD COMPONENTS

As in all biological systems the lipid molecules in food are often closely associated with neighboring nonlipid material such as proteins, carbohydrates, water, enzymes, salts, vitamins, and pro- and antioxidants.

286

When food is subjected to heat in the presence of air, reactions of the lipids, or their decomposition products, with such compounds will influence the oxidative reaction pathways and thus the final outcome of the treatment. For example, the basic groups in proteins may catalyze aldol condensation of carbonyls produced from lipid oxidation, resulting in the formation of brown pigments. Lipid hydroperoxides may induce oxidative changes in sulfur-containing proteins causing significant nutritional losses, and secondary oxidation products from lipids can initiate free radical reactions in proteins or form Schiff-base addition products with the epsilon amino groups of lysine (Fig. 4). A number of volatile compounds are known to result from thermal interaction of amino acids with fatty acid esters or triaclyglycerols. These include amides, acid nitriles, alkyl pyridines, and pyrroles (Lien and Nawar, 1974; Breitbart and Nawar, 1981; Henderson and Nawar, 1981). Similar compounds have been identified from such foods as fried chicken and broiled beef.

BIOLOGICAL EFFECTS

The possibility that consumption of heated and/or oxidized fats may produce adverse effects has been a major concern, and has stimulated extensive research. It is frustrating to attempt to draw decisive conclusions from the available literature on this subject since experimental conditions varied widely, and the interplay of the many pertinent variables has not always been considered. At times, the oils were heated continuously at excessive temperatures and without contact with food. Often, pure chemicals typical of those produced by thermal oxidation, or fractions of heated oils were fed to animals at unreasonably high levels in their diets. This has resulted in confusing and contradictory results inapplicable to normal frying situations.

In certain studies lipid peroxides were found to be toxic to rats in doses of 16 Meq O /Kg body weight. In other experiments, rats tolerated daily doses of 75 mg of concentrated peroxide for 6 weeks. Hydroxy-fatty acids (e.g. methyl ricinoleate) when fed to rats at 5% of diet for 4 weeks caused growth depression, reduced protein efficiency, elevated respiratory quotient, increased plasma triacylglycerols and accumulation of ricinoleate in depot fat. In contrast, other studies have shown that dihydroxystearic acid is poorly absorbed and not deposited in the tissues.

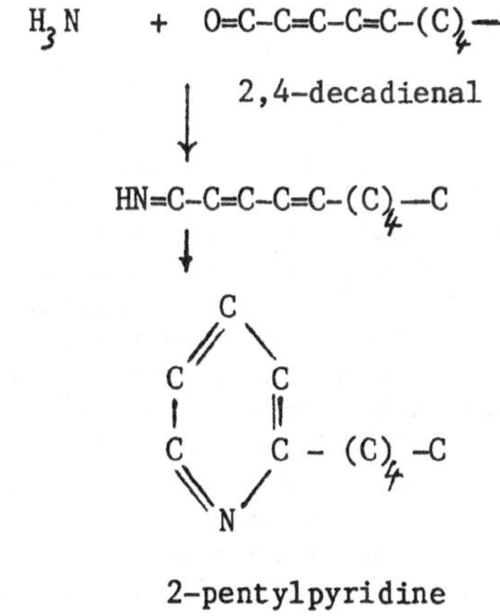

$$H_3N \quad + \quad O=C-C=C-C=C-(C)_4-C$$

2,4-decadienal

$$HN=C-C=C-C=C-(C)_4-C$$

2-pentylpyridine

Fig. 4. Reaction of aldehydes with primary amines

287

Aromatic compounds are known to cause acute toxicity when administered to rats in large doses. When material containing 60% aromatics, prepared by cyclization of fish oil and subsequent fractionation, was fed to rats at a level of 15%, the rats survived and adapted during an 18-week study. The aromatic compounds were excreted in the urine. When the same material was fed at a 9-30% level, these animals died within 5 weeks.

A number of studies were concerned with dimeric or polymeric compounds. Heated fats were fractionated and the non-urea-adductable fractions were reported to be toxic. These however were crude fractions which surely contained mixtures of many other compounds. In some studies, polar dimers were reported to be toxic, while the non-polar dimers were not. A prevalent view is that polymeric material of relatively high molecular weight is not absorbed in the body and thus is not toxic.

Of all products of thermal oxidation, the cyclic monomers pose the most concern in case frying oils are heated excessively. Iwaoka and Perkins (1978) reported that incorporation of cyclic monomers into the diet of rats caused low weight gains and low feed consumption in animals fed low levels of protein. A level of only 0.15% of cyclic fatty acids in diets containing 8% protein, produced fatty livers. In addition, decreased rates of lipogenesis have been observed in the liver of rats fed 8% protein and higher levels of cyclic fatty acids. An increased rate of lipogenesis occurred in the adipose tissue of these animals.

In most cases, feeding of heated fats that have become severely oxidized produces various detrimental effects in animals. However, when the experimental conditions approach those of normal frying, much milder symptoms are observed. Decreased feed intake, lower fat absorbability, decreased growth, and liver enlargement are among the various adverse effects reported in several animal feeding studies involving highly abused frying oils. In contrast, other investigators have reported that used frying fats produce no ill-effects.

In a' study by Billek (1979), sunflower oil used for industrial production of fish-fingers was taken at the end of a production period when the oil was usually discarded, and fractionated into a polar fraction (which presumably contained most of the oxidation and decomposition products), and a non-polar fraction (which consisted mostly of unaltered acylglycerols). The animal group fed the polar fraction exhibited growth retardation. The greatest increase in weight was observed with the unheated sunflower oil control while the group fed the non-polar fraction, or the complete used oil, showed somewhat less gain in weight. A rather extensive analysis using various techniques of clinical chemistry and hematology was also conducted. In most cases, no significant deviation from the normal values or difference among the four types of diets could be observed. On the other hand, values for serum glutamic pyruvic transaminase were significantly higher for the group fed the polar fraction, indicating a certain amount of liver damage. The polar fraction also produced an increase in serum glutamic oxaloacetic transaminase. In addition, the average weights of the livers and kidneys were significantly greater in animals receiving the polar fraction than they were in animals receiving other diets. Histological investigation of various organs showed no major irregularities. While the used oil at the time of discarding contained 30% decomposition products, the polar fraction contained 90%. Furthermore, the average daily intake of this fraction was 10g/Kg body weight while humans usually consume only about 0.1g frying oil/Kg of body weight. This author concluded that, in commercial practice, fring oils reach the point of deterioration and are usually discarded, long before they start to be toxic.

A number of investigators have reported tumors in animals fed severely heated fats. However, the temperatures used in most of these studies were 150–220°C above the temperature of normal frying. Various other studies have failed to show evidence of carcinogenic effects from normally-used frying fats. In a study of the mutagenicity of used frying fat, negative results were obtained when various fractions were subjected to the Salmonella/microsome mutagenicity test (Ames test). In a more recent study it was found that severely abusive frying conditions were necessary to produce appreciable levels of mutagenic activity in french fried potatoes or fish fillets (Taylor, et al., 1983).

Thermal oxidation of the lipids in food may have an effect on the nutritive value of its proteins. For example, the aldehydes produced from the lipid oxidation may react with the epsilon amino groups of lysine (Fig. 4) and thus compromise the availability of this essential amino acid.

Although it is evident that toxic compounds can be generated in fat by severe heating and/or oxidation, it appears reasonable to conclude that no significant hazard to health is to be expected from moderate ingestion of fried foods, provided high quality oils are used and recommended practices are followed.

ACKNOWLEDGMENT

The author wishes to thank Regina Whiteman and Carol Breton for assisting in the preparation of this manuscript.

REFERENCES

Artman, N., and Alexander, J.C., 1968, Characterization of some heated fat components, J. Amer. Chem. Soc., 45:643.

Billek, G., 1979, Heated oils - chemistry and nutritional aspects. Nutr. Metab., 24 (Suppl. 1): 200.

Breitbart, D., and Nawar, W.W., 1981, Thermal interaction of lysine and triglycerides. J. Agr. Food Chem., 29: 1194.

Chang, S.S., Peterson, R.J., and Ho, C.T., 1978, Chemical reactions involved in the deep-fat frying of foods, J. Am. Oil Chem. Soc., 55:718.

Crnjar, E.D., Witchwoot, A., and Nawar, W.W., 1981 Thermal oxidation of a series of saturated triacylglycerols, J. Agr. Food Chem. 29:39.

Frankel, E.N., Neff, W.E., Rohwedder, W.K., Khambay, B.P.S., Garwood, R.F., and Weedon, B.C.L., 1977a. Analysis of autoxidized fats by gas chromatography-mass spectrometry; III. Methyl linolenate, Lipids, 12:1055.

Frankel, E.N., Neff, W.E., Rohwedder, W.K., Khambay, B.P.S., Garwood, R.F., and Weedon, B.C.L., 1977b, Analysis of autoxidized fats by gas chromatography-mass spectrometry; I. Methyl oleate, Lipids, 12:901.

Frankel, E.N., Neff, W.E., Rohwedder, W.K., Khambay, B.P.S., Garwood, R.F., and Weedon, B.C.L., 1977c. Analysis of autoxidized fats by as chromatography-mass spectrometry; II. Methyl linoleate, Lipids, 12:908.

Henderson, S.K., and Nawar, W.W., 1981, Thermal interaction of linoleic acid and its esters with valine, J. Amer. Oil Chem. Soc., 58:632.

Horvat, R., McFadden, W., Ng, H., and Shepherd, A., 1966, Identification of 2,4,6-trialkyl-1,3,5-trioxanes from autoxidized methyl linoleate by mass spectrometry, J. Amer. Oil Chem. Soc., 43:350.

Iwaoka, W.T., and Perkins, E.G., 1978, Metabolism and lipogenic effects of the cyclic monomers of methyl linolenate in the rat, J. Amer. Oil Chem. Soc., 55:734.

Lien, Y.C., and Nawar, W.W., 1974, Thermal interaction of amino acids and triglycerides, valine and tricaproin, J. Food Sci., 39:917.

Matthews, R.F., Scanlon, R.A., and Libbey, L.M., 1971, Autoxidation products of 2,4-decadienal. J. Amer. Oil Chem. Soc. 48:745.

Michael, W., 1966a, Thermal reactions of methyl linoleate. II. The structure of aromatic C18 methyl esters, Lipids, 1:359.

Michael, W., 1966b, Thermal reactions of methyl linoleate. III. Characterization of C18 cyclic esters, Lipids, 1:365.

Perkins, E., 1976, Chemical, nutritional and metabolic studies of heated fats, Rev. Fr. Corps Gras, 23:257.

Paulose, M.M., and Chang, S.S., 1973, Chemical reactions involved in deep fat frying of foods: VI. Characterization of nonvolatile decomposition products of trilinolenin, J. Amer. Oil Chem. Soc., 50:147.

Selke, E., Rohwedder, W.K. and Dutton, H.J., 1977, Volatile components from triolein heated in air, J. Am. Oil Chem. Soc., 54:62.

Sen Gupta, A.K., 1966, Radical reactions on the thermal treatment of oleic methyl ester under exclusion of oxygen, Fette, Seifen, Anstrichmittel, 68:475.

Sen Gupta, A.K., 1967, Investigations on the structure of dimeric fatty acids, part 1: structure of dimeric methyl oleates, Fette, Seifen, Anstrichmittel, 69:907.

Swoboda, P.A.T. and Lea, C.H., 1965, The flavor volatiles of fats and fat-containing foods. II. A gas chromatographic investigation of volatile autoxidation products from sunflower oil, J. Sci. Fd. Agric. 16:680.

Taylor, S.L., Berg, C.M., Shoptaugh, N.H., and Traisman, E., 1983, Mutagen formation in deep-fat fried foods as a function of frying conditions, J. Amer. Oil Chem. Soc., 60:576.

RESIDUES AND CONTAMINANTS IN EDIBLE

FATS AND OILS

André Prevot

Institut des Corps Gras
Rue Monge - Parc industriel
33600 Pessac, France

INTRODUCTION

A contaminant is a substance which occurs especially without being
expected. The idea of toxicity or the idea of trouble for the consumer
is very often linked as far as this notion is concerned but, as it will
be seen, it is not always the case : some undesirable substances are
not at all toxic. We are particularly interested in contaminants which
could, if present at a high level, constitute a hazard to human health.

It is well known that contaminants occur naturally or come from
farming, transportation, storage and processing. Sometimes, a contaminant
may have various origins. That is the reason why a choice of classification is adopted, based on each type of contaminant, the main source of
contamination, the detection limit and, as often as possible,
bibliographical references That is to say which contaminant, where it
comes from, why it occurs, how it can be detected, how much of it is
found.

MINERAL OIL (Table I)

A seed contamination with mineral oil found in 1969 in Hungary was
due to jute sacks being used for shipping (1). Transportation in dirty
tankers causes contamination of seeds and crude oils. This is probably
more frequent than one would expect and can go undetected because of
difficulty in detection of small amounts. The IUPAC Working Group 3/79
dealing with the "Determination of mineral oil residues in oils and fats"
had a meeting in August 1983 (2). HENDRISKE commented on the results
obtained from a ring test organized by NNI. Two methods have been used :
one combining TLC and GLC, a second introducing first an isolation by
column chromatography. The first method has given a better recovery but
the second was more simple. In 1984, P. HENDRISKE reported on his study
of the AOCS procedure in which a light petroleum solution of the oil is
eluted through an alumina column. There was some difficulty with marine
diesel and petroleum owing to the presence of volatiles; however in a
refined oil this should not be a problem -they would be removed during
deodorisation of the oil. In 1985, P. HENDRISKE gave a report on recent
work undertaken in the Netherlands involving the analysis of soya and
rape seed oil "spiked" with marine diesel oil and palm oil "spiked" with
paraffin oil. Results from only two had been received. A. KARLESKIND

Table I

WHICH CONTAMINANT	WHERE WHEN	REASON WHY COMING FROM HOW MUCH	Method of detection / Detection limit	Reference
MINERAL OIL *	1969 Transportation and storage of rice in Hungary	Originates from jute sacks used for shipping	} unsaponifiable matter + TLC 100 ppm	CIELESKY (1)
	During transportation of seeds or crude oil	Dirty tanks	}	IUPAC. WG 3/79 (2)
	Processing	Heating fluid	} if GLC is used in addition : 10 ppm	Laboratory
TRICRESYL PHOSPHATE *	1930's USA 50.000 persons			
	1957 DURBAN		extraction with Acetonitrile + GLC with Phosphorus detector	SMITH et al (3)
	1959 MOROCO outbreak of paralysis 2000 cases	Lubricating oil sold as olive oil		
	1962 BOMBAY			
	1977-1978 SRI-LANKA 20 young female cases of neuropathy	Sesame oil. 5600 ppm	0.1 ppm	SENANAYAKE (4)

emphasised the importance of having a method which was sensitive to
10 ppm to detect contamination which was now occuring in a large number
of oil shipments. P. HENDRISKE stated that he was not yet in a position
to propose a method for study by the Commission ; low recoveries of
mineral oil added to palm oil had been obtained during the tests
conducted in the Netherlands.

TRICRESYLPHOSPHATE (Table I)

Half a dozen catastrophes have been identified as being due to
tricresylphosphate (TCP). It is now a well known fact that TCP produces
neurotoxic effects. For example in the 1930's TCP crippled 50.000 persons
in the USA. Since then several other epidemics have occured elsewhere,
the outbreaks in Morocco in 1959 (3), Bombay and Durban being some of the
better known. In most intances poisoning occured after consumption of
edible oils accidentally contaminated or voluntarily adulterated with
mineral oils containing TCP. Because TCP is a widely used industrial
chemical in lubricating oils it has always been a likely contaminant of
food.

In an epidemic in Sri Lanka in 1977-78, only young girls attaining
menarche or women soon after childbirth were affected (4). This unusual
feature was the result of local customs and traditions and provided the clue
to the diagnosis. Contamination probably occured during transport of the oil
in containers previously used for storing mineral oils.

Results concerning the removal of TCP in refining process have not been
found but there is little chance of a possibility of elimination by refining.
Fortunately such a contaminant is rare. Using the specific Nitrogen Phospho-
rus detector the detection limit should be as low as 0.1 ppm. The IUPAC has
not planned to standardize a method.

POLYCHLORO BIPHENYLS (PCBs) (5-12) (Table II)

A mass rice bran oil contamination was reported in western Japan in 1968
affecting more than one thousand victims. These were cases of direct consumption of high doses of polychlorinated biphenyls (2000 to 3000 ppm of PCBs).
The use of these substances was banned in Taïwan in 1972 but did not prevent
a recurrence: in March 1979, patients suffering from chloracne began appearing
on the west coast of Taïwan. By October the disease had reached epidemic
proportions, affecting over 1.100 people. The exposure factor common to these
patients was the consumption of rice oil produced by the same rice oil processing company situated in Changhua. Investigations of the manufacturing process
identified a leak in a heat exchanger, which resulted in oil contamination
with Kanecholr 400, a 48 per cent chlorinated biphenyl. Fortunately, such cases
are rare but traces of PCBs can be found in eddible fats and oils because of
the contamination of different ecosystems.
In most cases the clean up of the fat extract is made by liquid chromatography and the amounts are determined by gas-chromatography with an electron
capture detector or by GM-MS. In France, the amounts of PCBs in tallows are
very low : between 11 and 97 ppb (see next table).

SPANISH TOXIC OIL (13-19) (Table II)

Up to now, the toxic contaminant has not been identified. The chemicals
formed by heat removal of aniline used as denaturant as required by spanish
customs, the fatty anilides have been suspected but they have not been identified as being responsible for such accidents. Many experiments on many
animals such as rats, monkeys, rabbits, pigs have been done. The real cause
remains unknown. This is the main conclusion given by an expert group of the
WHO Danish, German, Italian... In France, Dr PASCAL of INRA is preparing a

Table II

WHICH CONTAMINANT	WHERE WHEN	REASON WHY COMING FROM HOW MUCH	Method of detection / Detection limit	Reference
POLYCHLORO BIPHENYLS (PCBs)	1968 FUKUOKA (Western JAPAN) rice bran oil 1000 people poisoned	2000 to 3000 ppm quaterphenyls	GC-MS	KAMPS et al (10) INAGAMI et al (5)
	1977 JAPAN cooking rice bran oil	1,1 % PCB an intake of 20 mg/patient		MORITA et al (9)
		PCB found in the blood of poisoned patients		KUWABARA (8)
	1979 TAICHUNG CHANGHUA (West coast of Taiwan) 1000 people rice bran oil	KANECHLOR 400 - PCBs and PCDF; 1 sample 405 ppm; 5 samples 54-99 ppm	GC-ECD and CC-MS	CHANG (11) CHEN (12)
UNKNOWN IN SPANISH TOXIC OIL	1981 SPAIN 20.000 cases 336 fatal cases	aniline (denaturant) + free fatty acids fatty anilides but they do not give the toxicity; 1 sample 480 ppm; 1 sample 38 ppm	p.dimethylaminocinnamaldehyde 20 ng; Thermal decomposition + fluorescence 1 ppm; GCL+ECD · 5ppb; Identification HPLC+MS + IR	VIOQUE et al (15); STAHL et al (16); JEURING (17); WHEALS (18)

report. 70 main references are available on this subject. The references given in the table mainly concern the detection methods. Contamination probably occured during transport of the oil in dirty containers.

POLYAROMATIC HYDROCARBONS (PAH) (20-29)

The fact that PAH occurs in different processes has aroused the interest of many authors since 1962. The main problem is the quantitative extraction from oil. Determinations are done mainly by HPLC + Fluorescence or by capillary GC + MS (detection limit 0.1 ppb). Usually the PAH mean content in current vegtable oils and margarines is low (less than 2ug/Kg.) It is essentially due to the preliminary contamination of the crude oils. Animal fats (butter, lard) show practically no signs of contamination and this confirms the results obtained concerning the transfer in the nutritional chain. The idea put forward that PAH was formed from carotenoïdes of palm oil has not been confirmed. In fact, it is not borne out by experimental evidence. During domestic use and in catering establishments the oils undergo severe temperatures and it is a well known fact that this can lead to the production of cyclisation and oxydation products, particularly during prolonged deep fat frying. BORIES has shown that PAH disappears during deep fat frying. The table shown below shows this. We notice that the PAH content of commercial oils are higher than after deep fat frying.

PAH AMOUNTS

	min	mean value	max
Fresly refined oils (all kind, 10 samples)	0,12ppb	0,5 ppb	1,35 ppb
Deep fat frying oil (from restaurants, 21 samples)	N.D.	0,12 ppb	0,55 ppb

Recently KOLAROVIC and TRAITLER using the GERTZ'S method by complexing with caffeine in formic acid solution, found some samples with an unusually high PAH content. For example, in one sample of soya-bean oil and peanut oil 30 ppb and 100 ppb of benzo[a]pyrene were found respectively. In the other vegetable oils tested : grapeseed, rapeseed, sunflower and cocoa butter the amounts were between less than 2 and 5 ppb. It is not possible to say up to now if these higher amounts found are due to the new method used and the possibility of artefact or whether they reflect the truth. Nervertheless the daily doses due to fats and oils are low.

A very good review on determination of polycyclic aromatic hydrocarbons in fats and oils has been done by David FIRESTONE of the Food and Drug Administration in the Working Group 3/78 of IUPAC "Determination of polycyclic aromatic hydrocarbons" in 1979 with 32 references (22).

During the last meetings of this IUPAC Working Group (2) G. OSTERMAN made a survey of several methods published recently. In 1984, he stated that ring tests had been conducted in Germany on foodstuffs and referred to the work being carried out by BCR on dried kale. But the determination of PAH in oils and fats was much more difficult ; in 1985, G. OSTERMAN, in a written note, said that the elaboration of a standard method was extremely laborious. He referred to the work being done in conjunction with the Community Bureau of Reference (BCR) in which participants were using their own methods and comparing the results of provided samples. Members of the Commission willing to participate in the work were invited to contact G. OSTERMAN ; the problem would also be taken up with the Food Chemistry Commission.

AFLATOXINS (Fig.1)

The polarity of the aflatoxin molecule gives it a bad solubility in oil. Usually, the amounts are about five times less in crude peanut oils than in nuts. A special extraction method from oil using acetonitrile must be used for the determination. A variant of the C.B. procedure was adopted by the IUPAC Working Group. During the refining, the neutralization with sodium hydroxyde destroyes 90 to 98 per cent of the present aflatoxin because of the lactone bond breaking up. The rest is completely absorbed on bleaching earth, even for an unusually high content of alfatoxin (30). That is the reason why aflatoxins must not be considered as contaminant in refined oils. Nevertheless, they could be contaminants in crude peanut oils. This must be kept in mind for consumers for whom non refining is considered as a dietetic quality It is the same for the next contaminants which are going to be considered : pesticides. In the french 'Inventaire National de la Qualité Alimentaire" among 30 samples of maïze oil, only one sample has been found with a positive amount of zearalenone (detection limit 50 ppb).

Only a few scientific investigations deal with mycotoxins in olive products as compared with other raw oil materials. Aspergillus flavus and A. ochraceus strains have been isolated but toxigenesis has always been weak on olives. However, aflatoxin B1 or Ochratoxin A have been found in commercial samples of edible olives or olive oil (30 b).

PESTICIDES (31-43) (Table III)

300 active substances are registred in France, weed-killers 56 per cent, fongicides 34 per cent, insecticides, acaricides 10 per cent. Usually, the more efficient ones degrates quicker. They have a good solubility in fats and oils but the refining process eliminates them. It is the reason why the amounts into chocolate, butter and olive oil are relatively higher than they are in refined oil (Table III) Improvements have recently been done. For example for dimethoate and phosphamidon (registred for use against Dacus oleae), no detectable residue were found in olive oil, even when the olives were sampled one day after spraying. Another example concerns the decamethrine used at a very low level of 10 g per hectare (100 times less than usually). In milk and dairy products the amounts decrease regularly (44). The amounts in animal fats were low in France in 1983 (44). The IUPAC Working Group 4 is closed since 1984 (2). The amounts in eddible fats and oils are lower than they are in other food ; in addition, the daily intake due to fats and oils becomes negligeable because the consumption is far lower (7 Kg per inhabitant in France) compared with other food (300 Kg) (40).

METAL TRACES AND UNDESIRABLE ELEMENTS (Table IV)(45-52)

Natural oligo-elements such as iron, copper, sodium, at the levels found in crude oil are rather necessary for the health of human beings but they are not good for the health of the oil because they are prooxydant catalysts. Fortunately, the refining process eliminates most of them and the refined oils can easily be stored for a period of one year. The ratio of undesirable elements such as arsenic, mercury, lead and cadmium are lowered too. The table shows that the amounts are usually below the detection limit. Some metal traces may come from packing ; this is the case of olive oil in tin can and seed oils in plastic bottles stabilized with organotin. But tin is not considered as a toxic element and the amounts found in oils are far lower than those in canned products. Nickel, an artifact of hydrogenation, must be removed from the oil for health, stability and safety considerations.

More often determinations are done by Flameless Atomic Absorption Spectroscopy. An improvement has been recently developed in Revue Française des Corps Gras by HOCQUELET (52) using diethyldithiocarbonates and special

Table III : Traces of pesticides in France - mean value in ppb (in bracket : percentage of positive samples) [42]

PESTICIDE	detect. limit	CHOCOLATE	PEANUT OIL	OLIVE OIL	SUNFLOWER OIL	BUTTER	MARGARINE	LARD
SAMPLE NUMBER :		12	5	8	5	9	5	5
α HCH	5 ppb	17 ppt (100 %)	27 ppb (40 %)	15 ppb (80 %)	0 %	24 ppb (77 %)		10 ppb (20 %)
β HCH	5 ppb	5 ppt (8 %)		8 ppb (80 %)	0 %	24 ppb (77 %)		
γ HCH	5 ppb	25 ppt (100 %)	20 ppb (40 %)	11 ppb (100 %)	0 %	30 ppb (77 %)	5 ppb (60 %)	29 ppb (40 %)
HCH	5 ppb		8 ppb (20 %)	6 ppb (40 %)	0 %	17 ppb (55 %)		
DP6 DP5	5 ppb	131 ppb (50 %)				217 ppb (44 %)	166 ppb (100 %)	
DDT	5 ppb	8 ppt (33 %)		18 ppb (100 %)	0 %			
DDD	5 ppb			7 ppb (60 %)				
DDE	5 ppb			6 ppb (100 %)		17 ppb (44 %)		
Heptachlor Epoxyde	5 ppb	12 ppb (8 %)		10 ppb (20 %)		30 ppb (55 %)		

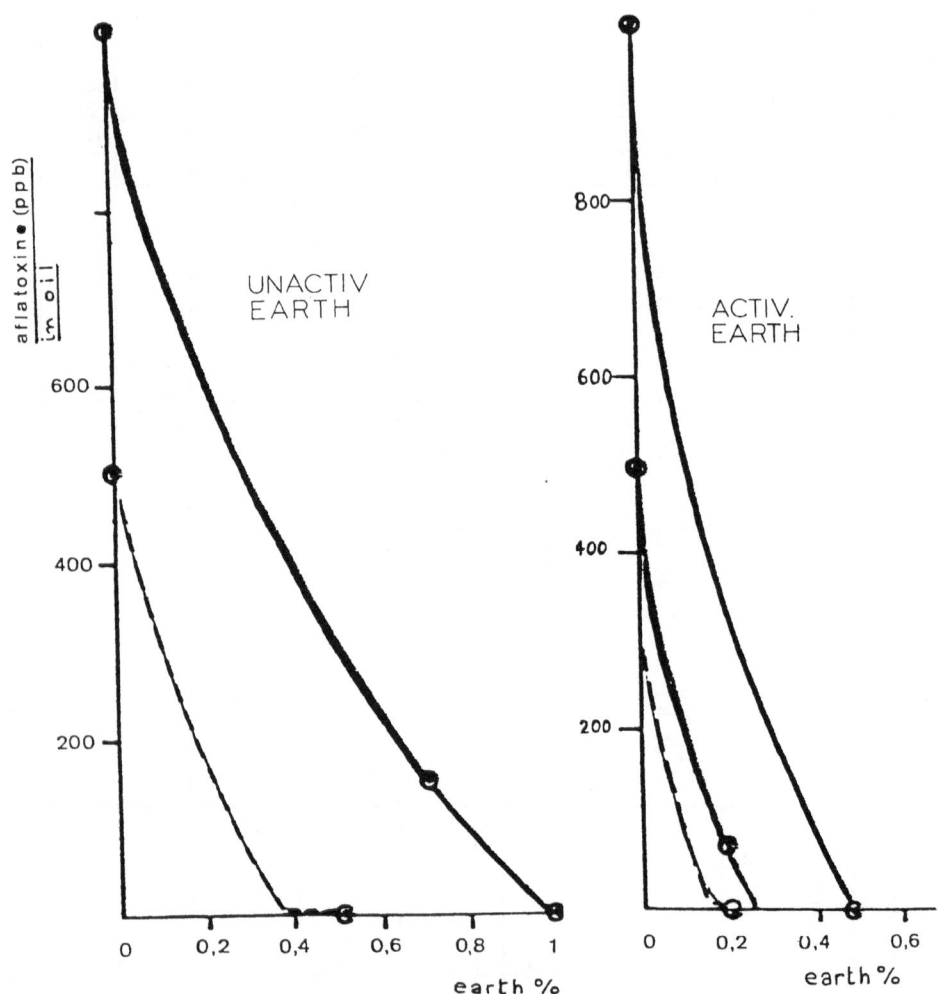

Figure 1.

TABLE IV : Heavy Metals (mean value)

METAL	detect. limit	SAMPLE	PEANUTS	CHOCOLATE	PEANUT OIL	OLIVE OIL	SUNFLOWER OIL	BUTTER	MARGARINE	LARD
		NUMBER	5	5	5	5	5	5	10	5
Hg	5 ppb	MEAN VALUE	56	5	<	<	<	<	10	<
		positive sample %	100 %	40 %	0	0	0	0	20 %	0
Cd	10 ppb	MEAN VALUE	72	<	<	<	<	15	20	<
		positive sample %	100 %	0	0	0	0	60 %	20 %	0
Pb	50 ppb	MEAN VALUE	<	103	<	0	0	101	<	0
		positive sample %	0	20 %	0	0	0	60 %	0	0

Source: Inventaire National de la Qualité Alimentaire 1983 ; Ministeries Environnement, Agriculture, Recherche, Industrie
Direction Prévention de la Production ; Mission de Controle des Produits – PARIS –

ashing temperature in flameless atomic absorption. He found these very
low amounts :

	Sn	As	Cd	Pb
rapeseed oil	10 ppb	3 ppb		5 ppb
peanut oil	7 ppb			
soyabean oil	1 ppb	3 ppb	0,5 ppb	20 ppb
sunflower oil		3 ppb	0,5 ppb	

At the 1985 IUPAC Meeting (2), P. HENDRISKE and J.P. WOLFF stated that
the procedure was not at the stage at which it could be standardised.

MIGRATION

I will be very short on this subject. As a matter of fact if at the end
of the 1960's many attacks were launched against monomeres of plastics such
as Vinyl Chloride Monomere (VCM), Acrylonitrile, Styrene. Such efforts were
made by plastic manufacturers so that this no longer presents a problem up
to now. For example, if we compare the VCM amounts found in 1970 and in 1977
we are now below the detection limit.

VCM AMOUNTS
(ppm)

Year of	Sample n°	Institut des Corps Gras (Paris)	Service fédéral de : l' hygiène publique : (Berne)
	1	105	118
	2	40	37
1974	3	40	23
	4	100	104
	5	110	144
	A	10	5
	B	10	5
	C	10	5
1977	D	10	5
	E	10	5
	F	10	5
	G	10	5
	H	10	5

At the 1983 IUPAC meeting (2), about the Working Group 13 "Determination
of undesirable plastic-based contaminants (other than polyethylene) in oils
and fats", it appeared that there were several reasons for abandoning the
investigation : I. lack of legal maximum level for monomers - II. weak value
of vinylchloride monomers in the polymer - III. practical absence of acrylonitrile in fatty materials packaging. This working group is now closed.
Nevertheless LEVIN said that a problem subsists for styrene monomers and one
must keep in mind the difficulties of determining a plastic monomer in fatty
materials.

CONTAMINATION OF AN OIL BY ANOTHER OIL OR FAT

The presence of another oil or fat in a pure oil or fat is not really a
contamination because the notion of toxicity is not present. It is more often
a problem of regulation eg. : tallow in lard, fish oil in palm oil, soyabean
oil in peanut oil, margarine in butter, seed oil in olive oil, cholesterol in
cholesterol-free margarine, olive foot in olive oil. Gas-liquid chromatography
of fatty

acid methyl esters or of sterols may give 1 per cent in the best of cases.
At the last IUPAC Meeting (Working Group 1/85) J.P. WOLFF and KAPOULAS
decided to organize a ring test based on HPLC of linoleic-triglycerides in
order to detect mainly cotton oil and sunflower oil in olive oil. The use of
other methods like argentation liquid chromatography is necessary in some
cases. The control of the absence or pork fat in tallow required in islamic
countries, needs the isolation of saturated glycerides after oxydation of
unsaturated glycerides, gel-silica column chromatography, hydrolysis of
saturated glycerides by means of the pancreatic lipase, isolation of mono-
glycerides by thin layer chromatography and glass capillary column gas liquid
chromatography of their fatty acids. This method enables the determination
of 2 per cent of pork fat in tallow.

CONCLUSION

This report is very extensive and each chapter could provide material for
a complete lecture in itself. The odds however are against this because we
have different causes, different methods, different amounts. What can we
conclude from all this ? The very serious accidents in Japan and Taïwan due
to tricresyl phosphate and polychlorobiphenyls must be kept in mind. That is
why severe controls are necessary just as for the other food products. The
main point being the cleanliness of the tanks : they must be checked very
conscientiously. If these unfortunate accidents are not taken into acount,
edible fats and oils are really contaminant free and very often, the amounts
are below the detection limit of the very sensitive methods available. If we
consider the relatively small amounts of fatty materials consumed, the daily
doses are quite negligible.

REFERENCES

Mineral oil
(1) Problems of transport and storage of rice in sacks impregnated with
     mineral oils. Y. CIELESZKY, K. SODS, J. ROMHANYL, Elelmez.Ipar 1969,
     23 (4), 106-10.
(2) IUPAC : minutes of the commissions meetings,1980-1983-1984-1985.
TCP
(3) Outbreak of Paralysis in Morocco due to orthocresyl phosphate poisoning,
     H.V. SMITH, J.M.K. SPALDING, Lancet, dec. 1959,1019-1021.
(4) Toxic polyneuropathy due to gingili oil contaminated with tri-cresyl
     phosphate affecting adolescent girls in Sri Lanka. N. SENANAYAKE,
     J. JEYARATMAN, Lancet jan. 1981, 88-89.
PCB
(5) Poisoning by chlorobiphenyls, K. INAGAMI, T.KOGA, Y.TOMITA, Shokuhin
     Eisagaku Zasshi 1969, 10(5), 312-17 (C.A. 72, 1970, 120116a).
(6) Toxic principle in a by product in rice oil rendering Toxic principle,
     K. KIYOMI, H. WATANABE, Y.MOCHIDA, S. HIRATSUKA, Nippon chikusan gakkai,
     Ho. 1970, 41(a) 439-44 (C.A. 76, 1972, 68956r).
(7) Simultaneous clean up of fish fat containing low levels of residues and
     separation of PCB from chlorinated pesticides by thin layer chromatogra-
     phy, M.L. HATTULA, Bull. Environn. Contam. Toxicol.12 (3) 1974, 331-37.
(8) Studies on polychlorinated biphenyls (PCBs) in the blood of patients
     poisoned by contaminated cooking oil. Part 3n K. KUWABARA, T. YAKUSHIJI,
     J. WATANABE, S. YOSHIDA, Y. YOSHIDA, H. MURATA, K. KOYAMA, N. KUNITA,
     Osaka -firutsu Koshu Eisei Kenkyusho Kenkyn Hokuku 1977, 8, 83-6 (C.A.
     89, 1978, 891262m).
(9) Detailed examination of polychlorinated dibenzofurans in PCB preparations
     and Kanemi Yusho Oil., M. MORITA, J. NAKAGAWA, K. AKIYAMA, S. MIMURA,
     N. ISONO, Bull. Environn. Contam. Toxicol. 18 (1) 1977, 67-73 (FTSA 11,
     1979, 01, 1 N 45).
(10) Polychlorinated quaterphenyls identified in Rice Oil Associated with
     Japanese "Yusho" Poisoning, L.R. KAMPS, W.J. TANNER, B. Mc MAHON, Bull.

Environm. Contam. Toxicol. 20, 1978, 589-91.

(11) PCB in rice oil, W.Y.B. CHANG, Nature 296, 1982, 192.

(12) Polychlorinated biphenyls and polychlorinated dibenzofurans in the toxic bran oil that caused PCB poisoning in Taïwan, P.H. CHEN, K.T. CHANG, Y.D. LU, Bull. Environm. Contam. Toxicol. 26,4, 1981(489-95(FSTA 15,1983 5N217).

(12bis)Chloro-biphenyls Poisonings -Gas-chromatographic Detection of Chloro-Biphenyls in the Rice Oil and Biological Materials-T. KOJIMA, H.FUKUMOTO, S. MADISUMI, Department of Legal Medecine, Kyushu University of School Medecine, Fukuoka- Jap. J. Leg. Med. 23 (5), 415-420, 1969.

Spanish Tox-Oil

(13) Toxic oil syndrome in Spain, W.N. ALDRIDGE, T.A. CONNORS, Fd Chem. Toxic 20, 1982, 989-92.

(14) Análysis conventional de las muestras disponible de aceites supuertamente toxicos, L.M. Ventura DIAZ et al, Grasas y Aceites 33 (2) 1982, 73-108

(15) Detection de anilina y anilidas grasas for cromatografia de capa fina en aceites desnaturalizadas con anilina, E. et A. VIQUE id 33(3) 1982,162

(15bis)Determination de Anilina y Aminas aromaticas totales en aceites causantes del "Sindrome Toxico",J.J. SANCHEZ SAEZ, P.CALVO ANTON, M.C. MIRAMAR BLAZQUEZ, Boletin de CeNAN, N° 6-7-8, 1981.

(16) Detection rapida y sensible de anilina y anilidas in "aceites comestibles", E. STAHL, A. VIOQUE, Grasas y Aceites 33 (4) 1982, 220-24.

(17) The gas-chromatographic determination of toxic residues in adulterated spanish olive oil, H.J. JEURING, W. WERWAAL, A. BRANDS, K. Van EDE, J. W. DORNSEIFFEN, Z. Lebensm. Unters. Forsch. 175 (1982) 85-87.

(18) Application of liquid chromatographic and spectroscopic methods for the characterisation of fatty anilides in contamined cooking oil, B.B. WHEALS, M.J. WHITEHOUSE, C.J. CURRY, J. Chromat. 238 (1982) 203-15.

(19) Rapport sommaire -groupe de travail sur le syndrome de toxicité de l'huile d'oeillette dénaturée W.H.O. ICP/RCE 905 (1)(S)15371, 21 Avril 1983.

PAH

(20) Collaborative study IUPAC of July 1974 on determination of PAH in sunflower oil, G. GRIMMER, IUPAC report, 1975.

(21) Schnellmethode zur Isolierung und Bestimmung von 3,4-Benzpyren in Lebensmitteln, C. GERTZ, Z. Lebensm. Unters. Forsch., 167, 1978, 233-37.

(22) Determination of Polycyclic Aromatic Hydrocarbons in fats and oils, D. FIRESTONE, Report of IUPAC work group 3/79 Appendix 23-79.A57, 1979.

(23) Analytical methodology and reported findings of Polycyclic Aromatic Hydrocarbons in foods, J.W. HOWARD, T. FAZIO, J. Ass. off. Anal. Chem., 63, (5); 1980, 1077-1104.

(24) Determination of Polycyclic Aromatic Hydrocarbons in fat products by High Pressure Liquid Chromatography, A. VAN HEDDEGHEM, A. HUYCHEBAERT, H. DEMOOR, Z. Lebensm. Unters. Forsch, 171, 1980, 9-13.

(25) Berbessertes Verfahren zur quantitativen Abtrennung von 3,4-Benzpyren in Lebensmitteln, C. GERTZ, ibid, 173, 1981, 208-12.

(26) Polyaromastische Kohleuwasserstoffe in rohen Olen und Fetten und irhe Entfernung durch Behandhung mit aktiviert kohle, Fette Seifen Anstrich. 63, 1981, 541-2.

(27) Incidences des façons culinaires sur la contamination des aliments par les hydrocarbures aromatiques polycycliques, G. BORIES, Cahiers de Diététique et Nutrition, 17, 1982, 9-16.

(28) Determination of Polycyclic Aromatic Hydrocarbons in vegetable oils by caffeine complexation and glass capillary gas chromatography, L. KOLAROVIC, H. TRAITLER, J. Chromat., 237, 1982, 263-72.

(29) Quantitative determination of Polycyclic Aromatic Hydrocarbons (PAH) in vegetable oils by GC/MS, H. TRAITLER, U. RICHLI, L. KOLAROVIC, Int. Mass. Spectrom. Ion Phys. 48, 1983, 331-4.

Aflatoxines

(30) Evolution et méthodes d'élimination des aflatoxines dans les produits oléagineux (huiles et tourteaux) A. PREVOT Rev. Franc. Corps Gras 21 1974, 35-47

(30bis) Possible mycotoxine contamination of olives and olive products :
Latest developments, A. TANTAOUI-ELARAKI, B. LE TUTOUR, Oléagineux
1985, 40, n° 8-9, p. 451-454.

Pesticides

(31) Index phytosatinaire, ACTA ed. PARIS 1981
(32 Nature des pesticides utilisés dans la culture des plantes oléagineuses,
dosage des pesticides ou de leurs métabolites dans les corps gras,
R. MESTRE, Rev. Franç. Corps Gras, 21, 1974, 145-153.
(33) Toxicité des résidus de pesticides dans les corps gras, G. WIEL, id.,
155-160.
(34) Evolution des pesticides au cours des opérations industrielles d'obtention et de raffinage des corps gras, J.P. WOLFF, id, 161-64.
(35) Rückstände von Pesticiden in fetthaltigen Lebensmitteln unter besonderer
Berücksichtigung der tierischen un pflanzlichen Fette un Ole, Von H.A.
MEEMKEN, Chemisches Landes-Untersurchungsamt Nordrhein-Westfalen, Münster
Fette Seifen Anstrisch., 1975, 19-25.
(36) Experimental data and critical review of the occurence of Hexachlorobenzene
in the Italian Environment, V.LEONI, S.U. D'ARCA, Sci. tot. Environm.,5,
1976, 253-72.
(37) Recherche quantitative des résidus d'hexachlorobenzène dans les matières
grasses concurremment à celles des résidus des autres pesticides et des
micropolluants organochlorés, R. MESTRES, FRANCOIS, S. ILLES, J. TOURTE
M. CAMPO, Trav. Soc. Franç. Pharm. Montpellier 56 (1) 1976, 43-58.
(38) Dosage des résidus de décaméthrine dans les produits végétaux, R.MESTRES,
Ch. CHEVALLIER,Cl. ESPINOZA, R. CORNET, id 38 (2) 1978, 183-92.
(39) Note sur l'analyse des résidus de décaméthrine, R. MESTRES, Cl. ESPINOZA
Ch. CHEVALLIER, G. MARTI, id, 34,(4), 1979, 329-36.
(40) Les résidus dans l'alimentation, R.MESTRES, M. HASCOET, compte-rendu,
Congrès sur la lutte contre les insectes en milieu tropical, Marseille,
13-16 Mars 1979, 839-58.
(41) Determination of organochlorine pesticides, IUPAC, Appendix 2-80, A-8.
(42) Inventaire National de la Pollution Alimentaire, Ministères Agriculture
Recherche Industrie, Direction, Prévention de la Pollution. Mission de
contrôle des Produits, Paris 1983.
(43) Organophosphorus Insecticides Residues in olives and olive oil, J.R.
FERREIRA, A.M. TAINHA, Pestic. Sci. 14 (1983) 167-72.
(44) Contamination des produits laitiers français par les résidus de composés
organochlorés, A.VENANT, L.RICHOU-BAC, Le Lait 61 (1981)619-33.

Metal traces

(45) Determination of sodium and potassium in oils and fats, A.PREVOT,Atomic
Absorption Newsletter, 5 (2) 1966, 13-16.
(46) Méthodes récentes de dosage des traces de métaux : incidence de ces
traces sur la stabilité des huiles, A. PREVOT, Rev. franç. Corps Gras,
18, (11), 1971, 655-68.
(47) Oxygen-Rich atmosphere for direct determination of copper in oils by
non-flame atomic absorption spectrometry, K. KUNDU, A. PREVOT, Anal.
Chem., 46, (1974) 1951-95.
(48) Comparison of three atomic absorption techniques for determining metals
in soyabean oil, L.T.BLACK, J. Amer. Oil Chem. Soc., 52 (1975) 88-91.
(49) Les progrès de l'absorption atomique et les corps gras, A. PREVOT, M.
GENTE-JAUNIAUX, O.MORIN, Rev. Franç. Corps Gras 24(8-9)(1977) 409-18.
(50) Données récentes sur la toxicité des contaminants métalliques apportés
à l'aliment conservé par son emballage, Cl. BOUDENE, Méd. et Nut., 15,
(1979), 425-30.
(51) Metal in soyabean oil,F.J. FLIDER, F.T. ORTHOEFER, J. Amer. Oil. Chem. Soc. 1981,
270-272.
(52) P. HOCQUELET, Rev. Franç. Corps Gras, Mars 1984, 117.

Migration

(53) Collaborative study on the methodology for specific migration studies,
W. KARCHER, G. HAESEN, B. LE GOFF, Commission of the European Communities,
(ed., 1983, EUR 8286 FN.

(54) Santé, Plastiques, Questions, Réponses : Syndicat Professionnel des Producteurs de matières plastique, ed. Sept. 1981, Paris.

(55) Influence of techniques on the quality of food products in the fat and oil industry, J.P. HELME, Rev. Franç. Corps Gras 27 (1980), 121-30.

Oil by another and diverse

(56) Detection of adulteration of olive oil by argentation thin Layer Chromatography, D.S. GALANOS, V.M. KAPOULAS, E.C. VOUDOURIS, J. Amer. Oil Chem. Soc., (1968) 825-29.

(57) Chromatographie en couche mince de l'insaponifiable de l'huile d'olive pour détecter sa falfification par les huiles de grignon, G. DIMOULAS, Rev. Franç. Corps Gras, 16 (11) 1969, 721-22.

(58) Etude de la fraction stérolique des margarines végétales, F. BLANCHARD, J. CASTANG, M. DERBESY, J. ESTIENNE, M. OLLE, M. SOLERE, Ann. Fals. Exp. Chim. 72, 1979, 25-37.

(59) Analyse des mélanges de graisses animales. Application au contrôle de l'absence de graisse de porc dans les suifs. L. RUGRAFF, A. KARLESKING, Rev. Franç. Corps Gras, 30, 1983, 323,31.

(60) La Pollution et les Produits de l'industrie des Corps Gras. J.P. HELME, M.Th. JUILLET, Rev. Franç. Corps Gras, Rev. fse, Diét. Vol.18, n° 70, 1974, pp 49-55.

(61) Contaminations et réactions secondaires dans la fabrication des corps gras, J.P. WOLFF, Nutr. Métab. 24 (Suppl. 1) : 187-99 (1980).

COMPLEX NATURAL PHOSPHOLIPID BLENDS

H. Traitler and A. Nikiforov

Nestlé Research Department, NESTEC LTD., CH-1800 Vevey
Switzerland
Institute of Organic Chemistry, Univ. Vienna, Währingerstr.38
1090 Vienna, Austria

SUMMARY

For the extensive use of complex natural phospholipid blends in nutritional industry and related research fields, the analytical monitoring of
this system is an essential requirement. The number of analytical methods
involved herein include the direct phospholipid class analysis by TLC, the
direct species characterisation within phospholipid classes by computer
averaged integrating (CAI) FAB-MS, CAI-FD-MS, high performance TLC and
H-NMR as well as methods of indirect phospholipid species characterisation
according to saturation and fatty acid (FA) chain lengths. The latter
methods are mainly based on enzymatic cleavage of phospholipids to the
corresponding diacylglycerols with subsequent characterisation by GC,
GC-MS, CAI-FD-MS, computer "compression" of the GC-MS data or on characterisation in terms of FA analysis after total hydrolysis.

The common feature of all these methods is the partial information on
complex, natural and other phospholipid blends covered by these analytical
procedures. The comparison and characterisation of such complex phospholipid systems achieved by using only one method must therefore always be
incomplete. For this reason, the multidimensional matrix-based descriptive
system for cross-correlation, storage and presentation of data from different analytical methods was developed and is now used in our laboratory.
The main features of this so-called "C,N.S-System" and typical examples of
application are presented further on.

RESULTS AND DISCUSSION

A phospholipid mixture may represent an extraordinary complex system.
Generally, most of the phospholipids can be described in terms of three
substituents R1, R2 and R3 of glycerolphosphatic-acid moiety when the
information on optical activity is omitted - as shown in Figure 1. With a
few exceptions such as lysocompounds or sphingomyelins, the residues R1
and R2 are fatty acid residues and the substitution with R3 determines the
phospholipid class. In a first approximation, this would mean that by
using a three axis descriptive system a mixture of phospholipids may be
defined. As the fatty acid residues may differ with respect to number and
position of double bonds, essentially more additional descriptive dimensions for definitive characterisation are theoretically necessary. When

the descriptive level is reduced only to number of double bonds with no respect to their position, a peusdo-three-dimensional descriptive system can be used – as all present information dimensions are not continuous (fatty acids with 16,5 carbon atoms or 3,7 double bonds do not exist). Two or more discontinuous dimensions characterising single R1 or R2 substituents may be incorporated sequentially only on one axis and the whole system can be displayed in a three-dimensional presentation. However, the most accessible information obtainable from analytical methods on phospholipid systems is separation in terms of strongly reduced characterising dimensions. In Fig. 1 some examples on this are shown. The species separation (R3) is mostly achieved by TLC and LC, while mass spectrometric methods such as GC-MS or FAB and FD include in total (R1+R2) – separation additional information in terms of double bond equivalents (DBA). This information may be to a reduced extent also be obtained from other analytical methods (TLC,GC). The GC analysis of hydrolysed sample delivers integrated distribution of fatty acids present.

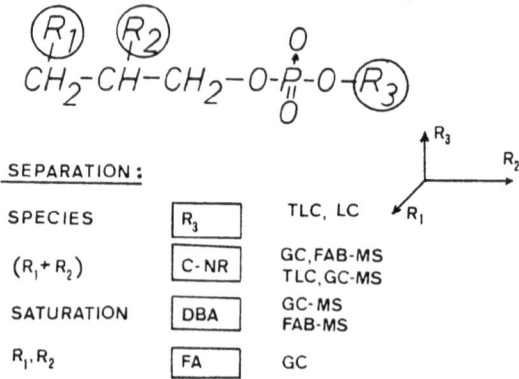

Figure 1 : Complementary phospholipid analyses

Fig. 2 shows an example of the so-called "FA-DATA BLOCK"-descriptive system with one species axis (R3 information) and two pseudo-monodimensional axis FA (corresponding to the R1 and R2 from previous figure). In perpendicular planes to the species axis, so-called species intensity areas, the complete species distribution is contained within one phospholipid class. In the figure, the FA axis are subdivided only according to the number of carbon atoms and the total number of double-bonds; however, any further subdivision may be added in. Every point of this Species/FA/FA matrix has an allocated intensity, which is displayed by a line as the pseudo forth dimension. Positional isomers of fatty acids are symmetrically displayed along the diagonal of every species intensity plane. The advantage of this type handling of phospholipid data is the relative simplicity of the matrix (complicated only by subsequent conversion of linear FA axis dimension in defined pseudolinearity of more dimensions placed sequentially in a single FA axis). The main disadvantage is the very complicated transformation of analytical results in this form and the nearly impossible reversed visible correlation to the analytical methods data. The main advantage of this matrix type is that the axis integration procedure explained later along, both FA axis brings a rapid correlation to the results of total fatty acid hydrolysis.

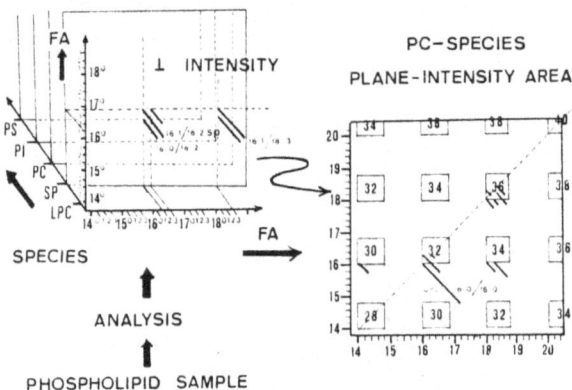

Figure 2 : The FA-Data Block

For this reason and to overcome the above-mentioned disadvantages of the FA-DATA BLOCK descriptive system, the alternative form of so-called "C,N,S"-systems was additionally developped.

The central establishment of the system is the C,N,S-data block for data storage and correlation of complex phospholipid systems in terms of "class" (C-axis), "saturation" (S-axis, number of unsaturated sites) and a "total number of carbon atoms in FA-residues" (N-axis) separation. In the three-dimensional C,N,S space to each X-cns point additionally via allocation vector, the intensity and the fatty acid residue distribution according to chain length and number of unsaturated sites of R1 substituents is allocated (two-additional dimensions) as displayed in Figure 3. The R2 substituent is determined via correlation of the N and S coordinate. In this respect the allocation vector allocated via fatty acid distribution of each X-cns data point corresponds to the pseudomonodimensional form of FA axis used in the FA-DATA BLOCK. The use of allocated fatty acid distribution essentially requires more storage place in the computer system and was not used in our 8-bit operated Apple II system. Principally additional extension of this allocated axis is possible. In the realisation of this extension one must only be aware that any additional pseudodimension of this axis multiplies the extension of the matrix system and thus the necessary minimal storage requirements. As the presented realisation was performed with an 8-bit operated Apple II computer (via Z-80 microprocessor) the mentioned limitation was necessary. The larger 68000-family microprocessor or 80286 microprocessor operated systems allow corresponding expansions.

307

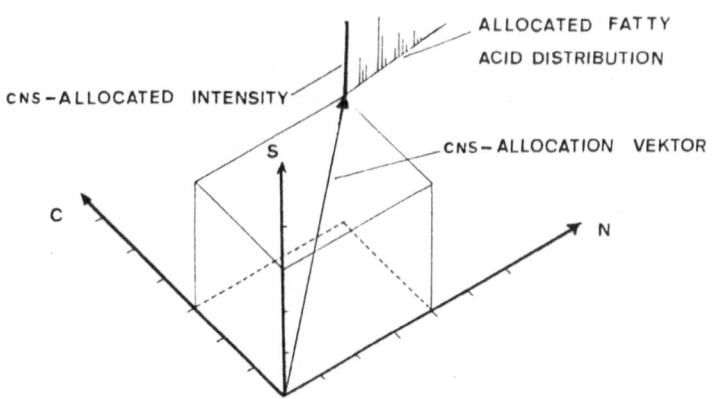

Figure 3 : The C,N,S-DATA BLOCK

The main advantage of the C,N,S-system is that the C,N and S axis separation corresponds to the analytical results of most of the common analytical procedures generally used. The C axis displays primarily the class separation by TLC or LC, the N axis reveals the analysis of corresponding diglycerides by GC, GC-MS or the results of direct analysis of single phospholipid classes by FAB, FAB-CAI etc., and both N and S axis together have the general dimension of the information delivered by the mass spectra.

For easy and quick correlation of C,N,S-data with results of the most common analytical methods the C,N,S-system is equipped with monitoring planes including the ground plane and with integration procedures for integrating the C,N,S-data along one or two axis to these monitoring planes. For correlation and cross-correlation of analytical data from different methods normalisation priority level directions are incorporated.

Monitoring planes including the main ground monitoring plane are designed for presentation and monitoring of partially summarized or integrated data. In the ground plane, the saturation information is suppressed and in the two-dimensional planes of C (class) and N (total number of carbon atoms in R1+R2) dimension and intensity mapping with intensity lines is obtained. These intensity lines obviously give no information on the distribution of saturation within single displayed species. This can be clearly seen in Figure 4. The evaluation of displayed data in the ground plane may be performed by further suppression in additional C or N monitoring planes.

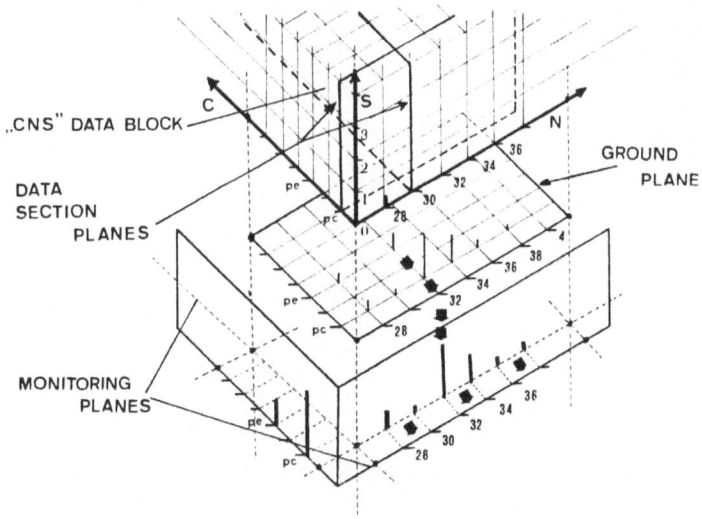

Figure 4 : C,N,S-system Description

In these planes only the class distribution of the sample (C monitoring plane) or the R1+R2 sum (N monitoring plane) is shown. For this reason, the X-cns data points in the C,N,S-data block must be partially integrated. This means that intensity values of all X-cns points having - as an example - the same c and n coordinates, but different s coordinates are summarized, stored and displayed in the ground monitoring plane at the c and n coordinates. This integration procedure may generally be carried out along any of the three presented axis and is called one axis integration and designed with a symbol 1AIs (in the case of ground plane). Index "s" means the suppressed evaluation or information. The one axis integration can thus produce class-saturation or saturation-R1+R2 intensity mapping of the sample. The one axis integration produces always a plane with additionally allocated intensity.

The next step in this direction is the so-called two axis integration. This way the data from already performed one-axis integration are submitted once more to this procedure. This means that in a data set of points from one axis integration which only have two coordinates, again as an example, in the data from ground plane with index X-c,n, the allocated intensity values of all points with the same c-coordinates but different n-coordinates are summarized under only one c-coordinate. As a result, the overall intensity distribution of the whole sample with respect to phospholipid classes, R1+R2 numbers (N) or saturation degree is then visible and stored in the monitoring plane.

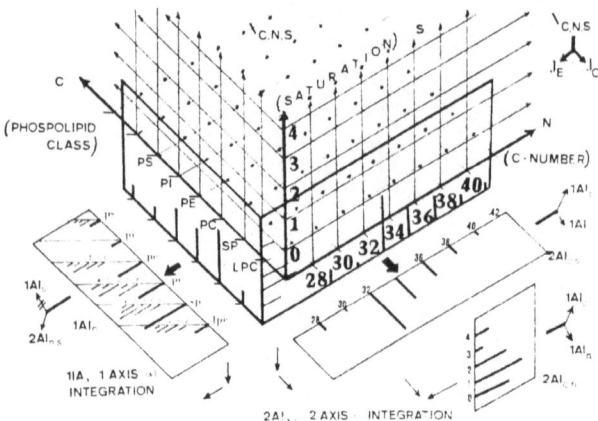

Figure 5 : Multiple Axis Integration

The data set from the two axis integration may of course be reevaluated and redevelopped to the data set of the one axis integration. The corresponding description may be seen from Figure 5.

Using the presented system the data from the C,N,S-data block can be integrated in the direction of any desired plane or axis and displayed. The overall or partial distribution of either the whole sample or single class elements in terms of classes, saturation and C-numbers can be correlated and shown. Thus not only the cross correlation of different methods can be achieved but via a computing procedure the experimental fatty acid total hydrolysis may also be verified and correlated. The essential requirement for the correlation of data from different methods is however the normalisation procedure and for this purpose the normalisation priority levels are established in the C,N,S-system. The first priority level is located in the class monitoring plane.

This means that the 100% of the sample (which results from hypothetical 3-axis integration) must result from the summation of intensity values of overall class distribution (C monitoring plane, 2 axis integration 2AIns). The second priority level of normalisation is defined along the N axis. All entries in the ground plane are normalised in the way that by summation along the n axis the class abundance contained in the C plane is produced. In an analog manner the third priority level exists in the direction of the s-axis, which is simply the entry level. The monitoring planes are important for correlation tasks. As molar abundances are principally entered, the combined dislocation of the R1+R2 sum (N number) in the N monitoring plane can be used for correlation of GC and GC-MS data etc.

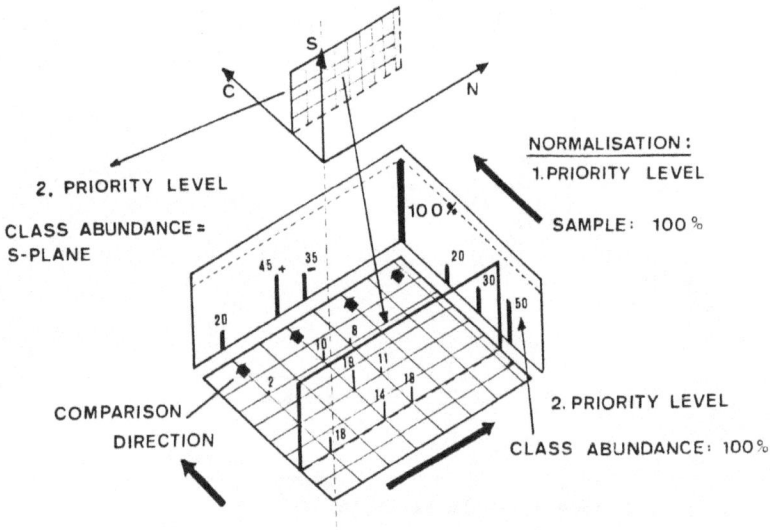

Figure 6 : Normalisation

By iterative matching of computed and experimentally found fatty acid total hydrolysis the conclusions on data planes not having been analysed may be drawn. For that purpose a hierarchic system of priority levels and directions for the normalisation of the fed data exists in the C,N,S-data block for the cross correlation compatibility and can be used for cross correlation of TLC,MS and GC data under implementation of fatty acid hydrolysis data.

Following the same application examples of systems for data storing evaluation already presented, a determination in complex natural phospholipid blends can be performed.

An example for presenting the data from the data block of egg yolk in gound plane (1 axis integration) along the saturation axis is shown in Figure 7. The minor components below 2% are not shown in this overview graphic display, but are presented in the magnified table print out. The main normalisation level represents the class distribution from TLC data. The summarized R1+R2 chain distribution was also received from HPTLC. The distribution of double-bond equivalents along the s-axis was obtained from FAB data. This complete data block was then correlated with overall N-distribution of corresponding diglycerides after enzymatic cleavage of the whole sample as determined by GC.

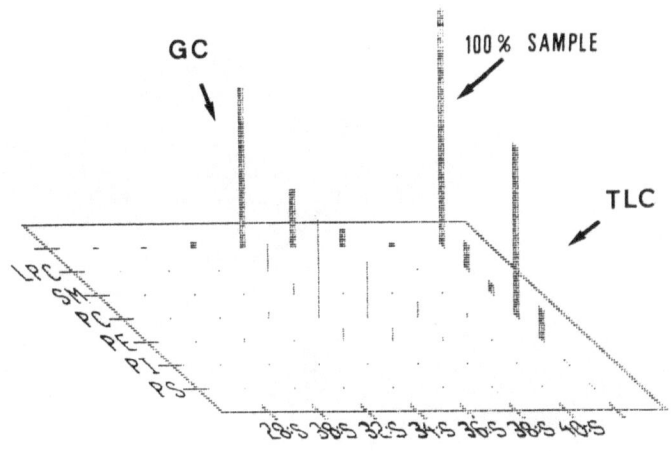

Figure 7 : Data Presentation

In the following figure the TLC densitogramme with data from GC are shown. Especially interesting in the presented plate is the C-number mixture containing C28, C30, C32, C34, C36 and C44 lecithin standards.

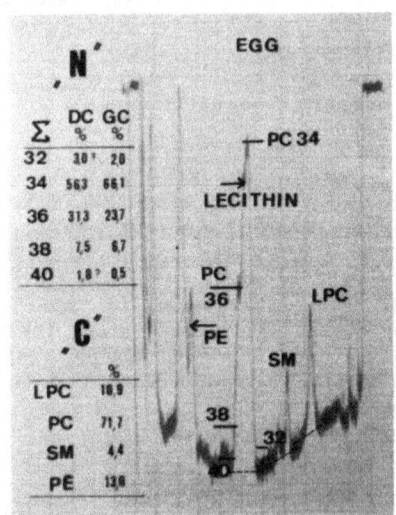

Figure 8 : TLC Densitogram

As we could show under certain circumstances only a C-number separation on TLC plates with no respect to the degree of unsaturation can be achieved.

In order to incorporate TLC results in this system rather complex investigations were necessary so as to ensure a reliable detection of saturated species next to the unsaturated ones. In the following it is shown that most of the used detection systems for phospholipids selectively react with unsaturated species and leave the saturated compound nearly invisible. Such results naturally affect the reliability of any correlation.

In Table 1 an example of the numerical print out of ground-, data- and monitoring planes of an egg yolk sample is shown. The first screen shows distribution of classes versus N-numbers in the ground plane. In the first line the N-monitoring plane is displayed, in the last column the C-monitoring one. The remaining field area is occupied by the data of ground plane. The next two screens show the distribution of unsaturation of lecithins and cefalins of this sample in the corresponding data planes. This is the content of the data plane which is perpendicular to the ground plane and to the S-axis. The next two screens display another print out of the C,N,S-data block : the overall composition of single phospholipid classes with respect to the number of unsaturated sites. This is the content of the data plane which is perpendicular to the ground plane and to the N-axis.

Table 1 : Screen Presentation

	28:S	30:S	32:S	34:S	36:S	38:S	40:S	
	0	0	2	67	24	7	1	100
LPC	0	0	0	11	0	0	0	11
SM	0	0	0	4	0	0	0	4
PC	0	0	2	41	23	6	1	72
PE	0	1	0	5	5	1	0	13
P1	0	0	0	0	0	0	0	0
PS	0	0	0	0	0	0	0	0

PC								
	0	1	2	3	4	5	6	7
23	0	0	0	0	0	0	0	0
30	0	0	0	0	0	0	0	0
32	2	2	0	0	0	0	0	0
34	7	15	11	5	0	0	0	0
36	2	10	8	3	5	2	0	0
33	0	0	0	0	0	0	0	0
40	0	0	0	0	0	0	0	0

PE								
	0	1	2	3	4	5	6	7
23	0	0	0	0	0	0	0	0
30	0	0	0	0	0	0	1	0
32	0	0	0	0	0	0	0	0
34	1	3	1	1	0	0	0	0
36	0	2	2	1	0	0	0	0
38	0	0	0	0	1	0	0	0
40	0	0	0	0	0	0	0	0

	34:0	34:1	34:2	34:3	34:4	34:5	34:6	34:7
	8	18	12	6	0	0	0	0
LPC	0	0	0	0	0	0	0	0
SM	0	0	0	0	0	0	0	0
PC	7	15	11	5	0	0	0	0
PE	1	3	1	1	0	0	0	0
P1	0	0	0	0	0	0	0	0
PS	0	0	0	0	0	0	0	0

	36:0	36:1	36:2	36:3	36:4	36:5	36:6	36:7
	2	12	10	4	5	2	0	0
LPC	0	0	0	0	0	0	0	0
SM	0	0	0	0	0	0	0	0
PC	2	10	8	3	5	2	0	0
PE	0	2	2	1	0	0	0	0
P1	0	0	0	0	0	0	0	0
PS	0	0	0	0	0	0	0	0

313

In the following table an example of the incorporation of the computative evaluation of the values from the total hydrolysis of the sample to the fatty acids is shown.

Table 2

CLASS : PC

------------------------MONO----------------------------

	16:0	18:0	18:1	18:2	18:3	20:4
MASS	256	284	282	280	278	304
%	34.9	11.4	32.8	12.2	0	1.9

------------------------DUAL----------------------------

	32:0	32:1	34:0	34:1	34:2	34:3	36:0
MASS	734	732	762	760	758	756	790
%	3.2	3.4	9.9	21.2	15	7.5	3.4

	36:1	36:2	36:3	36:4	36:5
MASS	788	786	784	782	780
%	13.5	10.7	4.3	6.4	2.8

In the section MONO the found experimental percental values from hydrolysis are fed in the computation system as well as present known species of dual combinations of fatty acids in diglyceride units (data from FAB-MS or GC-MS) are entered. In Table 3 the computation of possible combinations of present fatty acids to diglyceride combinations are listed. In the next step below the theoretical possible maximal and minimal values of fatty acids after hydrolysis of the sample are calculated (maximum value from possible multiple combinations only the ones with the particular fatty acid are considered; minimum value - if any other combinations than the particular one are possible thus the particular one is omitted). Using an iterative procedure of this kind, by fitting of calculated and experimentally found values a more detailed distribution within particular classes may be found.

Table 3

%	MONO	MIN	MAX	DISTRIBUTION	
16/0	34.9	30	33.2	32/0 --^{o}	16/0-16/0
18/0	11.4	22.6	22.6	32/1 --^{o}	
18/1	32.8	30.2	33.4	34/0 --^{o}	16/0-18/0
18/2	12.2	16.4	22.8	34/1 --^{o}	16/0-18/1
18/3	0	7.3	10.5	34/2 --^{o}	16/0-18/2
20/4	1.9	0	3.2	34/3 --^{o}	16/0-18/3
				36/0 --^{o}	18/0-18/0
				36/1 --^{o}	18/0-18/1
				36/2 --^{o}	18/0-18/2 18/1-18/1
				36/3 --^{o}	18/0-18/3 18/1-18/2
%	SOLL	IST		36/4 --^{o}	16/0-20/4 18/1-18/3 18/2-18/2
16/0	34.9	31.056		36/5 --^{o}	18/2-18/3
18/0	11.4	18.85			
18/1	32.8	24.831			
18/2	12.2	14.826			
18/3	0	7.281			
20/4	1.9	1.056			

In Fig. 9 the principle of the incorporation of this calculative
- so called MO–DU procedure in combination with the C,N,S data block is
shown.

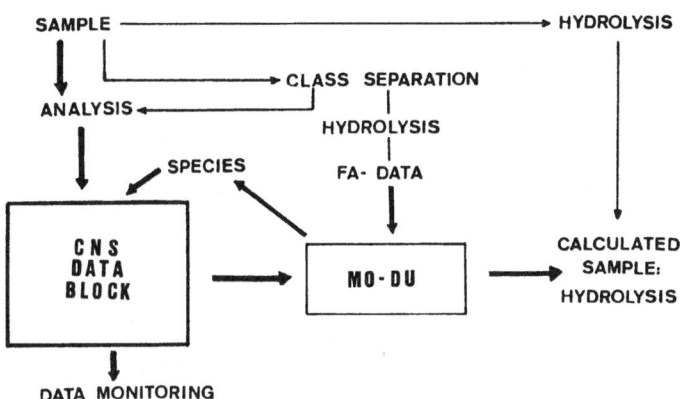

Figure 9 :

The realisation of this model in a Z80 operated CP/M system with
data blocks of 14kB per sample with a Pascal written source as well as
together with the principal possibility of extension and application of
the data block system to other compound classes like terpenes or steroids
may be an interesting approach in the characterisation of extremely
complicated mixtures as they are constituted by natural phospholipid
blends and may be only the first step in developing a new category of
compound characterisa- tion.

REFERENCES

1. L. Kolarovic and H. Traitler : J.H. Res. Chrom. 341, 8 (1985).
2. L. Kolarovic and H. Traitler : Ibid, 383, 8 (1985).
3. H. Traitler and A. Nikiforov. In : Chromatography and Mass Spectrometry
   in Nutrition and Food Science, 299, Elsevier Science Publishers, B.V.
   Amsterdam (1984).
4. A. Nikiforov, G. Dicher, He Meiyu and H. Traitler, Mh. Chem. 1985 in
   preparation.
5. A. Nikifovor, A. Lohninger, M. Specker and L. Linhardt : Eur. J. Mass
   Spectrom, 105, 2 (1982).
6. J.J. Myher, A. Kuksis, L. Marai and S.K.F. Young : Anal. Chem. 557, 50
   (1978).
7. A. Lohninger and A. Nikiforov : J. Chromatography, 185, 192 (1980).

Figure 4

RECOMMENDATIONS FROM THE ADVANCED NATO WORKSHOP

Nutrition Working Group

Selvino Italy
March, 1986

1.0. BACKGROUND

1.1. Health Concern Related to Nutrition

a) Heart disease and Stroke

In North America and Europe, one in four men have a heart attack or stroke before their age of retirement. In countries where awareness of the risk factors is present (e.g. the USA) and where intervention programmes have been in progress (e.g. in Norway and Finland) the trends in food consumption, the fall in mean blood cholesterol levels, and the downward movement in mortality from heart disease are consistent with the hypothesis of diet as a causative factor. In countries where there has been no action, mortality has risen or remained the same (e.g. the UK). Whilst this is an over-simplification of the present position, the evidence linking diet and heart disease is ample justification for recommendations on diet, to be persued.

b) Cancer

Breast and colon cancer roughly follow the incidence of heart disease from country to country and in migrating populations. Whilst the reason for the link between the incidence of cancer and heart disease can only be a matter of speculation, the present indications are that dietary advice based on the prevention of heart disease, is likely to be beneficial in cancer prevention.

c) The vulnerability during growth and development

The incidence of fetal growth retardation , perinatal mortality and handicap is substantially higher in low socio-economic sections of a population. Low birth weight is associated with low calorie intakes. Retardation of brain growth during early development can lead to permanent handicaps. It is evident that the period of growth and development of a biological system, is the period when it is most susceptible or vulnerable to nutritional deficits or distortions. This principle will apply equally to the vascular system hence the critical importance of maternal, infant and child nutrition.

Some 10 millions of children die each year from under or malnutrition. In these populations the incidence of low birth weight,
perinatal mortality and developmental distortions of nutritional
origin is high: 70% of the undernutrition is due to inadequate energy
intakes.

e) Transfer of food and agricultural policies
    Past history has seen both benefits and serious health problems
resulting from the introduction of foods from one culture into another.
For example, the introduction of alcohol, refined carbohydrates and
sugar rich foods to technically undeveloped communities has introduced
alcoholism, liver disease, dental caries, and intestinal disorders
where previously none existed. In certain NATO countries where the
incidence of heart disease was low, the introduction of Western type
foods is effectively introducing Western diseases.

1.2. Present knowledge on Essential Fatty Acids

a) Essential Fatty Acids
    Linoleic and alpha-linolenic acids together with their long
chain derivatives are essential components of the cell membrane structural lipids. Animals require essential fatty acids for growth, development and reproduction. The long chain derivatives of the essential
fatty acids provide precursors for a group of hormone like substances,
the eicosanoids, which are involved in regulation of cell function,
blood flow, homeostasis, and the immune system. The structural lipids
are quantitatively the most important structural membrane component
in the brain and nervous system and the second most important next
to protein in all other soft tissues. Two families (linoleic or n-6
and the alpha-linolenic or n-3 families) of polyunsaturated fatty
acids posses essential fatty acid activity. Both the parent and the
derivative fatty acids are used in cell membranes and there is increasing evidence that the n-3 family contributes to cell regulation.
The balance between the two families influences the function of the
immune system, the vascular system and thrombogenic mechanisms.
Attention should therefore be given to the balance between the n-6
and n-3 fatty acids in consideration of prevention of cardiovascular
disease and cancer.

b) Non Essential Fatty Acids
    Evidence has accumulated that the non essential fatty acids
fall into different groups with different biological properties.
A certain proportion of the 16,18 and 24 carbon chain length fatty
acids are used for membranes alongside the essential fatty acids
but at disproportionately high levels in the diet will interfere
with essential fatty acid incorporation. Trans and positional isomers
behave in a similar manner to the non-essential fatty acids. Whilst
there is no convincing evidence of specific toxic effects, the positional and trans-isomers do not possess essential fatty acid functions
and at high intakes will compete for utilisation with the essential
and saturated fatty acids. It is becoming increasingly clear, that
just as the different polyunsaturated fatty acids have different

and indipendent properties, so also the non-essential fatty acids are different depending on chain length and or position of the double bond.

2.0. RECOMMENDATIONS ON NUTRITION

2.1. Strategy adopted by the Workshop

A number of National and International Committees have discussed diet and heart disease at length and formulated recommendations. This NATO Workshop does not have comparable amounts of time to produce its own specific recommendations. However, it was considered possible to identify weaknesses in previous reports and recommend attention be given to these by future Committees. As the workshop included several experts in fatty acid metabolism and the nutritional implications, it was appropriate to identify previous gaps in the recommendations and guidelines as they pertain to the specialist aspects of dietary fats.

2.2. The principle underlying the recommendations

The following recommendations are based on aspects of previous National and International Reports, which have been omitted or inadequately expressed.

In principle, a request is made for clear statements on what is meant in terms of the different fatty acids, age groups and population backgrounds, in the interest of scientific accuracy and of stimulating clearer understanding of the nutritional and biological interactions with health and disease.

3.0. THE RECOMMENDATIONS

3.1. The nutritional value of the different fatty acids

3.1.1. The amount of dietary fats - The question which should be addressed is : "What is the optimum quantity of fat at different ages appropriate for human health and the prevention of cardiovascular disease and cancer".

3.1.2. Polyunsaturated Fatty Acids - The term is ambiguous as it can refer to fatty acids with or without essential fatty acid activity. The term "essential fatty acids" should be used to place them in their proper nutritional perspective.

3.1.3. Essential Fatty Acids - There is now enough evidence to ask for clear statements on the nature of the essential (polyunsaturated) fatty acids being specified. It is known that :

a) The n-6 family is not the same as the n-3 family
b) Linoleic is not the same as arachidonic acid
c) Alpha-linolenic is not the same as eicosapentanoic nor docosahexaenoic acids.

Consequently, recommendations on fatty acid intakes should say what is intended with regard to the specific essential fatty acids and their long chain derivatives.

3.1.4. Monoenes - Similarly it is clear that palmitoleic is not the same as oleic which is not the same as erucic nor nervonic acids. Consequently, reference to dietary monoenes should be precise as to what is meant. There is now growing evidence that oils rich in oleic acid may exert favourable effects on plasma lipid and lipoprotein levels. These effects should be further investigated.

3.1.5. Saturated Fatty Acids - The saturated fatty acids should be considered in the recognised categories of (i) short, (ii) medium, (iii) long and (iv) branched chain fatty acids. There is growing evidence that the physiological effects of these fatty acids are somewhat different depending on chain length and their position in the triglyceride molecule.

3.1.6. Trans and positional isomers - The contribution of the trans and positional isomers to the total fat intake has been neglected, yet they now amount to appreciable proportions of the fat intake in certain processed foods. If omitted, there can be a significant proportion of the fat intake left unaccounted. Labelling of such foods with their total isomer contents would be one method of drawing their quantitative presence to the attention of the consumer and stimulating discussion and research on their health implications.

3.1.7. Minor components - Tocopherols, tocotrienols, phytosterols, beta-carotene and related compounds may have beneficial effects. Some oils and fats are very rich sources and new technologies should be developed to spare such nutrients during processing.

3.2. Special age groups

3.2.1. General - The long term aim of reducing the saturated fat intake to 10% or below and maintaining a proper balance of both families of essential fatty acids, is in the interest of preventing cardio-vascular disease. As the optimum function of the cardiovascular system is relevant to all ages and both sexes, the implementation of appro-priate recommendations for both primary and secondary prevention would seem logical for all ages and both sexes. However, special needs may be apparent for different aspects of the life cycle.

3.2.2. Pregnancy and lactating mothers - The requirement for lipids are highlighted in pregnancy. Fat is stored as the energy reserve for fetal growth and lactation. Fetal growth is dependent on placental blood flow and hence the development and growth of the placental vascular system which mainly occurs before the fetal growth thrust. Quantitatively, lipids are of special significance during lactation. Sixty percent of the 600-800 calories delivered daily in human milk is as lipid. It might be wrong to recommend a low fat intake to lac-tating mothers yet previous recommendations have been directed globally at an adult population. Because of the special role of fats in human milk and its production and the importance of this period of development in the infant, it would seem proper to draw the attention of recommending committees to this oversight.

3.2.3. Primary prevention (primordial prevention) in children - Previous recommendations have focused on the adult population. However, during growth, developing systems are at their most vulnerable to nutritional distortion. With the evidence of early vascular changes and the rise in blood cholesterol and pressure in children there is a clear responsability for future recommendations to consider the question of maternal and infant nutrition as well as the principles applying during growth of the child. On the specific issue of milk formula substitutes for breast milk and infant feeds, there is no commercially available product which matches human milk with regard to its content of n-6 and n-3 fatty acids despite a recommendation to that effect by FAO/WHO in 1978. Where foods are designed to meet the whole or nearly all the infant's nutrient requirements, te balance and spectrum of essential fatty acids should match that of human milk.

3.2.4. The elderly - With increasing numbers of elderly people in the Western countries, the possibility that the elderly may have accentuated requirements for essential fatty acids and related nutrients, should be investigated. There is evidence that the elderly may have insufficient intakes of essential fatty acids and related nutrients. Because of the relevance of the essential fatty acids to the vascular and nervous systems and the special significance of these systems in old age, further documentation on the nutrient intake in the elderly is needed with a view to making specific recommendations for this age group.

3.3. International Issues

3.3.1. Need for increased fat intake of the appropriate type - In developing countries carbohydrate based diets may be associated with a high bulk of food which can limit intake of dietary energy with adverse effects on growth and development in children and a consequent high incidence of malnutrition. The addition of 12-14 g of oil to 100 g of boiled rice can double the energy density. In many developing countries a reduction in fat intake could result in an increase in malnutrition. The aim of reducing fat intake in Western populations should be coupled with an increased fat intake in developing countries where the energy density of the diet is undesirably low. At the same time, it would be important for such populations, to consider the need to meet the optimum essential fatty acid requirements.

3.3.2. Food policy

3.3.2.1. Export of food policy, heart disease and cancer - In view of the very serious nature of nutrition related disease in certain technically advanced countries, it would be undesirable for those countries to transfer their principles of food technology and agriculture to other countries in which cardiovascular disease, breast or colon cancer are not at this time a problem.

3.3.2.2. Scientific and political coordination - At this stage of history it is possible to prevent the introduction of these diseases in the low incidence countries. There is an opportunity, which will be short lived, for policy makers and medical scientists to cooperate to develop a code of practice and to educate governments on the need for early action. It is now possible to identify nutrition related disease within a culture, take steps to eradicate these and to build on the best of traditional practices instead of importing food structures that are identified with heart disease and other serious disorders which the technically advanced are now attempting to prevent.

Concern was expressed at the workshop about the very long time-lag between the appearance of scientific evidence and any action by those responsible for health and food production. Food and agricultural policies are still not linked with health. This is a serious inditement of the present system. Failure to act on existing knowledge has resulted in premature deaths and disability which could otherwise have been prevented. Failure to safeguard the food structures of low incidence countries both in and outside of NATO is creating disease which was not there before. There is an urgent need for collaboration between the decision makers and the scientific community to counteract the spread of nutrition related disease from one country to another and to ensure that food and agricultural policies are not just matters of economics and politics but are consistent with the higher objective of community health world wide.

3.4. Analytical Technology
There is an obvious gap in knowledge of (i) details of fatty composition of foods and tissue lipids and (ii) agreement between laboratories on reported composition of organ glyceride fatty acid composition.
It was agreed that recommendations should be made on methodology for fatty acid analysis with the objective of standardising information.

GENERAL ASPECTS IN TECHNOLOGY

The Round Table participants considered a series of issues directed to the role of technology in providing edible fat products acceptable to the consumer.

A. It was noted that in many countries consumers want products which respond to slimming considerations, that is, with reduced fat content while maintaining organoleptic quality. Most countries require margarine products to have a fat content above 80%. Diet products containing less than 40% fat are also generally permitted. A large share (22%)

of the US market is products having a fat content of about 60%, labeled as spreads. These products are not permitted in EEC countries.

In order to provide the consumer the opportunity to optimize dietary fat content, participants agreed to the following recommendations: that EEC regulations be modified to allow low-fat spreads having fat content in the range between 40% and 80%.

B. Products which provide dietary approaches to amelioration of disease and dysfunction are available. These include medium chain triglycerides, i.e. having fatty acids C_8 to C_{10} chain length from fractionated coconut oils. Geriatric stores provide products formulated from un-hydrogenated fats, but some concern was expressed about the storage stability of such products. Since most studies agree that there are no demonstrated ill-effects of consuming hydrogenated fats, such formulations specifying unhydrogenated fats address the desire for "natural" foods.

C. Products which supply needed calories for undernourished populations can best be provided by development of efficient small-scale technology to process the raw materials available in developing nations.

D. Products which provide fats with a balance between stability and beneficial nutritional components were discussed. It was noted that there is still much controversy about what are beneficial nutritional components. Technology is not standing still, it is a continuum, and at any point in time it is necessary to have the best application of the technology available. Research should be directed to a better knowledge of minor components of fats and oils. Useful components should be retained during processing of raw materials. Impact on minor components of the stabilization of fats by hydrogenation needs to be understood to a greater extent. In regards to new nutritional information relative to n-3 and n-6 long chain (C_{20} - C_{22}) fatty acids, much needs to be known concerning the stability of such compo-nents in consumer products. Thus, research continues to be needed concerning techniques of preservation and storage of fats.

E. Improvement of existing technologies is important, as well as the development of new technology. Processors should be concerned about raw material quality, pre- and post-processing oil quality, and fat product quality. It was noted that few raw material speci-fications reflect a concern for quality of the oil to be obtained. Analytical technology is needed to evaluate oil quality early in the processing chain and specifications for crude oil need to be improved to assure the pre-process quality of the oil.

During 1986 essential fatty acids (n-6 nd n-3) continued to hit the headlines with fish being particularly active. NATO held an Advanced Research Workshop to discuss them (and some remarkable results on palm oil from Gerard Hornstra) in Selvino during March.

At the Gala Dinner its organiser, Claudio Galli, allowed me ten minutes to try to get the delegates singing, hence the following which updates a version sung at an ICBL meeting and which I hope you will hum to yourself as you enjoy your Christmas and 1987.

Hugh Sinclair

International Nutrition Foundation
Sutton Courtenay , Oxon, England

CONTRIBUTORS

INDEX